MULTISENSOR DECISION AND ESTIMATION FUSION

The Kluwer International Series on
ASIAN STUDIES IN COMPUTER AND INFORMATION SCIENCE

Series Editor
Kai-Yuan Cai
Beijing University of Aeronautics and Astronautics, Beijing, CHINA

Also in the Series:

STABLE ADAPTIVE NEURAL NETWORK CONTROL, *by S. S. Ge, C.C. Hang, T. H. Lee and T. Zhang;* ISBN: 0-7923-7597-1

FULLY TUNED RADIAL BASIS FUNCTION NEURAL NETWORKS FOR FLIGHT CONTROL, *by N. Sundararajan, P. Saratchandran and Yan Li;* ISBN: 0-7923-7518-1

NONLINEAR CONTROL SYSTEMS AND POWER SYSTEM DYNAMICS *by Qiang Lu, Yuanzhang Sun and Shengwei Mei;* ISBN: 0-7923-7312-X

DATA MANAGEMENT AND INTERNET COMPUTING FOR IMAGE/PATTERN ANALYSIS, by David Zhang, Xiaobo Li and Zhiyong Liu; ISBN: 0-7923-7456-8

COMMON WAVEFORM ANALYSIS: ***A New and Practical Generalization of Fourier Analysis***, *by Yuchuan Wei and Qishan Zhang;* ISBN: 0-7923-7905-5

DOMAIN MODELING-BASED SOFTWARE ENGINEERING: ***A Formal Approach***, *by Ruqian Lu and Zhi Jin;* ISBN: 0-7923-7889-X

AUTOMATED BIOMETRICS: ***Technologies and Systems***, *by David D. Zhang*; ISBN: 0-7923-7856-3

FUZZY LOGIC AND SOFT COMPUTING, *by Guoqing Chen, Mingsheng Ying, Kai-Yuan Cai*; ISBN: 0-7923-8650-7

MULTISENSOR DECISION AND ESTIMATION FUSION

Yunmin Zhu
Sichuan University, P.R. China

KLUWER ACADEMIC PUBLISHERS
Boston / Dordrecht / London

Distributors for North, Central and South America:
Kluwer Academic Publishers
101 Philip Drive; Assinippi Park
Norwell, Massachusetts 02061 USA
Telephone (781) 871-6600
Fax (781) 681-9045
E-Mail: kluwer@wkap.com

Distributors for all other countries:
Kluwer Academic Publishers Group
Post Office Box 322
3300 AH Dordrecht, THE NETHERLANDS
Telephone 31 786 576 000
Fax 31 786 576 254
E-Mail: services@wkap.nl

Electronic Services <http://www.wkap.nl>

Library of Congress Cataloging-in-Publication Data

Zhu, Yunmin, 1944-
Multisensor decision and estimation fusion / Yunmin Zhu.
p. cm.-- (Kluwer international series on Asian studies in computer and information science ; 14)
Includes bibliographical references and index.
ISBN 1-4020-7258-9 (alk. paper)
1. Multisensor data fusion. 2. Multicriteria decision making. I. Title. II. Series.

TK7870 .Z43 2002
006.3--dc21 2002034044

Printed on acid-free paper. Printed in the United States of America.

To My Parents,

Hong You & Guoding Zhu,

My Wife, Shengna Gu,

My Daughter, Ji Zhu,

My Son, Yu Zhu

SERIES EDITOR'S ACKNOWLEDGMENTS

I am pleased to acknowledge the assistance to the editorial work by Beijing University of Aeronautics and Astronautics and the National Natural Science Foundation of China

Kai-Yuan Cai
Series Editor
Department of Automatic Control
Beijing University of Aeronautics and Astronautics
Beijing 100083
China

Contents

List of Figures

List of Tables

Preface

YUNMIN ZHU

In the past two decades, multisensor or multi-source information fusion techniques have attracted more and more attention in practice, where observations are processed in a distributed manner and decisions or estimates are made at the individual processors, and processed data (or compressed observations) are then transmitted to a fusion center where the final global decision or estimate is made. A system with multiple distributed sensors has many advantages over one with a single sensor. These include an increase in the capability, reliability, robustness and survivability of the system. Distributed decision or estimation fusion problems for cases with statistically independent observations or observation noises have received significant attention (see Varshney's book *Distributed Detection and Data Fusion*, New York: Springer-Verlag, 1997, Bar-Shalom's book *Multitarget-Multisensor Tracking: Advanced Applications*, vol. 1-3, Artech House, 1990, 1992, 2000). Problems with statistically dependent observations or observation noises are more difficult and have received much less study. In practice, however, one often sees decision or estimation fusion problems with statistically dependent observations or observation noises. For instance, when several sensors are used to detect a random signal in the presence of observation noise, the sensor observations could not be statistically independent when the signal is present.

This book provides a more complete treatment of the fundamentals of multisensor decision and estimation fusion in order to deal with general random observations or observation noises that are correlated across the sensors. Progress in two directions is presented. For the multisensor decision fusion problem, we give a necessary condition for the set of the optimal sensor rules given a fixed fusion rule. It must be a fixed point of an integral operator, which depends on the joint conditional probability densities of the sensor observations and on the fixed fusion rule, no matter whether or not the sensor observations are statistically independent. Then, we propose an efficient discretized iterative algorithm to search for the optimal sensor rules. We prove the finite convergence of this algorithm and that the solutions approach the solutions of the original continuous algorithm as a step-size parameter is taken to zero.

Much attention in this book is paid to the optimal fusion rule problem. There are two different optimization strategies in practice. The first is a two-level optimization strategy–optimizing the sensor rules first and then optimizing the fusion center rule given the sensor rules. When each sensor in practice must first make its own optimal decision based only on the local information the sensor can get, the optimal fusion rule given all sensor rules is then presented for very general multisensor decision system. The second is a global optimization problem. When one wants to pursue globally optimal performance of the decision system, one has to find a globally optimal fusion rule first, then get the optimal sensor rules under the optimal fusion rule. Since the number of possible fusion rules is usually very large, it is in general computationally intractable to exhaustively search for the globally optimal fusion rule from all possible rules. In the global optimization sense, we analyze how to classify all possible fusion rules into equivalent classes and give up those classes that are worse than one of other classes. After such analysis, we can reduce the number of the valuable fusion rules that need to be considered in the optimization. More importantly, we present some new findings about the optimal fusion rules. In particular, we show some highly unexpected properties of the optimal fusion rules. In some specific sensor communication patterns of the system, we prove that a fixed fusion rule proposed in the book is a unified version of all possible fusion rules, i.e., any possible fusion rule is a special case of the unified fusion rule provided that some of the sensor rules are restricted to special versions in advance. This unified fusion rule depends only on the number of bits transmitted by the sensors, and optimizing the sensor rules under this fusion rule can obtain globally optimal performance.

A specific example of these cases is there an l-sensor distributed binary decision system uses a total of $l-1$-bit decisions distributed over the first $l-1$ sensors, while the lth sensor uses 2^{l-1} bits in its decision. Moreover, in these cases, the fusion rule does not depend on the statistical properties of the observational data. Of course, to get optimal performance under this fusion rule, the choice of the sensor rules does depend on the statistical properties of the observational data. Further, we show that, in these cases, increasing the number of bits used in the lth sensor decision will not improve performance, even if the observations themselves are sent from this sensor to the fusion center. Using this unified/optimal fusion rule, one just needs to compute the optimal sensor rules to obtain overall optimal performance. These findings help us reduce huge computation. Many numerical results given in this book support the above statements. All of the above results can be extended to Neyman-Pearson testing and sequential testing, as well as to the multisensor multi-hypothesis network decision systems (parallel, tandem, and tree networks).

The progress in the second direction is concerned with the multisensor estimation fusion. For the multisensor point estimation fusion problem, a general version of the linear minimum error variance estimation fusion rule is developed. It has the Least Mean-Square (LMS) error among all linear unbiased estimation fusion rules. It is very general–it relies only on two assumptions: (I) the local estimators are unbiased and (II) the error covariance matrix C_k of all local estimates at each time k is known. Not only does it include existing fusion results as special cases, but it is also valid for many more general cases, including (A) coupled measurement noises across sensors; (B) sophisticated network structures or communication patterns; (C) different local dynamic models or estimator types; and (D) efficient fusion of asynchronous estimates.

First, we formulate the problem of distributed estimation fusion in a general setting, which is the key to the subsequent results in this book. In this setting, the fused estimator is a weighted sum of local estimates. We show that the set of weights is optimal if and only if it is a solution of a matrix quadratic optimization problem subject to a linear equality constraint. Secondly, we present a general solution to the above optimization problem, which depends only on the covariance matrix C_k, and prove that the final fused estimate is unique whether or not the solution of the optimal fusion weight is unique or not. We also discuss the generality and usefulness of the fusion formulae developed. Finally, we provide an off-line recursion for C_k for a class of multisensor linear systems with coupled measurement noises. In addition, more importantly, we will give a necessary and sufficient condition for the two types, with and without a priori information on the parameter, of distributed estimation fusion being identical. We also proposed two fusion formulas of the recursive Least Squares algorithm with a forgetting factor which is efficient for state estimate of dynamic systems without knowledge of the noise covariances. The optimality of the multisensor Kalman filtering with feedback is also rigorously analyzed.

The multisensor interval estimation fusion problem is also discussed. According to two types of the different available messages that the fusion center gets from sensors, two different interval estimation fusion methods are presented. Concerning the first fusion method, the fusion center can receive sensor statistics and know their joint distribution, then optimally fuse them to get a final interval estimate by using two conventional interval estimation methods:the pivotal quantity method and the inverting a hypothesis testing method. As for the second fusion method, the outputs of every sensor are its interval estimate and confidence degree. Under the assumption of independent sensor estimates, a combination rule to yield final interval outputs and their confidence degrees is proposed. In terms of different optimization criteria, then, using the combined interval outputs and their confidence degrees, the final optimal interval estimate can be derived. Besides, when the fusion center has extra information on the sensor estimates, a conditional confidence degrees can be derived. As an application, we will deal with the fault-tolerant interval estimation fusion problem.

This book can be used as a graduate-level textbook. It is assumed that the reader has been exposed to elementary decision and estimation theory. The book will also serve as a useful reference for practicing engineers and researchers. It is organized into seven chapters. Chapter 1 provides necessary background knowledge for a good understanding of the multisensor decision fusion. Then, we start in Chapter 2 with the two sensor binary decision system, to present a fundamental framework, iterative algorithms, and convergence analysis for the distributed Bayes decision systems. The key result in this chapter is a necessary condition that the optimal sensor rules must satisfy given a fusion rule. We do not assume the independence between the sensor observations. We then extend the previous results to the multisensor Neyman-Pearson, and sequential decision systems and the multi-hypothesis network decision systems in Chapters 3 and 4. Chapter 5 is dedicated to the optimal fusion rule problem. In Chapters 6 and 7, we discuss the multisensor distributed point estimation fusion problem and the interval estimation fusion problems, respectively.

Acknowledgments

I would like to express my deep appreciation to Dr. Zhi-Quan Luo and Kon Max Wong, as well as to Dr. X. Rong Li, for their support of my visits to the Communications Research Laboratory, McMaster University, and University of New Orleans, respectively. A part of the work was performed there with them. I would also like to thank Professor H. Kushner very much for his many valuable comments and suggestions on this book. He spent much time in reading the early version of the manuscript and corrected many inaccuracies in it. I am very grateful to Professor P. Varshney and Dr. Kai-Yuan Cai for their continued interest in my research and encouragement throughout this period. Thanks to Dr. Jie Zhou for his great help in typing this book and in preparation of some of the figures. Finally, I would like to thank the National Natural Science Foundation and the National Key Project of China for their constant supports.

I

DECISION FUSION

Chapter 1

INTRODUCTION

1.1 CONVENTIONAL STATISTICAL DECISION

In practice, one is very often faced with a decision-making problem, i.e., selecting a course of action among several possibilities. For instance, in a digital communication system, one of several possible symbols is transmitted over a channel, we need to determine the symbol that was transmitted based on the received noisy observations. In a radar detection system, a decision based on the radar return is to be made regarding the presence or absence of a target. In a medical diagnosis problem, based on an electrocardiogram, one needs to determine if the patient has a heart attack. In a target recognition problem, one needs to identify the type of aircraft being detected based on some observed aircraft features. In all of the above practical applications, the common issue is to make a decision from several possibilities based on available noisy observations, where the truly present phenomenon cannot be observed directly. The branch of statistics dealing with these type of problems is known as statistical decision or hypothesis testing.

In preparation for the development of the multisensor statistical decision theory and methods, a brief introduction of the conventional statistical decision will now be given.

1.1.1 BASIC MODEL OF STATISTICAL DECISION

The basic components of a simple binary statistical decision problem are shown in the following Fig. 1.1.1.

The first component is the physical source, which generates one of two possible outputs. More generally, the source generates one of m possibilities as an output. The source output cannot be observed directly.

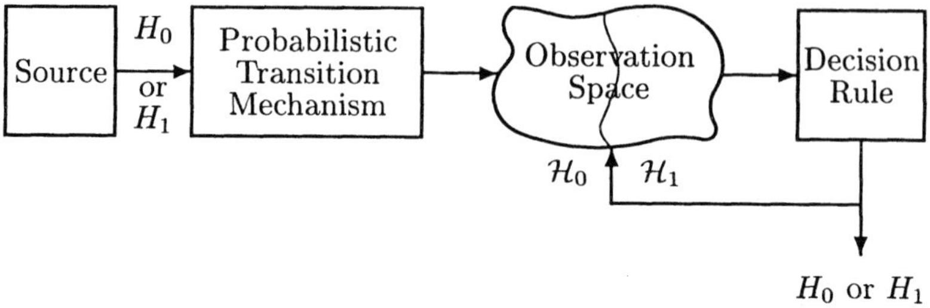

Figure 1.1.1. Basic model of statistical decision.

The second component is a observation probabilistic transition mechanism, which generates an indirect observation of the source output as well as knows the conditional probabilistic characteristic given any source output of all possible possibilities. In other words, although the observation cannot directly tell us what possibility happened, the observations really contain some information on the source output via the conditional probabilistic characteristic of the observation. One just uses the differences among all possible conditional probabilistic characteristics to find a method to optimally make a decision in terms of a certain criterion.

The third component is an observation space that one has access to. The actual distribution of points in the observation space has been determined by the source output but is unknown by the decision maker.

The last component is a decision rule, which is determined based on the observation, the knowledge of all possible conditional distributions given the source output, and a certain performance criterion. When any point in the observation space is received, only a single output can be declared according to the decision rule, this rule is called , *the deterministic decision rule*, which partitions the observation space into two (or m) disjoint decision regions as shown in Fig. 1.1.1. Otherwise, for some points in the observation space, more than one outputs can be declared randomly with their own probabilities. This rule is called *the randomized decision rule.* One can view the deterministic decision rule as a special case of the randomized decision rule. Obviously, based on the same criterion, the optimal randomized decision rule outperforms the optimal deterministic decision rule. However, it is very hard in most cases to get a solution for the optimal randomized decision rule. In this book, we will concentrate ourselves mostly on the deterministic decision rules. The randomized decision rule will be considered only in Neyman-Pearson decision in Subsection 1.3.2 and the multisensor distributed decision with the randomized fusion rules in Section 2.6.

1.1.2 HYPOTHESIS TESTING

In statistics, one likes more to call statistical decision *Hypothesis Testing*. Hypotheses are all possible outputs from the source. Testing is a procedure to make a decision that a decision maker declares which output true.

When there are only two different outputs from the source, the test is called binary decision. When there are more than two outputs from the source, one calls the test multiple hypothesis testing or simply m-ary hypothesis testing.

If one differentiates hypothesis tests in terms of criteria, there are *Bayes test, Neyman-Pearson test*, and *Minimax test* among others. In the Bayes test, the hypotheses happen randomly. A priori probabilities of all hypotheses and all decision costs related to all possible true phenomena are known. A Bayes decision rule minimizes the average (expected) decision cost. The Bayes criterion requires the knowledge of a priori probabilities which may not be readily available. In such situations, the Minimax criterion is a feasible alternative. Under this criterion, one uses the Bayes decision rule corresponding to the least favorable prior probability assignment, i.e., one minimizes the maximum decision risk. In many practical situations, not only are the prior probabilities unknown, but the decision cost assignments are also difficult to determine. In this case, the Neyman-Pearson criterion for a binary decision is employed. According to Neyman-Pearson decision rule, one constrains a decision error probability under a hypothesis to an acceptable value and maximize a right decision probability under another hypothesis or equivalently minimizes a decision error probability under another hypothesis.

In the above three types of the hypothesis testings, the sample sizes of the observations are all fixed. In many practical situations, however, observations are collected sequentially and more information becomes available as time progresses. When we are not satisfied with the decision performance, we naturally would like to observe more data. In this case, we may wish to process the observations sequentially and make a final decision, i.e., terminate the testing as soon as all obtained observations fall in a decision region which can guarantee a required decision performance. Otherwise a new observation is needed and we continue the testing. Although the number of observations is not fixed at each test, the average number of observations can be smaller than that of the testing with fixed number of observations. This type of testings is called *Sequential Testing*.

In addition to the above hypothesis testings, there are some other types of hypothesis tests, for example, the composite hypothesis test. In this case, there are a family of conditional probability distributions given a source output. the composite hypothesis test deals with this type of decision problems. In this book, we focus only on the developments for Bayes test, Neyman-Pearson test, and the sequential test in multisensor case.

1.2 MULTISENSOR STATISTICAL DECISION FUSION SUMMARY

1.2.1 BRIEF INTRODUCTION TO MULTISENSOR DATA FUSION

In recent years, multisensor data fusion and information fusion have received significant attention for both military and non-military applications. Group decision-making structure are found in many real world situations. Data fusion techniques combine data from multiple sensors which observe a common phenomenon. They can achieve improved accuracies, system survivability, and more specific inferences than could be achieved by the use of a single sensor alone. The concept of multisensor data fusion is hardly new. Human and animals have evolved the capability to use multiple senses: sight, touch, smell, sound, and taste, to improve their ability to survive. Historically, data fusion methods were developed primarily for military applications. However, in recent years, these methods have been applied to civilian applications. Application areas are widespread. They include: automated target recognition, guidance for autonomous vehicles, battlefield surveillance, financial institutions, air traffic control, oil exploration, medical diagnosis, robotics, optical engineering, and security inspection systems.

Techniques to combine or fuse information are drawn from a diverse set of fields ranging from more traditional to more modern disciplines: digital signal processing, statistical decision and estimation, control theory, optimization theory and numerical methods, artificial intelligence, fuzzy set theory, evidence inference, neural networks, etc.

In a multisensor decision fusion system, there is one more basic component that should be added to Fig.1.1.1: receiver (see Fig. 1.2.1), which may consist

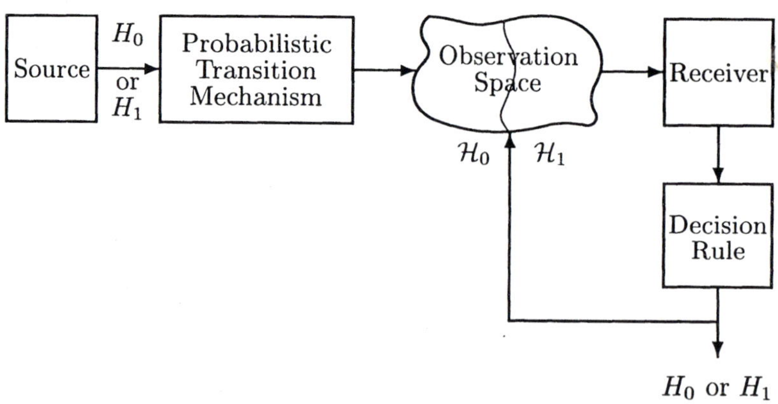

Figure 1.2.1. Basic model of multisensor statistical decision.

of geographically distributed or functionally different multiple sensors and a fusion center. The sensors make local decisions based on their own received information or just preprocess (compress) it for global optimization purpose and then transmit the processed information to the fusion center (for example, see Figs. 4.1.1–4.1.4). Besides, the decision rule in this case is usually called the fusion rule according that the fusion center fuse the all received information and make a final decision.

If there are no constraints on communication channel and processor bandwidth, complete observations can be transmitted to a central processor for data processing. In this case, all sensors act as simple data observers without any local data processing. Data processing performs only at the fusion center. The entire multisensor system can be viewed as a single sensor system in nature, and conventional optimal methods can be implemented. Such a multisensor system is called the *centralized system.* Clearly, the centralized system has the best performance since complete observations are used.

In many practical applications, in particular, when many sensors are dispersed over a wide geographic area, there may be limitations on the amount of communication among sensors and between sensors and the fusion center, or limitations on the amount of computations and data load. Besides, for local purpose or system survivability, one may require to know local decision or estimation first. In these cases, a preprocessing is performed at the individual sensors and a compressed version of sensor data is transmitted to the fusion center. Accordingly, we call this system the *decentralized system* or *distributed system.* So called data fusion primarily is performed for this system, which we will be concerned with in this book.

1.2.2 SOME BASIC ISSUES

To design a high quality multisensor data processing system, a variety of issues need to be examined. The most basic issues of them are:

I. How much favor people have on the various system objectives?

If the survivability of the system is the most important, or the local decisions/estimations are necessary for people, the sensors must make a locally optimal decision/estimation based on their own information and own criterion. Then the processed information in terms of the local criterion is transmitted to the fusion center to process finally. If the performance of the system is the most important, the sensors just compress their observations to appropriately long binary codes or quantized observations according to the limitations of communication bandwidth and computation capability, then the fusion center uses the received information compressed by the sensors to make final decision/estimation in terms of global optimization criterion. Obviously, the two different system objectives above determine two different optimization strategies. The former uses two-level optimization strategy while the latter uses global optimization only. Hence, in the latter case the compressions of observations at sensors are not determined by the local objectives and must be adjusted by the global optimization criterion. Since one can view the local sensor decisions as the observations of the fusion center in the two-level optimization case, the optimal fusion problem in this case can be dealt with as the conventional decision problem. In fact, local sensors can use their observations twice for the local and global purposes, respectively. They first make locally optimal decisions/estimation for their own purposes by using the conventional decision method. In the meantime, they compress observations in the global optimization sense, then send the compressed data to the fusion center for the global purpose. In this way, the local and global requirements are both fulfilled. Therefore, the globally optimal decision fusion problems are much more significant and most part of this book considers this type of multisensor decision problems.

II. How much communication bandwidth and computational capability for the system are available?

The answer to the issue determines the total amount of communications and computations of the system as well as how to optimally allocate the amount of communications among sensors and between each sensor and the fusion center. There has been little study for the latter so far, some study will be presented in this book.

III. What distributed decision/estimation topology and proper communication pattern are selected for the system network structure?

Indeed, this issue is related to the above two issues. It is a primary concern to design the system. The most popular one is the parallel network structure, where only communications happen between each sensor and the fusion center, no communication exists among sensors. We will indicate that for the global optimization purpose, in most cases, the parallel structure has the best performance under the same limitation of quantity of communication. Another frequently used network structure is the serial or tandem structure. In this system, the first sensor observes the phenomenon, processes the data, and transmits it to the next sensor. Based on the incoming processed information from the first sensor and its own observation, the second sensor does its data processing and transmits the processed information to the next sensor. This procedure continues until the final sensor of the tandem network which yields the global inference. For the global optimization purpose, this network structure may not be a good choice in most cases since the information received by the last sensor is compressed too much. However, for the system survivability, the tandem structure is a better choice than the parallel structure. In this system, the performance is getting better at the next sensor. If the fusion center fails, the remainder of the tandem network becomes a smaller tandem systems, while the parallel system becomes a number of single sensors. A hybrid network structure combining the parallel and tandem structures is the tree network structure. More generally, one may introduce feedback from the fusion center to individual sensors. The feedback gives the local sensor more information and improves the system performance. However, the drawbacks of feedback costs communication bandwidth and cause time delay of the final decision. In a word, one need to optimally select system networks and assign proper communication pattern under variety of practical constraints and requirements.

1.2.3 THE PREVIOUS STUDIES OF DECISION FUSION

Now we like to give a brief introduction to the previous studies of decision fusion. Although the earliest decision problem of group decision makers was studied about four decades ago [87], a pioneering work on the distributed decision that is referenced most frequently in the literature is Tenney and Sandell's paper in 1981 [77]. In their paper, using the minimum average cost criterion, optimal sensor rules to compress observation to a binary code under a given fusion rule was discussed. They found that in case of conditionally independent sensor observations, the optimal sensor rules are of the familiar likelihood ratio test. Although the set of optimal thresholds need to be found from all solutions of a group of coupled nonlinear equations, and the required computation is not simple in general, the form of optimal sensor rules is well known at least. Results are presented for two sensors, but are easily extended to the case of multiple sensors. If the sensor observations are dependent mutually, the above results are not true in general[1]. However, the assumption of conditionally independent sensor observations cannot be fulfilled in many practical applications. For example, if a signal to be detected is random, the sensor observations are certainly dependent mutually even if the sensor noises are independent of each other whenever the signal presents. While [77] produced necessary conditions for the single-bit sensor decision case, similar equations for multiple-bit sensor decision cases were provided in [43]. Then, there were many studies to extend Tenney and Sandell's results to a variety of hypothesis tests, such as the distributed Neyman-Pearson decision [42, 29], the distributed sequential decision [24, 47, 86], the distributed composite hypothesis test [27], the distribute minimax test [85]. Since there are various communication patterns among the sensors to choose, which is a novel issue not encountered by single sensor system, a variety of the multisensor network decision systems are discussed [75, 76, 80].

[1]Some special examples where the optimal sensor rules are still the likelihood ratio test will be given later in this chapter.

A very important work [82] on the computational complexity of distributed decision problems was contributed by Tsitsiklis and Athans in 1985.

In most of the literature to date, a common assumption is the independence of observations among sensors. Varshney's book [84] in 1997 collected and summarized well the previous primary research results. Problems with statistically dependent observations have received much less study, mainly because they have been shown to be difficult. It was explained in [77] that finding solutions to these equations is difficult, due to the coupling. Some progress was made for weak signal cases. Closed form expressions for the optimal sensor detectors for weak signal cases were found in [9]. A Gauss-Seidel approach was suggested in [73] to numerically find the functional forms of the sensor rules for cases where signals may not be weak. Discretizing the sensor optimization problem is not addressed in [73]. In [98], using the fixed point theory, for general observations a necessary condition for the optimal sensor rules, given a fusion rule, was found, i.e., the optimal sensor rules must be a fixed point of an integral operator that is dependent on the conditional probability densities and the given fusion rule. Motivated by this result, an efficient discretized Gauss-Seidel iterative algorithm to search for optimal sensor rules was proposed. The finite convergence of this algorithm was proved, and the discretized solutions approach the solutions of the original continuous algorithm as a step-size parameter goes to zero was also shown to be true. On the other hand, studies of fusion rules for distributed signal detection systems have also been appeared. The form of the optimal fusion rule for cases with independent observations was produced in [13]. Further studies of fusion rules for independent observations have also appeared [86, 92]. General studies of fusion rules for dependent observations have appeared less frequently, but a limited number of studies can be found [29, 45]. Still in [98], some new findings about optimal fusion rules was presented. In particular, we show some highly unexpected properties of fusion rules. We show that a specific fixed fusion rule is a unified version of general fusion rule, it depends only on the number of bits used in the sensor rule, and can be used to obtain overall optimal performance in some specific cases since any possible fusion rule is a special case of the unified fusion rule provided that some of the sensor rules are restricted to be special versions in advance. A specific example of these cases is that an l-sensor distributed binary decision system uses a total of $l-1$-bit decisions distributed over the first $l-1$ sensors, while the last sensor uses 2^{l-1} bits in its decision. This result gives an easily implementable practical solution for the case where the last sensor is also the fusion center, as in the tandem network decision system. Moreover, in these cases, the fusion rule does not depend on the statistical properties of the observational data. Of course, to get optimal performance under this fusion rule, the choice of the sensor rules does depend on the statistical properties of the observational data. Further, we show that, in these cases, increasing the number of bits used in the last sensor decision will not improve performance, even if the original observations themselves are sent from this sensor to the fusion center. Using this unified/optimal fusion rule, one just need to compute the optimal sensor rules to obtain overall optimal performance. These findings help us reduce huge computation. Many numerical results given in this book support the above statements. All of the above results can be extended to Neyman-Pearson testing and sequential testing, as well as multisensor multi-hypothesis network decision systems (parallel, tandem, and tree networks) [99]–[103], and form the basic framework of the decision fusion part of the book.

1.3 THREE CONVENTIONAL SINGLE SENSOR DECISIONS

In the multisensor decision problems there are many new issues. But the major optimization criteria are the same as those used for the conventional single sensor decision problem. They are the Bayes criterion, the Neyman-Pearson criterion, and the minimum average sampling size criterion for the sequential decisions. We will now discuss these in more detail.

1.3.1 BAYES DECISION

Let us consider the simple binary decision problem with the two hypotheses denoted by $\mathbf{H}_0$ and $\mathbf{H}_1$. For Bayes decision problem, the hypotheses $\mathbf{H}_0$ and $\mathbf{H}_1$ are assumed to occur randomly, and the prior probabilities P_0 and P_1 of the two possibilities are known. Let $\mathcal{Y}$ be the entire observation space, the observation be denoted by $\mathbf{y} \in \mathcal{Y}$, and the conditional densities under the two hypotheses be $p(\mathbf{y}|\mathbf{H}_i), i = 0, 1$. There is a cost associated with each decision. Let $C_{ij}, i = 0$, 1, $j = 0$, 1, represent the cost of declaring $\mathbf{H}_i$ true when $\mathbf{H}_j$ is actually true. Two of these correspond to correct decisions and two to incorrect decisions. Suppose that we declare $\mathbf{H}_i$ if $\mathbf{y} \in \mathcal{H}_i$, a decision region to be determined. Hence, for the deterministic decision rule, $\mathcal{H}_0 \cup \mathcal{H}_1 = \mathcal{Y}$ and $\mathcal{H}_0 \cap \mathcal{H}_1 = \emptyset$, the null set. In particular, the optimization problem is minimization of the expectation of the decision cost–the following cost functional:

$$\begin{aligned} C(d, H) &= \textstyle\sum_{i=0}^{1} \sum_{j=0}^{1} C_{ij} P(d = \mathbf{H}_i, H = \mathbf{H}_j) \\ &= \textstyle\sum_{i=0}^{1} \sum_{j=0}^{1} C_{ij} P_j P(d = \mathbf{H}_i | \mathbf{H}_j) \\ &= \textstyle\sum_{i=0}^{1} \sum_{j=0}^{1} C_{ij} P_j \displaystyle\int_{\mathcal{H}_i} p(\mathbf{y}|\mathbf{H}_j) d\mathbf{y}, \end{aligned} \tag{1.3.1}$$

where d is the decision variable, and H is the hypothesis variable. We can expand (1.3.1) as

$$\begin{aligned} C(d,H) = {} & P_0C_{00}\int_{\mathcal{H}_0} p(\mathbf{y}|\mathbf{H}_0)d\mathbf{y} + P_0C_{10}\int_{\mathcal{Y}-\mathcal{H}_0} p(\mathbf{y}|\mathbf{H}_0)d\mathbf{y} \\ & + P_1C_{01}\int_{\mathcal{H}_0} p(\mathbf{y}|\mathbf{H}_1)d\mathbf{y} + P_0C_{11}\int_{\mathcal{Y}-\mathcal{H}_0} p(\mathbf{y}|\mathbf{H}_1)d\mathbf{y}. \end{aligned} \tag{1.3.2}$$

Noting that

$$\int_{\mathcal{Y}} p(\mathbf{y}|\mathbf{H}_i)d\mathbf{y} = 1, \quad i = 0,1$$

and collecting terms, (1.3.2) reduces to

$$\begin{aligned} C(d,H) = {} & P_0C_{10} + P_1C_{11} \\ & + \int_{\mathcal{H}_0} [P_1(C_{01} - C_{11})p(\mathbf{y}|\mathbf{H}_1) - P_0(C_{10} - C_{00})p(\mathbf{y}|\mathbf{H}_0)]d\mathbf{y}. \end{aligned} \tag{1.3.3}$$

The first two terms on the right hand side are constant. Therefore, the cost can be minimized by letting $\mathcal{H}_0$ be the set of the points of $\mathcal{Y}$ for which the integrand of (1.3.3) is non-positive. Assuming that $C_{01} > C_{11}$ and $C_{10} > C_{00}$, the decision rule which minimizes the cost is defined by the likelihood ratio test (LRT)

$$\frac{p(\mathbf{y}|\mathbf{H}_1)}{p(\mathbf{y}|\mathbf{H}_0)} \underset{d=\mathbf{H}_0}{\overset{d=\mathbf{H}_1}{\gtrless}} \frac{P_0(C_{10} - C_{00})}{P_1(C_{01} - C_{11})}. \tag{1.3.4}$$

The quantity on the left hand side of (1.3.4) is called the likelihood ratio and is denoted by $LR(\mathbf{y})$. The right hand side is called the threshold of the test and is denoted by η, i.e.,

$$LR(\mathbf{y}) \underset{d=\mathbf{H}_0}{\overset{d=\mathbf{H}_1}{\gtrless}} \eta. \tag{1.3.5}$$

Since the natural logarithm is a monotonically increasing function and the two sides of (1.3.5) are positive, an equivalent decision rule is

$$\log LR(\mathbf{y}) \underset{d=\mathbf{H}_0}{\overset{d=\mathbf{H}_1}{\gtrless}} \log \eta. \tag{1.3.6}$$

When the cost coefficients are hardly determined, commonly, one sets $C_{00} = C_{11} = 0$ and $C_{01} = C_{10} = 1$. Then, the cost of a correct decision is zero and the cost of an error is unity. In this case,

$$\begin{aligned} C(d,H) &= P(d = \mathbf{H}_0, H = \mathbf{H}_1) + P(d = \mathbf{H}_1, H = \mathbf{H}_0) \\ &= P_0\int_{\mathcal{H}_1} p(\mathbf{y}|\mathbf{H}_0)d\mathbf{y} + P_1\int_{\mathcal{H}_0} p(\mathbf{y}|\mathbf{H}_1)d\mathbf{y} \\ &= P(\text{decision error}). \end{aligned} \tag{1.3.7}$$

The Bayes cost now simply reduces to the probability of error in the decision. The two types of errors ($d = \mathbf{H}_1, H = \mathbf{H}_0$) and ($d = \mathbf{H}_0, H = \mathbf{H}_1$) in decision making are called the error of the first kind and of the second kind, respectively. In radar terminology, they are known as a false alarm and a miss, respectively. The associated two conditional probabilities $P(d = \mathbf{H}_0|\mathbf{H}_1)$ and $P(d = \mathbf{H}_1|\mathbf{H}_0)$ are called as probability of a false alarm and of a miss, respectively. Then,

$$P_f = P(d = \mathbf{H}_1|\mathbf{H}_0) = \int_{\mathcal{H}_1} p(\mathbf{y}|\mathbf{H}_0)d\mathbf{y}, \tag{1.3.8}$$

and

$$P_m = P(d = \mathbf{H}_0|\mathbf{H}_1) = \int_{\mathcal{H}_0} p(\mathbf{y}|\mathbf{H}_1)d\mathbf{y}. \tag{1.3.9}$$

We also define the probability of detection P_d as follows:

$$P_d = P(d = \mathbf{H}_1|\mathbf{H}_1) = \int_{\mathcal{H}_1} p(\mathbf{y}|\mathbf{H}_1)d\mathbf{y} = 1 - P_m. \tag{1.3.10}$$

1.3.2 NEYMAN-PEARSON DECISION

In many physical situations it is hard to assign realistic costs or a priori probabilities as required by the Bayes criterion. A simple procedure to bypass this difficulty is to work with the conditional probabilities P_f and P_d (or P_m). In general, we would like to make P_f as small as possible and P_d as large as possible. For most problems of practical importance these are conflicting objectives. Thus, a reasonable criterion is to constrain one of the probabilities to an accepted value and then maximize (or minimize) other. This is formalized as follows.

Neyman-Pearson Criterion. Given $\alpha > 0$, constrain $P_f \leq \alpha$ and design a test to maximize P_d (or to minimize P_m) under this constraint.

The resulting test is called Neyman-Pearson test.

A standard method of solution uses Lagrange multipliers. Letting λ denote the multiplier, the augmented objective function J is defined as follows:

$$\begin{aligned} J &= P_m + \lambda(P_f - \alpha) \\ &= \int_{\mathcal{H}_0} p(\mathbf{y}|\mathbf{H}_1)d\mathbf{y} + \lambda[\int_{\mathcal{H}_1} p(\mathbf{y}|\mathbf{H}_0)d\mathbf{y} - \alpha] \\ &= \int_{\mathcal{H}_0} p(\mathbf{y}|\mathbf{H}_1)d\mathbf{y} + \lambda[1 - \int_{\mathcal{H}_0} p(\mathbf{y}|\mathbf{H}_0)d\mathbf{y} - \alpha] \\ &= \lambda(1-\alpha) + \int_{\mathcal{H}_0} [p(\mathbf{y}|\mathbf{H}_1) - \lambda p(\mathbf{y}|\mathbf{H}_0)]d\mathbf{y}. \end{aligned} \tag{1.3.11}$$

Clearly, when $\lambda \le 0$, taking $\mathcal{H}_0 = \emptyset$ makes J minimum since the integrand is non-negative. Therefore, we only need to consider positive λ. For fixed λ, the first term in the right hand side of (1.3.11) is constant and the second term is minimized if we employ the following likelihood ratio test:

$$LR(\mathbf{y}) = \frac{p(\mathbf{y}|\mathbf{H}_1)}{p(\mathbf{y}|\mathbf{H}_0)} \underset{d\,=\,\mathbf{H}_0}{\overset{d\,=\,\mathbf{H}_1}{\gtrless}} \lambda. \tag{1.3.12}$$

In the deterministic decision, to obtain the largest P_d under the constraint $P_f \le \alpha$, the optimal value of the Lagrange multiplier λ for the test is obviously chosen to be

$$\min\{\lambda : \int_{\mathcal{H}_1(\lambda)} p(\mathbf{y}|\mathbf{H}_0)d\mathbf{y} \le \alpha\}. \tag{1.3.13}$$

If for the λ^* given by (1.3.13),

$$P(d = \mathbf{H}_1|\mathbf{H}_0) = P(LR(\mathbf{y}) > \lambda^*|\mathbf{H}_0) < \alpha, \tag{1.3.14}$$

and

$$P(LR(\mathbf{y}) \ge \lambda^*|\mathbf{H}_0) > \alpha,$$

this obviously implies $P(LR(\mathbf{y}) = \lambda^*|\mathbf{H}_0) \ne 0$. Then P_f is not able to attain the level α. A more general formulation of the Neyman-Pearson test allows a randomized decision rule that can solve this problem. This decision rule is defined as

$$\begin{aligned} &P(d = \mathbf{H}_1|LR(\mathbf{y}) < \lambda^*, \mathbf{H}_i) = 0, \\ &P(d = \mathbf{H}_1|LR(\mathbf{y}) = \lambda^*, \mathbf{H}_i) = \gamma, \quad i = 0, 1, \\ &P(d = \mathbf{H}_1|LR(\mathbf{y}) > \lambda^*, \mathbf{H}_i) = 1, \end{aligned} \tag{1.3.15}$$

where

$$\gamma = \frac{\alpha - P(LR(\mathbf{y}) > \lambda^*|\mathbf{H}_0)}{P(LR(\mathbf{y}) = \lambda^*|\mathbf{H}_0)}.$$

Clearly, the decision rule (1.3.15) is randomized on the set $\{\mathbf{y} : LR(\mathbf{y}) = \lambda\}$. It is easy to verify that under this decision rule,

$$\begin{aligned} &P(d = \mathbf{H}_1|\mathbf{H}_0) \\ &= P(d = \mathbf{H}_1, LR(\mathbf{y}) > \lambda^*|\mathbf{H}_0) + P(d = \mathbf{H}_1, LR(\mathbf{y}) = \lambda^*|\mathbf{H}_0) \\ &= P(d = \mathbf{H}_1|LR(\mathbf{y}) > \lambda^*, \mathbf{H}_0)P(LR(\mathbf{y}) > \lambda^*|\mathbf{H}_0) + \gamma P(LR(\mathbf{y}) = \lambda^*|\mathbf{H}_0) \\ &= \alpha. \end{aligned} \tag{1.3.16}$$

In the meantime, the detection probability P_d given by the randomized rule (1.3.15), denoted by $P_d(1.3.15)$, is larger than P_d given by the deterministic rule, denoted by $P_d(1.3.14)$, since by Eqs. (1.3.15) and (1.3.14)

$$\begin{aligned} P_d(1.3.15) &= P(d = \mathbf{H}_1, LR(\mathbf{y}) = \lambda^*|\mathbf{H}_1) + P(d = \mathbf{H}_1, LR(\mathbf{y}) = \lambda^*|\mathbf{H}_1) \\ &= P(LR(\mathbf{y}) > \lambda^*|\mathbf{H}_1) + \gamma P(LR(\mathbf{y}) = \lambda^*|\mathbf{H}_1) \\ &\geq P(LR(\mathbf{y}) > \lambda^*|\mathbf{H}_1) = P_d(1.3.14). \end{aligned} \tag{1.3.17}$$

1.3.3 SEQUENTIAL DECISION

The Neyman-Pearson test started with the constraint $P_f \leq \alpha$. In many practical situations, we might also want $P_m \leq \beta$ for some given $\beta > 0$. However, with the given sample size, the Neyman-Pearson test may yield $P_f \leq \alpha$ and $P_m > \beta$. To reach our goal, the sample size need to be increased. If we employ a test with a fixed large sample size, $P_f \leq \alpha$ and $P_m \leq \beta$ may be satisfied simultaneously. But the required sample size might be excessively large. If data are collected sequentially, then more information becomes available as time goes on, we may wish to process the data sequentially and make a final decision as soon as possible without exceeding the maximum allowable error probabilities α and β. To solve the above problem, Wald developed the Sequential Probability Ratio Test (SPRT) half a century ago (cf. [90, 91]). In the sequential decision process, the sample size is not fixed with probability one in advance. At each stage, namely the kth stage, the kth observation space is After each observation, the decision maker checks in which one of the three regions: $\mathbf{H}_0$ decision region $\mathcal{H}_0$, $\mathbf{H}_1$ decision region $\mathcal{H}_1$ and the continuing observation region $\mathcal{C}$ the observed data are. The decision maker either decides for one of the two hypotheses or decides to take another observation. The main advantage of sequential decision is that it requires, on an average, fewer observations to satisfy $P_f \leq \alpha$ and $P_m \leq \beta$ than the fixed-sample-size method.

1.3.3.1 WALD'S SPRT

Wald's SPRT compares the likelihood ratio at each time with two time-invariant thresholds (i.e., stopping bounds) η_0 and η_1 so as to decide on the corresponding hypothesis or to collect additional data (for details, see Eq. (1.3.19) below). For any available specific pair (η_0, η_1) with the associated two actual decision error probabilities $P_{f(SPRT)}$ and $P_{m(SPRT)}$ given by (1.3.19), Wald's SPRT is optimal in the sense that the two conditional expectations $E(T|\mathbf{H}_0)$ and $E(T|\mathbf{H}_1)$ of the stopping time T are both minimum with respect to all sequential tests with the same two error probabilities $P_{f(SPRT)}$ and $P_{m(SPRT)}$ as Wald's SPRT [35, 91].

Let $\mathbf{y}_1, \mathbf{y}_2, \ldots, \mathbf{y}_k, \ldots$ represent the sequence of independent observations under each hypothesis, where $\mathbf{y}_k \in \Re^n$ denotes the kth observation received sequentially. Denote

$$Y_k = (\mathbf{y}_1, \mathbf{y}_2, \ldots, \mathbf{y}_k).$$

The (joint) likelihood ratio at time k is defined as

$$LR(Y_k) = \frac{p(Y_k|\mathbf{H}_1)}{p(Y_k|\mathbf{H}_0)}. \tag{1.3.18}$$

Wald's SPRT is a sequential likelihood ratio test, given by

$$\begin{cases} \text{If } LR(Y_k) \geq \eta_1, \text{ decide } \mathbf{H}_1; \\ \text{if } LR(Y_k) \leq \eta_0, \text{ decide } \mathbf{H}_0; \\ \text{otherwise, take the } (k+1)\text{th observation } \mathbf{y}_{k+1}. \end{cases} \tag{1.3.19}$$

Thus, the stopping time for Wald's SPRT is defined as

$$\begin{aligned} T &= \min\{k : LR(Y_k) \notin (\eta_0, \eta_1)\} \\ &= \infty, LR(Y_k) \in (\eta_0, \eta_1) \quad \forall k. \end{aligned} \tag{1.3.20}$$

The next step is to compute the two thresholds η_0 and η_1 in terms of the specified values of P_f and P_m. For notational simplicity, assume that there exist η_0 and η_1 such that

$$P_f = P(LR(Y_T) \geq \eta_1|\mathbf{H}_0) = \alpha \tag{1.3.21}$$

and

$$P_m = P(LR(Y_T) \leq \eta_0|\mathbf{H}_1) = \beta. \tag{1.3.22}$$

It is still difficult in general to find the values of η_0 and η_1 so that (1.3.21) and (1.3.22) hold exactly. Hence, Wald proposed the following approximate solution.

Firstly, under a regularity condition he proved that

$$P(T < \infty|\mathbf{H}_i) = 1, \quad i = 0, 1. \tag{1.3.23}$$

Define the sequence of sets $\{B_k\}$ by

$$B_k = \{Y_k : LR(Y_i) \in (\eta_0, \eta_1), i \leq k-1; \quad LR(Y_k) \geq \eta_1\}. \tag{1.3.24}$$

Using (1.3.20), (1.3.21), (1.3.23) and (1.3.24), we have

$$\begin{aligned} \alpha &= \textstyle\sum_{k=1}^{\infty} \int_{B_k} p(Y_k|\mathbf{H}_0)dY_k \\ &= \textstyle\sum_{k=1}^{\infty} \int_{B_k} LR^{-1}(Y_k)p(Y_k|\mathbf{H}_1)dY_k \\ &\leq \frac{1}{\eta_1} \textstyle\sum_{k=1}^{\infty} \int_{B_k} p(Y_k|\mathbf{H}_1)dY_k \\ &= \frac{1}{\eta_1} P_d = \frac{1-\beta}{\eta_1}. \end{aligned} \tag{1.3.25}$$

Similarly,

$$\beta \leq \eta_0(1-\alpha). \tag{1.3.26}$$

The above two inequalities cannot be equalities because it is impossible in general for $LR(Y_k)$ to equal exactly η_1 (or η_0) with probability one when it falls outside the interval (η_0, η_1), by the definition of B_k in Eq. (1.3.24). To get η_0 and η_1, Wald used the following two approximations

$$\eta_0 \approx \frac{\beta}{1-\alpha}, \quad \eta_1 \approx \frac{1-\beta}{\alpha}. \tag{1.3.27}$$

Based on these approximations, the approximations of $E(T|\mathbf{H}_0)$ and $E(T|\mathbf{H}_1)$ are dependent only on α and β and the two conditional expectations of the data. In particular,

$$E(T|\mathbf{H}_0) \approx \frac{\alpha \log \frac{1-\beta}{\alpha} + (1-\alpha) \log \frac{\beta}{1-\alpha}}{E(\log LR(\mathbf{y}_1)|\mathbf{H}_0)}, \tag{1.3.28}$$

$$E(T|\mathbf{H}_1) \approx \frac{(1-\beta) \log \frac{1-\beta}{\alpha} + \beta \log \frac{\beta}{1-\alpha}}{E(\log LR(\mathbf{y}_1)|\mathbf{H}_0)}. \tag{1.3.29}$$

These attractive features make Wald's SPRT very convenient and easy to implement in practice. However, this approximation is too crude and the resulting errors are not insignificant for applications, i.e., the approximate constant stopping bounds $\eta_0 = \beta/(1-\alpha)$ and $\eta_1 = (1-\beta)/\alpha$ may be too conservative in the sense that $\alpha - P_f$ and/or $\beta - P_m$ is not necessarily very small, depending on the distributions involved. For instance, in [20], pp. 348, an example of Wald's SPRT is presented where the actual error probability P_f is significantly less than its maximum allowable value α. Besides, although Wald's SPRT has a finite stopping time T (i.e., it will terminate) with probability one, for a specific sample, T could be very large. This is a major criticism of Wald's SPRT. In applications, it would be much more desirable to have a sequential procedure with a bounded stopping time. Obviously, these drawbacks are due to the fact that the method does not use the given conditional probability densities.

To remove these weak points in Wald's SPRT, we propose a new method of SPRT-type sequential decision procedure, maintaining the Neyman-Pearson framework of controlling decision error probabilities. Its underlying idea differs from those of existing methods, such as [35, 93]. Instead of designing stopping bounds, the new method uses the conditional probability densities and focuses on properly allocating the type I (P_f) and type II (P_m) error probabilities for each time so that after a finite time L the continuing observation region becomes empty and the two total error probabilities P_f and P_m are guaranteed to approximate their maximum allowable values α and β, respectively. The procedure is based on uniform error allocation. Although this allocation is not optimal, it has three major advantages: (a) under mild conditions we give a constructive proof that the procedure has a finite and computable upper bound on the stopping time; (b) we also prove that the procedure is capable of making good use of the allowable error probabilities — its actual error probabilities can be arbitrarily close to (but not above) their maximum allowable values; and (c) we demonstrate via examples that the new procedure has smaller expected stopping times than the Wald's SPRT as well as and has the popular constant stopping bounds. At each stage, the new method needs to control two given error decision probabilities. It does reduce to two Neyman-Pearson decisions on the product space of the last continuing observation region of Y_{k-1} and the current observation space of $\mathbf{y_k}$. Therefore, we can easily derive the distributed Neyman-Pearson type sequential decision rule whenever we have a multisensor distributed Neyman-Pearson decision.

The proposed method allocates the maximum allowable error probabilities for each time *uniformly* in the sense that for some given M, N,

$$\max P(\text{decide } \mathbf{H}_1 \text{ at the } k\text{th stage}|\mathbf{H}_0) := \Delta\alpha \triangleq \alpha/N, \quad \forall k \leq N,$$

and

$$\max P(\text{decide } \mathbf{H}_0 \text{ at the } j\text{th stage}|\mathbf{H}_1) := \Delta\beta \triangleq \beta/M, \quad \forall j \leq M.$$

Now the crucial question is: is there a pair $\{M, N\}$ and reasonable regularity conditions such that at a finite time $L = \max\{M, N\}$ the continuing observation region becomes empty? If the answer is affirmative, then we can find $\min L$ from all the $\{M, N\}$ pairs satisfying the above constraints. Indeed, we will give a constructive proof of the existence of such finite $\{M, N\}$ pairs under mild conditions on the likelihood ratio and α, β. More specifically, we propose an algorithm for finding such pairs and rigorously prove that the algorithm is guaranteed to find a finite pair. We also suggest an algorithm for finding $\min L$. As a result, the proposed SPRT-type procedure has a computable finite upper bound on its stopping time. Furthermore, we prove that both error probabilities of the procedure can be made arbitrarily close to (but not above) their maximum allowable values simultaneously.

A variety of constant and variable stopping bounds (open intervals $(\eta_0(k), \eta_1(k))$ have been proposed (see, e.g., [35, 93]). A key issue here is how to account for the relevant information in the conditional distributions. The proposed method of error allocation provides an appealing means of determining proper stopping bounds that are sophisticated enough to make good use of the distributional information, particularly if a bounded stopping time is desired. For many commonly encountered decision problems, the proposed simple procedure based on uniform error allocation takes better advantage of the allowable error probabilities and thus has smaller conditional expected stopping times than the SPRT with either the constant stopping bounds suggested by Wald or the improved approximation by Siegmund [71].

There are four possibilities for a sequential procedure with an upper bound $L = \max\{M, N\}$ on its stopping time: (A) $P_f \leq \alpha$ and $P_m \leq \beta$ are guaranteed and $L < \infty$; (B) $L < \infty$ but $P_f \leq \alpha$ and $P_m \leq \beta$ are not guaranteed; (C) $P_f \leq \alpha$ and $P_m \leq \beta$ are guaranteed but $L = \infty$ (i.e., no bound on the stopping time); and (D) $P_f \leq \alpha$ and $P_m \leq \beta$ are not guaranteed and $L = \infty$. A procedure in class D is of no interest unless it possesses some other attractive properties. Wald's SPRT belongs to class C. All procedures with a bounded stopping time previously proposed, such as a truncated SPRT, belong to class B. We are mainly interested in class A. In fact, *the most important contribution* here is a constructive proof of the existence of class A.

Numerical examples are provided, which compare Wald's SPRT and Siegmund's modified SPRT [71] with our proposed procedure. The results show that our procedure not only has a finite upper bound on the stopping time, but also outperforms the other two SPRTs in terms of the average stopping time.

For simplicity of presentation, we assume throughout the section that the distributions involved are absolutely continuous with respect to Lebesgue measure (i.e., the probabilities density functions exist). Otherwise, the method has to be modified to handle the possible discrete points.

1.3.3.2 SEQUENTIAL DECISION BASED ON ALLOCATING ERROR PROBABILITIES

Since $\mathbf{y}_1, \mathbf{y}_2, \ldots, \mathbf{y}_k, \ldots$ are independent observations under each hypothesis, the (joint) likelihood ratio at time k can be written as

$$\begin{aligned} LR(Y_k) &= \frac{p(Y_k|\mathbf{H}_1)}{p(Y_k|\mathbf{H}_0)} = \prod_{i=1}^{k} \frac{p(\mathbf{y}_i|\mathbf{H}_1)}{p(\mathbf{y}_i|\mathbf{H}_0)} = \frac{p(\mathbf{y}_k|\mathbf{H}_1)}{p(\mathbf{y}_k|\mathbf{H}_0)} \prod_{i=1}^{k-1} \frac{p(\mathbf{y}_i|\mathbf{H}_1)}{p(\mathbf{y}_i|\mathbf{H}_0)} \\ &= LR(\mathbf{y}_k) LR(Y_{k-1}), \end{aligned} \tag{1.3.30}$$

where $LR(\mathbf{y}_k)$ is the marginal likelihood ratio. Denote the continuing observation region and the $\mathbf{H}_0$ and $\mathbf{H}_1$ decision regions at the kth time by $\mathcal{C}(k)$, $\mathcal{H}_0(k)$, and $\mathcal{H}_1(k)$, respectively. The general sequential decision rule is formulated by

$$\begin{cases} \text{If } Y_k \in \mathcal{H}_1(k), \text{ decide } \mathbf{H}_1; \\ \text{if } Y_k \in \mathcal{H}_0(k), \text{ decide } \mathbf{H}_0; \\ \text{otherwise, i.e., } Y_k \in \mathcal{C}(k), \text{ take the } (k+1)\text{th observation } \mathbf{y}_{k+1}. \end{cases} \tag{1.3.31}$$

The stopping time is

$$T = \begin{cases} \min\{k : Y_k \in \mathcal{H}_0(k) \quad \text{or} \quad Y_k \in \mathcal{H}_1(k)\}, \\ \infty, \quad \text{if all } Y_k \in \mathcal{C}(k). \end{cases} \tag{1.3.32}$$

To overcome the limitations of Wald's SPRT, most existing efforts are directed toward the design of variable stopping bounds $(\eta_0(k), \eta_1(k))$ directly so that for some K, $\eta_0(K) = \eta_1(K)$. Although it is relatively easy to guarantee a bounded stopping time with this approach, it is difficult to effectively control and manage the associated decision error probabilities.

The method that we are proposing here uses a different approach. Since controlling error probabilities is part of the goal and the stopping bounds merely serve as a means for achieving the goal, why don't we directly design a proper allocation of the maximum allowable error probabilities for each time? Once such an allocation is fixed, the corresponding stopping bounds can be easily (at least in principle) be determined by invoking the Neyman-Pearson lemma, which is capable of utilizing the distribution information optimally. Clearly, our uniform allocation of the error probabilities is not an optimal allocation. However, the numerical examples given below show that the new method outperforms Wald's SPRT very often.

We now provide a general description of this method of error allocation for the aforementioned Class A. The fundamental tasks are to find a finite upper bound L for the stopping time and to apportion/allocate the error probabilities; that is, design two sequences $\Delta\alpha(k) \geq 0$ and $\Delta\beta(k) \geq 0$ of maximum allowable error probabilities for each time k such that

$$\sum_{k=1}^{L} \Delta\alpha(k) \approx \alpha, \qquad \sum_{k=1}^{L} \Delta\beta(k) \approx \beta, \tag{1.3.33}$$

where

$$\begin{aligned} e_1(k) &:= P(Y_{k-1} \in \mathcal{C}(k-1), Y_k \in \mathcal{H}_1(k)|\mathbf{H}_0) \leq \Delta\alpha(k), \\ e_2(k) &:= P(Y_{k-1} \in \mathcal{C}(k-1), Y_k \in \mathcal{H}_0(k)|\mathbf{H}_1) \leq \Delta\beta(k), \end{aligned} \quad \forall k \leq L. \tag{1.3.34}$$

Clearly, it is required that $\mathcal{C}(L) = \emptyset$.

The inequalities in (1.3.34), rather than equalities, are due to the possibility that the region at time k is not large enough to have both $\Delta\alpha(k) = e_1(k)$ and $\Delta\beta(k) = e_2(k)$ simultaneously.

Assume now that k is a time at which the region is large enough to have both $\Delta\alpha(k) = e_1(k)$ and $\Delta\beta(k) = e_2(k)$. Then, the two decision thresholds $\eta_0(k)$ and $\eta_1(k)$ can be determined from

$$\begin{aligned} \Delta\alpha(k) &= P(Y_{k-1} \in \mathcal{C}(k-1), Y_k \in \mathcal{H}_1(k)|\mathbf{H}_0), \\ \Delta\beta(k) &= P(Y_{k-1} \in \mathcal{C}(k-1), Y_k \in \mathcal{H}_0(k)|\mathbf{H}_1) \end{aligned} \tag{1.3.35}$$

by the Neyman-Pearson lemma, based on the joint likelihood ratio $LR(Y_k)$. Specifically, $\eta_0(k)$ and $\eta_1(k)$ are the solutions of the following equations, respectively,

$$\begin{aligned}\int_{Y_{k-1}\in\mathcal{C}(k-1),LR(Y_k)\geq\eta_1(k)} p(Y_k|\mathbf{H}_0)dY_k &= \Delta\alpha(k),\\ \int_{Y_{k-1}\in\mathcal{C}(k-1),LR(Y_k)\leq\eta_0(k)} p(Y_k|\mathbf{H}_1)dY_k &= \Delta\beta(k).\end{aligned} \tag{1.3.36}$$

The stopping bounds $\eta_0(k)$ and $\eta_1(k)$ so determined are optimal if the allocation of the error probabilities is optimal. However, these two equations cannot be solved easily in general. Of course, we may have recourse to numerical solutions, which, however, requires heavy computation if k is large.

To reduce computation, we may use the suboptimal stopping bounds determined by the marginal likelihood ratios; that is, the solutions of the following equations:

$$\begin{aligned}\int_{LR(\mathbf{y}_k)\geq\eta_1(k)} p(\mathbf{y}_k|\mathbf{H}_0)d\mathbf{y}_k &= \Delta\alpha(k)/P(\mathcal{C}(k-1)|\mathbf{H}_0),\\ \int_{LR(\mathbf{y}_k)\leq\eta_0(k)} p(\mathbf{y}_k|\mathbf{H}_1)d\mathbf{y}_k &= \Delta\beta(k)/P(\mathcal{C}(k-1)|\mathbf{H}_1),\end{aligned} \tag{1.3.37}$$

where

$$P(\mathcal{C}(k)|\mathbf{H}_0) = P(Y_k \in \mathcal{C}(k)|\mathbf{H}_0), \quad P(\mathcal{C}(k)|\mathbf{H}_1) = P(Y_k \in \mathcal{C}(k)|\mathbf{H}_1). \tag{1.3.38}$$

Both of these two equations involve an n-fold integral, where n is the dimension of $\mathbf{y}_k$. They can be solved numerically.

Three related key issues with this method are: (a) is there a finite upper bound L on the stopping time? (b) how do we find such a bound if it exists? (c) how do we design a good error allocation scheme? In the next two sections, we consider a particular error allocation scheme that is among the simplest, show the existence of finite bounds on its stopping time, and present an algorithm for finding such bounds.

1.3.3.3 A NEYMAN-PEARSON PROCEDURE WITH UNIFORM ERROR ALLOCATION

The basic idea of a simple implementation of the above new method is to allocate the maximum allowable error probabilities *uniformly* over all time; that is, for some M, N, let $\Delta\alpha(k) = \Delta\alpha \triangleq \frac{\alpha}{N}$ and $\Delta\beta(k) = \Delta\beta \triangleq \frac{\beta}{M}$. In actual implementation, it turns out that this idea has to be adjusted slightly. Specifically, we use

$$P(Y_{k-1} \in \mathcal{C}(k-1), Y_k \in \mathcal{H}_1(k)|\mathbf{H}_0) = \Delta\alpha, \quad \forall k < N, \tag{1.3.39}$$

$$P(Y_{k-1} \in \mathcal{C}(k-1), Y_k \in \mathcal{H}_0(k)|\mathbf{H}_1) = \Delta\beta, \quad \forall k < M, \tag{1.3.40}$$

to determine the regions $\mathcal{H}_1(k)$, $\mathcal{H}_0(k)$, and $\mathcal{C}(k)$ by the Neyman-Pearson criterion, which form a partition of the region $\mathcal{C}(k-1) \times \Re^n$. At the final time $L = \max\{N, M\}$, we require that

$$P(Y_{L-1} \in \mathcal{C}(L-1), Y_L \in \mathcal{H}_1(L)|\mathbf{H}_0) = e_1 \leq \Delta\alpha, \tag{1.3.41}$$

$$P(Y_{L-1} \in \mathcal{C}(L-1), Y_L \in \mathcal{H}_0(L)|\mathbf{H}_1) = e_2 \leq \Delta\beta, \tag{1.3.42}$$

$$\mathcal{C}(L) = \emptyset, \tag{1.3.43}$$

simultaneously, where the inequalities, rather than equalities, hold generally because there is no guarantee at the final time L that the final decision region $\mathcal{C}(L-1) \times \Re^n$ can be partitioned into $\mathcal{H}_1(L)$ and $\mathcal{H}_0(L)$ exactly with allocated decision error probabilities $\Delta\alpha$ and $\Delta\beta$, respectively. Clearly, M and N can be different. Without loss of generality, suppose that $L = M \geq N$. Then, this implies that at each time k, $N \leq k < M$, we partition $\mathcal{C}(k) \times \Re^n$ into two disjoint parts $\mathcal{H}_0(k)$ and $\mathcal{C}(k)$. In other words, we do not decide on $\mathbf{H}_1$ ($\mathcal{H}_1(k) = \emptyset$).

Note that the actual error probabilities associated with this procedure are $P_f = (N-1)\Delta\alpha + e_1$ and $P_m = (M-1)\Delta\beta + e_2$ if $L = \max\{M, N\}$ is indeed the first time the continuing observation region becomes empty. They are bounded by $\alpha - \Delta\alpha \leq P_f \leq \alpha$ and $\beta - \Delta\beta \leq P_m \leq \beta$. However, by a simple adjustment of the procedure, it is shown below (see Corollary 1 of Theorem 1) that P_f and P_m can be made arbitrarily close to α and β, respectively: $\alpha - \epsilon_1 \leq P_f \leq \alpha$ and $\beta - \epsilon_2 \leq P_m \leq \beta$ for any $\epsilon_1, \epsilon_2 > 0$.

Evidently, the key issue here is the existence of a pair of finite $\{N, M\}$ such that (1.3.39)–(1.3.43) hold, and how to find such a pair.

Assume that the marginal, instead of joint, likelihood ratios are used to determine the stopping bounds. Then, in order to have a finite bound L on the stopping time and guaranteed error probabilities, the quantities $\Delta\alpha$, $\Delta\beta$, L, and the stopping bounds $\{\eta_0(1), \ldots, \eta_0(L)\}$ and $\{\eta_1(1), \ldots, \eta_1(L)\}$ have to satisfy

$$\begin{aligned} P_f &\leq \sum_{k=1}^{N} \int_{\mathcal{C}(k-1)} p(Y_{k-1}|\mathbf{H}_0) dY_{k-1} \int_{LR(\mathbf{y}_k) \geq \eta_1(k)} p(\mathbf{y}_k|\mathbf{H}_0) d\mathbf{y}_k \\ &= (N-1)\Delta\alpha + \epsilon_1 \leq \alpha, \end{aligned} \tag{1.3.44}$$

$$\begin{aligned} P_m &\leq \sum_{k=1}^{M} \int_{\mathcal{C}(k-1)} p(Y_{k-1}|\mathbf{H}_1) dY_{k-1} \int_{LR(\mathbf{y}_k) \leq \eta_0(k)} p(\mathbf{y}_k|\mathbf{H}_0) d\mathbf{y}_k \\ &= (M-1)\Delta\beta + \epsilon_2 \leq \beta, \end{aligned} \tag{1.3.45}$$

and

$$\mathcal{C}(L) = \emptyset, \tag{1.3.46}$$

where $0 < \epsilon_1 \leq \Delta\alpha$ and $0 < \epsilon_2 \leq \Delta\beta$. Of course, using the computationally much more demanding joint likelihood ratios in general yields better results than those for the marginal likelihood ratios. Note that the possible occurrences of case **I** or **II**, presented at the end of this section, can yield $P_f \neq (N-1)\Delta\alpha+\epsilon_1$ or $P_m \neq (M-1)\Delta\beta+\epsilon_2$.

Next we show how to find $\{M, N\}$ or $\{\Delta\alpha, \Delta\beta\}$ so that Eqs. (1.3.44)–(1.3.46) hold simultaneously. Then in the next section, we prove the existence of finite $\{M, N\}$.

We can use any guess for the initial value of $\{M, N\}$. Correspondingly, we have an initial pair of initial $\{\Delta\alpha, \Delta\beta\}$. Given $\{M, N\}$ (and thus $\{\Delta\alpha, \Delta\beta\}$), we can determine the stopping bounds $\{\eta_0(1), \ldots, \eta_0(M)\}$ and $\{\eta_1(1), \ldots, \eta_1(N)\}$ by Neyman-Pearson decision procedure as follows.

At the initial time,

$$\int_{LR(\mathbf{y}_1)\geq\eta_1(1)} p(\mathbf{y}_1|\mathbf{H}_0)d\mathbf{y}_1 = \Delta\alpha, \tag{1.3.47}$$

$$\int_{LR(\mathbf{y}_1)\leq\eta_0(1)} p(\mathbf{y}_1|\mathbf{H}_1)d\mathbf{y}_1 = \Delta\beta. \tag{1.3.48}$$

Note that the sequence $\{\mathbf{y}_i\}$ is conditionally independent. For $\forall\ i < N$, at the ith stage, using the Neyman-Pearson test with the constraint $P_f = \Delta\alpha$, we have

$$\begin{aligned}&\int_{\substack{\eta_0(j)<LR(\mathbf{y}_j)<\eta_1(j),\forall j\leq i-1\\ LR(\mathbf{y}_i)\geq\eta_1(i)}} p(Y_i|\mathbf{H}_0)dY_i\\ &= P(\mathcal{C}(i-1)|\mathbf{H}_0)\int_{LR(\mathbf{y}_i)\geq\eta_1(i)} p(\mathbf{y}_i|\mathbf{H}_0)d\mathbf{y}_i = \Delta\alpha,\end{aligned} \tag{1.3.49}$$

where the probability of the continuing observation regions is also determined by the marginal likelihood ratios and thus

$$P(\mathcal{C}(i-1)|\mathbf{H}_0) \triangleq \int_{\eta_0(j)<LR(\mathbf{y}_j)<\eta_1(j),\forall j\leq i-1} p(Y_{i-1}|\mathbf{H}_0)dY_{i-1}. \tag{1.3.50}$$

Similarly, for $\forall\ i < M$, using the Neyman-Pearson test with constraint $P_m = \Delta\beta$ yields

$$\begin{aligned}&\int_{\substack{\eta_0(j)<LR(\mathbf{y}_j)<\eta_1(j),\forall j\leq i-1\\ LR(\mathbf{y}_i)\leq\eta_0(i)}} p(Y_i|\mathbf{H}_1)dY_i\\ &= P(\mathcal{C}(i-1)|\mathbf{H}_1)\int_{LR(\mathbf{y}_i)\leq\eta_0(i)} p(\mathbf{y}_i|\mathbf{H}_1)d\mathbf{y}_i = \Delta\beta,\end{aligned} \tag{1.3.51}$$

$$P(\mathcal{C}(i-1)|\mathbf{H}_1) \triangleq \int_{\eta_0(j)<LR(\mathbf{y}_j)<\eta_1(j),\forall j\leq i-1} p(Y_{i-1}|\mathbf{H}_1)dY_{i-1}. \tag{1.3.52}$$

In the above, if $M > N$, let $\eta_1(i) = \infty$ for $i = N, N+1, \ldots, M-1$; if $M < N$, let $\eta_0(i) = 0$ for $i = M, M+1, \ldots, N-1$; that is, for every time between the Nth and the $(M-1)$th time if $N < M$ or between the Mth and the $(N-1)$th time if $N < M$, the two stopping bounds become actually only one.

There are only the following five possibilities for this search process.

I. For some $i < \min\{M, N\}$, $\eta_0(i) \geq \eta_1(i)$ (i.e., $\mathcal{C}(i) = \emptyset$) holds. In this case, the error probabilities are bounded by

$$(i-1)\Delta\alpha \leq P_f \leq i\Delta\alpha, \quad (i-1)\Delta\beta \leq P_m \leq i\Delta\beta, \tag{1.3.53}$$

which could be significantly smaller than their maximum allowable values α and β therefore possibly courses significantly longer expected stopping times. Alternatively, we may decrease $\{M, N\}$ properly (say, by one) and repeat the above process. However, it is possible that any reduction in M and/or N will lead to $\mathcal{C}(L) \neq \emptyset$, where L is the maximum of the reduced M and N. In this case, we stop and use the M and N values prior to the reduction.

II. For some i, $\min\{M, N\} \leq i < \max\{M, N\}$, $\mathcal{C}(i) = \emptyset$ holds. I.e., when $M \geq N$, $\eta_0(i) = \infty$, or when $N \geq M$, $\eta_1(i) = 0$. Then, for case **I**, we may either terminate the search, which leads to a lower P_f or P_m, or decrease M or N, whichever is larger, to reduce the expected stopping times.

Clearly, for the above two cases, we are guaranteed to eventually find the desired $\{M, N\}$ since the number of choices is finite.

III. The above two cases do not occur but $P(\mathcal{C}(L-1)|\mathbf{H}_0) \leq \Delta\alpha$. Then let $\mathcal{H}_1(L) = \mathcal{C}(L-1) \times \Re^n$ and $\mathcal{H}_0(L) = \emptyset$, or define $\mathcal{H}_1(L)$ and $\mathcal{H}_0(L)$ in any way that guarantees $\mathcal{C}(L) = \emptyset$ and $P(\mathcal{H}_0(L)|\mathbf{H}_1) \leq \Delta\beta$. Similarly, when $P(\mathcal{C}(L-1)|\mathbf{H}_1) \leq \Delta\beta$, we can easily define $\mathcal{H}_0(L)$ and $\mathcal{H}_1(L)$ to guarantee that $\mathcal{C}(L) = \emptyset$. Thus, $\{M, N\}$ is a good pair.

If the above three cases do not occur, then there are only the following two possibilities:

IV. If $\eta_0(L) \geq \eta_1(L)$ holds, then we can easily define $\mathcal{H}_0(L)$ and $\mathcal{H}_1(L)$ to guarantee that $\mathcal{C}(L) = \emptyset$ as in case **III.** If $\eta_0(L) < \eta_1(L)$ but at least one of the two conditional probabilities of $\mathcal{C}(L)$ equals zero, say, $P(\mathcal{C}(L)|\mathbf{H}_0) = 0$, then let

$$\mathcal{H}_1(L) = \{Y_L : Y_{L-1} \in \mathcal{C}(L-1), LR(\mathbf{y}_L) > \eta_0(L)\},$$
$$\mathcal{H}_0(L) = \{Y_L : Y_{L-1} \in \mathcal{C}(L-1), LR(\mathbf{y}_L) \leq \eta_0(L)\}.$$

We still have $\mathcal{C}(L) = \emptyset$ and thus $\{M, N\}$ is a good pair.

V. If $\eta_0(L) < \eta_1(L)$, $P(\mathcal{C}(L)|\mathbf{H}_1) \neq 0$, and $P(\mathcal{C}(L)|\mathbf{H}_0) \neq 0$, then we should increase M or N and repeat the above process.

It turns out that Possibility **V** does not occur implies the finiteness of $\{M, N\}$.

1.3.3.4 FINITE UPPER BOUND ON STOPPING TIME

We now proceed to prove that the search process proposed above will terminate after a finite number of modifications of $\{M, N\}$.

Theorem 1.3.1. Assume that the observations $\{\mathbf{y}_n\}$ form an i.i.d. vector sequence conditional on each hypothesis. Let η_0 and η_0 denote the infima of the $\bar{\eta}_0$ and $\bar{\eta}_1$ for which the following two equations hold.

$$\int_{LR(\mathbf{y}_1)\leq\bar{\eta}_0} p(\mathbf{y}_1|\mathbf{H}_1)d\mathbf{y}_1 = \hat{\beta} > 0, \tag{1.3.54}$$

$$\int_{LR(\mathbf{y}_1)^{-1}\leq\bar{\eta}_1} p(\mathbf{y}_1|\mathbf{H}_0)d\mathbf{y}_1 = \hat{\alpha} > 0. \tag{1.3.55}$$

Obviously, both η_0 and η_1 are decreasing as $\hat{\beta}$ and $\hat{\alpha}$ go to zero, respectively. Define

$$\lambda_0 \triangleq \lim_{\hat{\beta}\to 0} \eta_0 \geq 0,$$

$$\lambda_1 \triangleq \lim_{\hat{\alpha}\to 0} \eta_1 \geq 0.$$

Assume that either

$$\lambda_0 < \frac{\alpha\beta}{1-\alpha}, \quad \text{or} \quad \lambda_1 < \frac{\alpha\beta}{1-\beta}. \tag{1.3.56}$$

Then, the sequential procedure of uniform error allocation of Subsection 1.3.3.3, starting with either $N = 1$ or $M = 1$, has a finite upper bound on its stopping time almost surely.

Proof. It suffices to show that the search algorithm presented at the end of Subsection 1.3.3.3 leads to a pair $\{M, N\}$ after a finite number of modifications of $\{M, N\}$, such that $\mathcal{C}(L) = \emptyset$, $P_f \leq \alpha$, and $P_m \leq \beta$, where $L = \max\{M, N\}$.

We only need to consider case **V** with $N = 1$ or $M = 1$. It suffices to prove that for a fixed N, we can find a finite M such that Eqs. (1.3.44)–(1.3.46) hold. Without loss of generality, assume $M > N$.

According to our algorithm, we have three sets of regions $\{\mathcal{H}_1(1), \ldots, \mathcal{H}_1(N-1), \mathcal{H}_1(M)\}$, $\{\mathcal{H}_0(1), \ldots, \mathcal{H}_0(M)\}$, and $\{\mathcal{C}(1), \ldots, \mathcal{C}(M)\}$. They are given by, at the first time,

$$\mathcal{H}_1(1) = \{\mathbf{y}_1 : LR(\mathbf{y}_1) \geq \eta_1(1), \int_{LR(\mathbf{y}_1)\geq\eta_1(1)} p(\mathbf{y}_1|\mathbf{H}_0)d\mathbf{y}_1 = \Delta\alpha\}, \tag{1.3.57}$$

$$\mathcal{H}_0(1) = \{\mathbf{y}_1 : LR(\mathbf{y}_1) \leq \eta_0(1), \int_{LR(\mathbf{y}_1)\leq\eta_0(1)} p(\mathbf{y}_1|\mathbf{H}_1)d\mathbf{y}_1 = \Delta\beta\}, \tag{1.3.58}$$

$$\mathcal{C}(1) = \{\mathbf{y}_1 : \eta_0(1) < LR(\mathbf{y}_1) < \eta_1(1)\}, \tag{1.3.59}$$

and at the ith time ($\forall i, 2 \leq i \leq M$),

$$\begin{aligned}\mathcal{H}_1(i) = \{&Y_i : Y_{i-1} \in \mathcal{C}(i-1), LR(\mathbf{y}_i) \geq \eta_1(i),\\ &P(\mathcal{C}(i-1)|\mathbf{H}_0)\int_{LR(\mathbf{y}_i)\geq\eta_1(i)} p(\mathbf{y}_i|\mathbf{H}_0)d\mathbf{y}_i = \Delta\alpha\},\end{aligned} \tag{1.3.60}$$

$$\begin{aligned}\mathcal{H}_0(i) = \{&Y_i : Y_{i-1} \in \mathcal{C}(i-1), LR(\mathbf{y}_i) \leq \eta_0(i),\\ &P(\mathcal{C}(i-1)|\mathbf{H}_1)\int_{LR(\mathbf{y}_i)\leq\eta_0(i)} p(\mathbf{y}_i|\mathbf{H}_1)d\mathbf{y}_i = \Delta\beta\},\end{aligned} \tag{1.3.61}$$

$$\mathcal{C}(i) = \{Y_i : Y_{i-1} \in \mathcal{C}(i-1), \eta_0(i) < LR(\mathbf{y}_i) < \eta_1(i)\}. \tag{1.3.62}$$

Denote the two correct-decision probabilities at the ith time by

$$\beta_i \triangleq P(\mathcal{H}_1(i)|\mathbf{H}_1) = \int_{\mathcal{H}_1(i)} p(Y_i|\mathbf{H}_1)dY_i, \tag{1.3.63}$$

$$\alpha_i \triangleq P(\mathcal{H}_0(i)|\mathbf{H}_0) = \int_{\mathcal{H}_0(i)} p(Y_i|\mathbf{H}_0)dY_i. \tag{1.3.64}$$

We can compute the conditional probabilities of the continuing observation regions by Eqs. (1.3.60)–(1.3.62):

$$P(\mathcal{C}(1)|\mathbf{H}_0) = 1 - [P(\mathcal{H}_1(1)|\mathbf{H}_0) + P(\mathcal{H}_0(1)|\mathbf{H}_0)] = 1 - \Delta\alpha - \alpha_1, \tag{1.3.65}$$

$$\vdots$$

$$\begin{aligned}S_M \triangleq P(\mathcal{C}(M)|\mathbf{H}_0) &= 1 - (\Delta\alpha + \alpha_1) - \cdots - (\Delta\alpha + \alpha_M)\\ &= 1 - \alpha - \alpha_1 - \cdots - \alpha_M,\end{aligned} \tag{1.3.66}$$

$$P(\mathcal{C}(1)|\mathbf{H}_1) = 1 - \Delta\beta - \beta_1, \tag{1.3.67}$$

$$\vdots$$

$$P(\mathcal{C}(M)|\mathbf{H}_1) = 1 - \beta - \beta_1 - \cdots - \beta_{N-1} - \beta_M. \tag{1.3.68}$$

To complete the proof of the theorem, it suffices to prove that there exists an M such that $\eta_0(M) \geq \eta_1(M)$, or equivalently, that

$$S_M \leq 0. \tag{1.3.69}$$

Since $\{\mathbf{y}_i\}$ is a conditional i.i.d. sequence and $P(\mathcal{C}(i)|\mathbf{H}_0)$ and $P(\mathcal{C}(i)|\mathbf{H}_1)$ are both decreasing as i increases, it follows from Eq. (1.3.61) that $\eta_0(i)$ increases with i. Hence, we have, for $\forall i \leq M$,

$$P(\mathcal{C}(i-1)|\mathbf{H}_0) \geq P(\mathcal{C}(M-1)|\mathbf{H}_0) = S_M + \Delta\alpha + \alpha_M < 1, \tag{1.3.70}$$

$$\bar{\alpha}_i \triangleq \int_{LR(\mathbf{y}_i) \leq \eta_0(i)} p(\mathbf{y}_i|\mathbf{H}_0) d\mathbf{y}_i \geq \int_{LR(\mathbf{y}_1) \leq \eta_0(1)} p(\mathbf{y}_1|\mathbf{H}_0) d\mathbf{y}_1 = \alpha_1. \tag{1.3.71}$$

Thus, we have, for $\forall i > 1$,

$$\alpha_i = P(\mathcal{C}(i-1)|\mathbf{H}_0)\bar{\alpha}_i \geq (S_M + \Delta\alpha)\alpha_1. \tag{1.3.72}$$

It follows from the above inequality and the definition of S_M in (1.3.66) that

$$\begin{aligned} S_M &\leq 1 - \alpha \underbrace{- \alpha_1(S_M + \Delta\alpha) - \cdots - \alpha_1(S_M + \Delta\alpha)}_{M-1} \\ &= 1 - \alpha - (M-1)\alpha_1(S_M + \Delta\alpha). \end{aligned} \tag{1.3.73}$$

Hence,

$$S_M \leq \frac{1-\alpha}{(M-1)\alpha_1 + 1} - \frac{(M-1)\alpha_1\Delta\alpha}{(M-1)\alpha_1 + 1}. \tag{1.3.74}$$

On the other hand,

$$\begin{aligned} \beta = M\Delta\beta &= M \int_{LR(\mathbf{y}_1) \leq \eta_0(1)} p(\mathbf{y}_1|\mathbf{H}_1) d\mathbf{y}_1 \\ &\leq M \int_{LR(\mathbf{y}_1) \leq \eta_0(1)} \eta_0(1) p(\mathbf{y}_1|\mathbf{H}_0) d\mathbf{y}_1 = M\eta_0(1)\alpha_1. \end{aligned}$$

It turns out that

$$M\alpha_1 \geq \frac{\beta}{\eta_0(1)}. \tag{1.3.75}$$

By the assumptions of the theorem, $\lim_{M\to\infty} \Delta\beta = 0$ implies that

$$\limsup_{M\to\infty} \eta_0(1) = \lambda_0.$$

By Eq. (1.3.75), if $\lambda_0 = 0$, then we have

$$\liminf_{M\to\infty} M\alpha_1 = \infty. \tag{1.3.76}$$

Then, from (1.3.74), we have

$$\lim_{M\to\infty} S_M \leq -\Delta\alpha. \tag{1.3.77}$$

If $\lambda_0 > 0$, then we have

$$\liminf_{M\to\infty} M\alpha_1 \geq \frac{\beta}{\lambda_0}. \tag{1.3.78}$$

Substituting (1.3.78) into (1.3.74), the limit superior of the right hand side of (1.3.74) is not more than

$$\frac{1-\alpha+\Delta\alpha}{\frac{\beta}{\lambda_0}+1} - \Delta\alpha. \tag{1.3.79}$$

Therefore, the right hand side of (1.3.74) would be less than zero whenever

$$\lambda_0 < \frac{\Delta\alpha\beta}{1-\alpha}. \tag{1.3.80}$$

Hence, we should make the right hand side of the last inequality as large as possible. If $N = 1$, choose $\Delta\alpha = \alpha$. Then we know that $S_M < 0$ holds for some large enough M if

$$\lambda_0 < \frac{\alpha\beta}{1-\alpha}. \tag{1.3.81}$$

If assume $N > M$, similarly, for $M = 1$, we have $S_N < 0$ for large enough N in the procedure similar to if

$$\lambda_1 < \frac{\alpha\beta}{1-\beta}. \tag{1.3.82}$$

It turns out that there must be a finite pair of $\{M, N\}$ such that at least one of the inequalities $S_M < 0$ (N fixed) and $S_N < 0$ (M fixed) holds. The theorem thus follows. ■

Obviously, the theorem holds even if the procedure was not started with $N = 1$ or $M = 1$. we only need to modify the bounds on λ_0 and λ_1 in (1.3.56) accordingly.

Remark 1.3.1. At each time, say the ith time, the proposed procedure with the marginal likelihood ratio used partitions the region $\mathcal{C}(i-1) \times \Re^n$ into three pieces: $\mathcal{C}(i-1) \times \{\mathbf{y}_i : LR(\mathbf{y}_i) \leq \eta_0(i)\}$, $\mathcal{C}(i-1) \times \{\mathbf{y}_i : LR(\mathbf{y}_i) \geq \eta_1(i)\}$, and $\mathcal{C}(i-1) \times \{\mathbf{y}_i : \eta_0(i) < LR(\mathbf{y}_i) < \eta_1(i)\}$. Obviously, this is not an optimal partition. If we use the joint likelihood ratio $LR(Y_i)$ to partition $\mathcal{C}(i-1) \times \Re^n$, then due to the optimality of the Neyman-Pearson decision rule, $P(\mathcal{C}(i)|\mathbf{H}_j)$ would generally be smaller and thus would go to zero faster (see also the numerical results below). It follows that an almost sure finite upper bound on the stopping time must exist if the optimal partition is used in the above algorithm.

Remark 1.3.2. Clearly, when $\lambda_0 = 0$ or $\lambda_1 = 0$, from (1.3.76) and (1.3.77) such a finite $\{M, N\}$ pair must exist no matter how small $\alpha > 0$ and $\beta > 0$ are. Although it is easy to construct a specific example of two conditional probability densities for which both $\lambda_0 > 0$ and $\lambda_1 > 0$, for most distributions used in practice, $\lambda_0 = 0$ or $\lambda_1 = 0$ is true.

Using the above theorem and Remark 1.3.2, it is easy to derive the following theoretically significant corollary.

Corollary 1.3.1. Suppose that $\lambda_i = 0$, $i = 0$ or 1, in the above theorem. Consider a nontrivial decision problem for which there is no test of size one with bounded error probabilities $P_f \leq \alpha, P_m \leq \beta$. Then, for any $\epsilon_1, \epsilon_2 > 0$, we can always construct a modified version of the procedure proposed above not only with an almost sure finite upper bound on the stopping time, but also with

$$\alpha - \epsilon_1 \leq P_f \leq \alpha, \qquad \beta - \epsilon_2 \leq P_m \leq \beta. \tag{1.3.83}$$

Proof. We allocate the two decision error probabilities for the first time to be $\alpha - \epsilon_1$ and $\beta - \epsilon_2$, respectively. The corresponding thresholds are determined by

$$\int_{LR(\mathbf{y}_1) \geq \eta_1(1)} p(\mathbf{y}_1|\mathbf{H}_0) d\mathbf{y}_1 = \alpha - \epsilon_1, \tag{1.3.84}$$

$$\int_{LR(\mathbf{y}_1) \leq \eta_0(1)} p(\mathbf{y}_1|\mathbf{H}_1) d\mathbf{y}_1 = \beta - \epsilon_2. \tag{1.3.85}$$

Since the problem is nontrivial, we have

$$P(\mathcal{C}(1)|\mathbf{H}_0) = \delta_0 > 0, \quad P(\mathcal{C}(1)|\mathbf{H}_1) = \delta_1 > 0. \tag{1.3.86}$$

Then, we can treat the remaining decision problem from the second time on as a sequential decision problem with two maximum allowable error probabilities $\tilde{\alpha} = \epsilon_1/\delta_0$ and $\tilde{\beta} = \epsilon_2/\delta_1$. By Theorem 1 and Remark 2, the proposed procedure always has an almost sure finite upper bound on the stopping time, and the error probabilities are bounded by $P_f \leq \epsilon_1/\delta_0$ and $P_m \leq \epsilon_2/\delta_1$. Combining the two parts, the corollary follows. ■

1.3.3.5 NUMERICAL EXAMPLES

In this subsection, we consider the problem of detecting a one-dimensional signal s in noise ν. In the first three examples, the detection of deterministic as well as Gaussian and Reyleigh signals in Gaussian noises is considered, and the results of a comparison between Wald's SPRT and the procedure proposed here are given. In the last example, we consider the sequential testing problem where the two conditional distributions are Gaussian, and which satisfies the common-variance assumption of Siegmund's approximation [71]. Then we give results comparing Wald's SPRT, Siegmund's modified SPRT and the procedure proposed here. Note that Siegmund's modified SPRT is not applicable to the first three examples because the variances are not equal under the two hypotheses.

The signal and noises are

$$\mathbf{H}_1 : \mathbf{y}_i = s_i + \nu_i,$$
$$\mathbf{H}_0 : \mathbf{y}_i = \nu_i,$$

where $\{s_i\}$, $\{\nu_i\}$ are mutually independent i.i.d. sequences. In the first two examples, we consider the detection of deterministic and Gaussian signal in Gaussian noise, respectively:

$$s_i \sim N(m,\ \sigma_s^2), \quad \nu_i \sim N(0,\ \sigma_\nu^2).$$

The two conditional Probability Density Function (PDF) given $\mathbf{H}_0$ and $\mathbf{H}_1$ are

$$p(\mathbf{y}_i|\mathbf{H}_0) = N(\mathbf{y}_i; 0,\ \sigma_\nu^2), \quad p(\mathbf{y}_i|\mathbf{H}_1) = N(\mathbf{y}_i; m,\ \sigma_s^2 + \sigma_\nu^2).$$

Then we consider the detection of a Rayleigh distributed signal in Gaussian noise for the third example.

In all of the following numerical examples, we take the discretization stepsize $\Delta = 0.01$ and use Matlab software to compute the integrals. We give numerical values of $P(\mathcal{C}(k)|\mathbf{H}_i)$, $E(T|\mathbf{H}_i)$, $P(\mathcal{H}_j(k)|\mathbf{H}_i)$ $(i \neq j)$, P_f, P_m, $\eta_0(k)$, $\eta_1(k)$, for Wald's SPRT (denoted by subscript W) with the two approximate constant stopping bounds and for the proposed uniform error allocation procedure (denoted by subscript N) with the optimal (denoted by superscript opt) and suboptimal partitions of the regions $\mathcal{H}_j(k)$ and $\mathcal{C}_j$ using both the joint and marginal likelihood ratios.

Example 1.3.1. $m = 3$, $\sigma_s^2 = 0$, i.e., the signal s_i is a deterministic constant; $\sigma_\nu^2 = 1.5$. Hence

$$p(\mathbf{y}_i|\mathbf{H}_0) = N(\mathbf{y}_i; 0,\ 1.5), \quad p(\mathbf{y}_i|\mathbf{H}_1) = N(\mathbf{y}_i; 3,\ 1.5).$$

The maximum allowable values are $\alpha = \beta = 0.05$. The final numerical results are $M = 4$, $N = 4$, $P_f = 0.0493$, $P_m = 0.0494$, $E(T|\mathbf{H}_0) = 1.8066$, and $E(T|\mathbf{H}_1) = 1.8187$ for the new procedure with the suboptimal partition by using the marginal likelihood ratio $LR(\mathbf{y}_k)$, and $M = 3$, $N = 3$, $P_f = 0.0458$, $P_m = 0.0429$, $E(T^{\text{opt}}|\mathbf{H}_0) = 1.5714$ and $E(T^{\text{opt}}|\mathbf{H}_1) = 1.6332$ for the new procedure with the optimal partition by using the joint likelihood ratio $LR(Y_k)$. More detailed results are given in Table 1.3.1.

Table 1.3.1. Performance comparisons of Wald's and the proposed SPRTs for Example 1.3.1 with $\alpha = \beta = 0.05$.

k	1	2	3	4	$E(T\|\mathbf{H}_i)$
$P_W(C(k)\|\mathbf{H}_1)$	0.6956	0.3540	0.1444	0.1252	≥ 2.3192
$P_W(C(k)\|\mathbf{H}_0)$	0.6503	0.3470	0.1572	0.1338	≥ 2.2883
$P_N(C(k)\|\mathbf{H}_1)$	0.4929	0.2465	0.0793	0.0000	1.8187
$P_N(C(k)\|\mathbf{H}_0)$	0.4928	0.2365	0.0773	0.0000	1.8066
$P_N^{\text{opt}}(C(k)\|\mathbf{H}_1)$	0.4833	0.1499	0.0000		1.6332
$P_N^{\text{opt}}(C(k)\|\mathbf{H}_0)$	0.4301	0.1413	0.0000		1.5714
$\eta_1(k)_N$	7.3900	7.3000	4.1090	1.0030	
$\eta_0(k)_N$	0.1350	0.1450	0.2440	1.0030	
$\eta_1(k)_N^{\text{opt}}$	5.6600	5.0600	1.2640		
$\eta_0(k)_N^{\text{opt}}$	0.2300	0.2490	1.2640		
$\eta_1(k)_W$	19.000	19.000	19.000	19.000	...
$\eta_0(k)_W$	0.0526	0.0526	0.0526	0.0526	...
$P_W(\mathcal{H}_0(k)\|\mathbf{H}_1)$	0.0080	0.0037	0.0024	0.0022	...
$P_W(\mathcal{H}_1(k)\|\mathbf{H}_0)$	0.0055	0.0041	0.0032	0.0026	...
$P_N(\mathcal{H}_0(k)\|\mathbf{H}_1)$	0.0125	0.0125	0.0125	0.0119	$P_m = 0.0494$
$P_N(\mathcal{H}_1(k)\|\mathbf{H}_0)$	0.0125	0.0125	0.0125	0.0118	$P_f = 0.0493$
$P_N(\mathcal{H}_0^{\text{opt}}(k)\|\mathbf{H}_1)$	0.0167	0.0167	0.0095		$P_m = 0.0429$
$P_N(\mathcal{H}_1^{\text{opt}}(k)\|\mathbf{H}_0)$	0.0167	0.0167	0.0124		$P_f = 0.0458$

Now we set $\alpha = \beta = 0.1$. The final numerical results are $M = 2$, $N = 3$, $P_f = 0.0976$, $P_m = 0.0664$, $E(T|\mathbf{H}_0) = 1.5599$, and $E(T|\mathbf{H}_1) = 1.4184$ for the new procedure with the suboptimal partition, and $M = 2$, $N = 2$, $P_f = 0.0911$, $P_m = 0.0996$, $E(T^{\text{opt}}|\mathbf{H}_0) = 1.2941$, and $E(T^{\text{opt}}|\mathbf{H}_1) = 1.3446$ for the new procedure with the optimal partition. More detailed results are given in Table 1.3.2.

Example 1.3.2. $m = 3$, $\sigma_s^2 = 0.5$, $\sigma_\nu^2 = 1$. Hence

$$p(\mathbf{y}_i|\mathbf{H}_0) = N(\mathbf{y}_i; 0,\ 1), \quad p(\mathbf{y}_i|\mathbf{H}_1) = N(\mathbf{y}_i; 3,\ 1.5).$$

We first set $\alpha = \beta = 0.05$. The final numerical results are $M = 2$, $N = 3$, $P_f = 0.0459$, $P_m = 0.0365$, $E(T|\mathbf{H}_0) = 1.9021$, and $E(T|\mathbf{H}_1) = 1.2647$ for the new procedure with the suboptimal partition, and $M = 2$, $N = 2$, $P_f =$

Table 1.3.2. Performance comparisons of Wald's and the proposed SPRTs for Example 1.3.1 with $\alpha = \beta = 0.1$.

k	1	2	3	$E(T\|\mathbf{H}_i)$
$P_W(C(k)\|\mathbf{H}_1)$	0.4838	0.1972	0.0812	≥ 1.7622
$P_W(C(k)\|\mathbf{H}_0)$	0.5308	0.2011	0.0737	≥ 1.8056
$P_N(C(k)\|\mathbf{H}_1)$	0.3397	0.0787	0.0000	1.4184
$P_N(C(k)\|\mathbf{H}_0)$	0.2966	0.2633	0.0000	1.5599
$P_N^{\text{opt}}(C(k)\|\mathbf{H}_1)$	0.3446	0.0000		1.3446
$P_N^{\text{opt}}(C(k)\|\mathbf{H}_0)$	0.2941	0.0000		1.2941
$\eta_1(k)_N$	4.1100	1.7500	1.4910	
$\eta_0(k)_N$	0.3350	0.0000	1.4910	
$\eta_1^{\text{opt}}(k)_N$	3.3210	2.2250		
$\eta_0^{\text{opt}}(k)_N$	0.3930	2.2250		
$\eta_1(k)_W$	9.000	9.000	9.000	...
$\eta_0(k)_W$	0.1111	0.1111	0.1111	...
$P_W(\mathcal{H}_0(k)\|\mathbf{H}_1)$	0.0224	0.0091	0.0032	...
$P_W(\mathcal{H}_1(k)\|\mathbf{H}_0)$	0.0162	0.0101	0.0044	...
$P_N(\mathcal{H}_0(k)\|\mathbf{H}_1)$	0.0500	0.0000	0.0164	$P_m = 0.0664$
$P_N(\mathcal{H}_1(k)\|\mathbf{H}_0)$	0.0333	0.0333	0.0310	$P_f = 0.0976$
$P_N^{\text{opt}}(\mathcal{H}_0(k)\|\mathbf{H}_1)$	0.0500	0.0496		$P_m = 0.0996$
$P_N^{\text{opt}}(\mathcal{H}_1(k)\|\mathbf{H}_0)$	0.0500	0.0411		$P_f = 0.0911$

0.0442, $P_m = 0.0474$, $E(T^{\text{opt}}|\mathbf{H}_0) = 1.3952$, and $E(T^{\text{opt}}|\mathbf{H}_1) = 1.1862$ for the new procedure with the optimal partition. More detailed results are given in Table 1.3.3.

Now we set $\alpha = \beta = 0.1$. The final numerical results are $M = 2$, $N = 2$, $P_f = 0.0994$, $P_m = 0.0627$, $E(T|\mathbf{H}_0) = 1.3020$ and $E(T|\mathbf{H}_1) = 1.1431$ for the new procedure with the suboptimal partition. More detailed results are given in Table 1.3.4.

Example 1.3.3. In this example, we want to detect a Rayleigh distributed signal in Gaussian noise, where

$$s_i \sim R(2), \quad \nu_i \sim N(0,\ 1).$$

Let $\alpha = \beta = 0.1$. The final numerical results are $M = 3$, $N = 3$, $P_f = 0.0908$, $P_m = 0.0994$, $E(T|\mathbf{H}_0) = 2.2990$, and $E(T|\mathbf{H}_1) = 1.4304$ for the new procedure with the suboptimal partition. More detailed results are given in Table 1.3.5.

Table 1.3.3. Performance comparisons of Wald's and the proposed SPRTs for Example 1.3.2 with $\alpha = \beta = 0.05$.

k	*1*	2	*3*	$E(T\|\mathbf{H}_i)$
$P_W(C(k)\|\mathbf{H}_1)$	0.3865	0.1320	0.0399	≥ 1.5584
$P_W(C(k)\|\mathbf{H}_0)$	0.7227	0.2031	0.0654	≥ 1.9912
$P_N(C(k)\|\mathbf{H}_1)$	0.2149	0.0498	0.0000	1.2647
$P_N(C(k)\|\mathbf{H}_0)$	0.4594	0.4427	0.0000	1.9021
$P_N^{\text{opt}}(C(k)\|\mathbf{H}_1)$	0.1862	0.0000		1.1862
$P_N^{\text{opt}}(C(k)\|\mathbf{H}_0)$	0.3952	0.0000		1.3952
$\eta_1(k)_N$	3.2500	3.2490	3.2400	
$\eta_0(k)_N$	0.1000	0.0000	3.2400	
$\eta_1^{\text{opt}}(k)_N$	1.5520	1.0520		
$\eta_0^{\text{opt}}(k)_N$	0.1600	1.0520		
$\eta_1(k)_W$	19.000	19.000	19.000	$\cdots$
$\eta_0(k)_W$	0.0526	0.0526	0.0526	$\cdots$
$P_W(\mathcal{H}_0(k)\|\mathbf{H}_1)$	0.0080	0.0077	0.0020	$\cdots$
$P_W(\mathcal{H}_1(k)\|\mathbf{H}_0)$	0.0044	0.0014	0.0001	$\cdots$
$P_N(\mathcal{H}_0(k)\|\mathbf{H}_1)$	0.0250	0.0000	0.0115	$P_m = 0.0365$
$P_N(\mathcal{H}_1(k)\|\mathbf{H}_0)$	0.0167	0.0167	0.0127	$P_f = 0.0459$
$P_N^{\text{opt}}(\mathcal{H}_0(k)\|\mathbf{H}_1)$	0.0250	0.0224		$P_m = 0.0474$
$P_N^{\text{opt}}(\mathcal{H}_1(k)\|\mathbf{H}_0)$	0.0250	0.0192		$P_f = 0.0442$

Table 1.3.4. Performance comparisons of Wald's and the proposed SPRTs for Example 1.3.2 with $\alpha = \beta = 0.1$.

k	*1*	2	$E(T\|\mathbf{H}_i)$
$P_W(C(k)\|\mathbf{H}_0)$	0.4068	0.0926	≥ 1.4994
$P_W(C(k)\|\mathbf{H}_1)$	0.2658	0.0741	≥ 1.3399
$P_N(C(k)\|\mathbf{H}_0)$	0.3020	0.0000	1.3020
$P_N(C(k)\|\mathbf{H}_1)$	0.1431	0.0000	1.1431
$\eta_1(k)_N$	2.0000	0.4510	
$\eta_0(k)_N$	0.1520	0.4510	
$\eta_1(k)_W$	9.0000	9.0000	$\cdots$
$\eta_0(k)_W$	0.1111	0.1111	$\cdots$
$P_W(\mathcal{H}_0(k)\|\mathbf{H}_1)$	0.0306	0.0109	$\cdots$
$P_W(\mathcal{H}_1(k)\|\mathbf{H}_0)$	0.0134	0.0024	$\cdots$
$P_N(\mathcal{H}_0(k)\|\mathbf{H}_1)$	0.0500	0.0127	$P_m = 0.0627$
$P_N(\mathcal{H}_1(k)\|\mathbf{H}_0)$	0.0500	0.0494	$P_f = 0.0994$

Table 1.3.5. Performance comparisons of Wald's and the proposed SPRTs for Example 1.3.3 with $\alpha = \beta = 0.1$.

k	*1*	2	*3*	$E(T\|\mathbf{H}_i)$
$P_W(C(k)\|\mathbf{H}_0)$	0.9857	0.4828	0.2356	≥ 2.7041
$P_W(C(k)\|\mathbf{H}_1)$	0.4939	0.2118	0.1245	≥ 1.8302
$P_N(C(k)\|\mathbf{H}_0)$	0.8301	0.4689	0.0000	2.2990
$P_N(C(k)\|\mathbf{H}_1)$	0.3381	0.0923	0.0000	1.4304
$\eta_1(k)_N$	2.0000	1.7080	1.7070	
$\eta_0(k)_N$	0.2260	0.2500	1.7070	
$\eta_1(k)_W$	9.0000	9.0000	9.0000	...
$\eta_0(k)_W$	0.1111	0.1111	0.1111	...
$P_W(\mathcal{H}_0(k)\|\mathbf{H}_1)$	0.0070	0.0028	0.0010	...
$P_W(\mathcal{H}_1(k)\|\mathbf{H}_0)$	0.0073	0.0033	0.0012	...
$P_N(\mathcal{H}_0(k)\|\mathbf{H}_1)$	0.0333	0.0333	0.0328	$P_m = 0.0994$
$P_N(\mathcal{H}_1(k)\|\mathbf{H}_0)$	0.0333	0.0333	0.0242	$P_f = 0.0908$

In all tables above, using the data and the following formula

$$E(T|\mathbf{H}_i) = \sum_{k=1}^{L} k(P(\mathcal{C}(k-1)|\mathbf{H}_i) - P(\mathcal{C}(k)|\mathbf{H}_i))$$

$$P(\mathcal{C}(0)|\mathbf{H}_i) = 1, \quad P(\mathcal{C}(L)|\mathbf{H}_i) = 0,$$

it is easy to compute $E(T|\mathbf{H}_0)$ and $E(T|\mathbf{H}_1)$.

Although due to intractable (high dimensional numerical integral) computation, we can just show $P_W(\mathcal{C}(j)|\mathbf{H}_i)$ up to $j = 4$ other than $j \to \infty$, it can still be seen that $P_W(\mathcal{C}(j)|\mathbf{H}_i)$ decreases much more slowly than $P_N(\mathcal{C}(j)|\mathbf{H}_i)$ as j increases and that $E_N(T|\mathbf{H}_i) < E_W(T|\mathbf{H}_i)$. In addition, it can be seen that the new procedure is less conservative than Wald's SPRT in that it made better use of the allowable error probabilities α and β. On the other hand, the results also show that uniform allocation of error probabilities over time in the new procedure failed to sufficiently take advantage of the two maximum allowable parameters in some cases, such as when M or N are small (see Examples 1.3.1 and 1.3.2 with $\alpha = \beta = 0.1$).

Example 1.3.4. Now we compare the performances of Wald's SPRT, Siegmund's modified SPRT, and our proposed method. Suppose that the two conditional probability densities are

$$p(\mathbf{y}_i|\mathbf{H}_0) = N(\mathbf{y}_i; -0.8,\ 1), \quad p(\mathbf{y}_i|\mathbf{H}_1) = N(\mathbf{y}_i; 0.8,\ 1).$$

In Siegmund's approximation [71], the maximum allowable error probabilities are calculated from the two given thresholds. We set these two thresholds for the likelihood ratio (note: these thresholds are for the likelihood ratio, rather than the log-likelihood ratio as in Siegmund [71]) to be $\eta_0 = 1$ and $\eta_1 = 20.0855$. The corresponding allowable error probabilities are $\alpha = 0.0721$ and $\beta = 0.2851$. Then, using the two conditional probability densities and these values of α and β, we compare Wald's SPRT, Siegmund's SPRT, and the procedure proposed here. Note that Wald's two thresholds $\eta_0 = 0.3086$ and $\eta_1 = 9.3942$ are different from Siegmund's. The detailed results are given in Table 1.3.6.

Table 1.3.6. Performance comparisons of three SPRTs for Example 1.3.4 with $\alpha = 0.0721, \beta = 0.2851$.

k	*1*	2	3	$E(T\|\mathbf{H}_i)$
$P_W(C(k)\|\mathbf{H}_1)$	0.6622	0.3304	0.1571	≥ 2.1497
$P_W(C(k)\|\mathbf{H}_0)$	0.4774	0.2136	0.0985	≥ 1.7895
$P_S(C(k)\|\mathbf{H}_1)$	0.6261	0.3018	0.1309	≥ 2.0588
$P_S(C(k)\|\mathbf{H}_0)$	0.1909	0.0701	0.0290	≥ 1.2900
$P_N(C(k)\|\mathbf{H}_1)$	0.3834	0.0000		1.3834
$P_N(C(k)\|\mathbf{H}_0)$	0.2754	0.0000		1.2754
$\eta_1(k)_W$	9.3942	9.3942	9.3942	...
$\eta_0(k)_W$	0.3086	0.3086	0.3086	...
$\eta_1(k)_S$	20.086	20.086	20.086	...
$\eta_0(k)_S$	1.0000	1.0000	1.0000	...
$\eta_1(k)_N$	2.3750	1.4021		
$\eta_0(k)_N$	0.7305	1.4021		
$P_W(\mathcal{H}_1(k)\|\mathbf{H}_0)$	0.0146	0.0125	0.0064	...
$P_W(\mathcal{H}_0(k)\|\mathbf{H}_1)$	0.0569	0.0255	0.0111	...
$P_S(\mathcal{H}_1(k)\|\mathbf{H}_0)$	0.0039	0.0055	0.0029	...
$P_S(\mathcal{H}_0(k)\|\mathbf{H}_1)$	0.2295	0.0424	0.0144	...
$P_N(\mathcal{H}_1(k)\|\mathbf{H}_0)$	0.0380	0.0323		$P_f = 0.0703$
$P_N(\mathcal{H}_0(k)\|\mathbf{H}_1)$	0.1326	0.1235		$P_m = 0.2561$

From Table 1.3.6, it can be seen that the two conditional probabilities of continuing observation regions of Siegmund's SPRT, denoted by $P_S(C(k)|\mathbf{H}_i)$, $i = 0, 1$, decrease faster than those for Wald's SPRT as k is increasing, but still not as fast as those of the procedure proposed here. The average stopping time of the proposed procedure is smaller than those for Wald's SPRT and Siegmund's SPRT. Note that a major limitation of Siegmund's SPRT is that it is only applicable to testing between two Gaussian distributions with a common variance (i.e., only the expected values are different, see [71]) .

The above examples indicate the following. Wald's simple thresholds are too conservative because the information on the conditional distributions under the two hypotheses is not accounted for; Siegmund's approximation improves the performance slightly but is still too conservative because the (Gaussian) distributional information is used only crudely. The inefficiency of the thresholds in these examples demonstrates the difficulty of developing constant thresholds for sequential tests. Although the proposed method is a simple uniform error allocation scheme and the thresholds were determined by the marginal rather than joint likelihood ratio to save computation, it is able to make good use of the distributional information and yield significantly better results, on top of the advantage that it has bounded stopping times.

Using this new method, a multisensor decentralized sequential decision will be presented in Subsection 3.8. How to optimally allocate the allowable decision error probabilities for each time for the new method is still an open problem.

Chapter 2

TWO SENSOR BINARY DECISION

From this chapter through Chapter 4, in order to use the global optimization strategy, we suppose that the fusion rule is fixed. We concentrate on investigating optimal sensor rules under this fusion rule. For notational simplicity, we start with the simplest multisensor decentralized decision system, namely, the two sensor binary decision system. We focus on Bayes decision problem. However, the basic idea and methods which we derive for this simplest case can be extended to the general multisensor m-ary distributed decision system as well as to the Neyman-Pearson decision and the sequential decision problem.

2.1 INTRODUCTION

Consider the situation where a fusion center, in cooperation with two sensors, wishes to determine which one of the two possible hypotheses is true. In this chapter, we derive a set of fixed-point type equations that the optimal sensor rules g^1 and g^2 must satisfy. We then discuss the existence and uniqueness of solutions to these optimality equations. Based on the discretization of the optimality equations, we obtain a finitely convergent method for the computation of a pair of optimal sensor rules. Finally, we report some encouraging numerical results. It should be noted that the results in this chapter do not assume the conditional independence of the data vectors given the hypothesis. The independence assumption, though unrealistic in many practical situations, was made extensively in the previous research on this problem.

2.1.1 PROBLEM FORMULATION

Let there be two hypotheses $\mathbf{H}_0$, $\mathbf{H}_1$. Consider a situation where a fusion center, in cooperation with two local sensors, wishes to determine which one of these two hypothesis is true. The two sensors obtain data vectors $\mathbf{y}_1 \in \Re^{n_1}$

and $\mathbf{y}_2 \in \Re^{n_1}$, respectively. The conditional joint distributions of $(\mathbf{y}_1, \mathbf{y}_2)$ given each hypothesis are assumed to be known and given by $p(\mathbf{y}_1, \mathbf{y}_2|\mathbf{H}_0)$, $p(\mathbf{y}_1, \mathbf{y}_2|\mathbf{H}_1)$. We then define two real valued functions $g^i(\mathbf{y}_i)$, $i = 1, 2$, which serve to divide their own observation spaces into two parts:

$$\mathcal{H}_0^i = \{\mathbf{y}_i \in \Re^{n_i} : g^i(\mathbf{y}_i) \le 0\}, \quad \mathcal{H}_1^i = \{\mathbf{y}_i \in \Re^{n_i} : g^i(\mathbf{y}_i) > 0\}, \ i = 1, 2. \tag{2.1.1}$$

Then the two sensors transmit their compressed binary messages $d_1(\mathbf{y}_1)$ and $d_2(\mathbf{y}_2)$ which are given by

$$g^1(\mathbf{y}_1) \underset{d_1=0}{\overset{d_1=1}{\gtrless}} 0 \quad \text{and} \quad g^2(\mathbf{y}_2) \underset{d_2=0}{\overset{d_2=1}{\gtrless}} 0, \tag{2.1.2}$$

respectively, to the fusion center. Hence, g^1 and g^2 are also called the *sensor compression functions*. Clearly, the difference $LR(\mathbf{y}_i) - \lambda_i$ of the likelihood ratio function and the threshold is a specific data compression function in the single sensor case

$$LR(\mathbf{y}_i) - \lambda_i \underset{d=0}{\overset{d=1}{\gtrless}} 0.$$

On the basis of the two received binary messages $d_1(\mathbf{y}_1)$ and $d_2(\mathbf{y}_2)$, the fusion center then makes final decision $F \in \{0, 1\}$ with the objective of optimizing a certain system performance.

The final binary decision F of the fusion center is to fuse the two binary messages d_1, d_2 according to a *fusion rule* F:

$$F = F(d_1, d_2) : \{0, 1\} \times \{0, 1\} \mapsto \{0, 1\}.$$

The goal in this section is to minimize the Bayes cost functional

$$\begin{aligned} C(g^1, g^2; F) = {} & C_{00}P_0P(F = 0|\mathbf{H}_0) + C_{01}P_1P(F = 0|\mathbf{H}_1) \\ & + C_{10}P_0P(F = 1|\mathbf{H}_0) + C_{11}P_1P(F = 1|\mathbf{H}_1), \end{aligned} \tag{2.1.3}$$

where C_{ij}, P_0, and P_1 were defined in Subsection 1.3.1 above.

Substituting the definition of F and d_1, d_2 into (2.1.3) and simplifying, we see that

$$\begin{aligned} & C(g^1, g^2; F) \\ & = P_0C_{10} + P_1C_{11} + \int_{F(d_1,d_2)=0} \{[P_1(C_{01} - C_{11})p(\mathbf{y}_1, \mathbf{y}_2|\mathbf{H}_1)] \\ & \qquad - [P_0(C_{10} - C_{00})p(\mathbf{y}_1, \mathbf{y}_2|\mathbf{H}_0)]\}\, d\mathbf{y}_1 d\mathbf{y}_2 \\ & = c + \int_{F(d_1,d_2)=0} \{ap(\mathbf{y}_1, \mathbf{y}_2|\mathbf{H}_1) - bp(\mathbf{y}_1, \mathbf{y}_2|\mathbf{H}_0)\}\, d\mathbf{y}_1 d\mathbf{y}_2, \end{aligned} \tag{2.1.4}$$

where

$$c = P_0C_{10} + P_1C_{11}, \quad a = P_1(C_{01} - C_{11}), \quad b = P_0(C_{10} - C_{00}).$$

Thus, our goal is to select optimal sensor rules (g^1, g^2) together with a fusion rule F which will jointly minimize the cost functional $C(g^1, g^2; F)$. Therefore, we need to solve an "infinite" dimensional optimization problem.

As a Boolean function defined over $\{0,1\} \times \{0,1\}$, the fusion rule $F(d_1, d_2)$ has 2^4 different forms. However, among all the 16 possibilities we only consider two interesting ones: the "OR" rule and the "AND" rule. A more general method that can deal with any possible fusion rule will be presented in the next chapter.

The AND rule is defined by

$$F(d_1, d_2) = \begin{cases} 1, & \text{if } d_1 = d_2 = 1, \\ 0, & \text{otherwise.} \end{cases} \tag{2.1.5}$$

The OR rule is given by

$$F(d_1, d_2) = \begin{cases} 0, & \text{if } d_1 = d_2 = 0, \\ 1, & \text{otherwise.} \end{cases} \tag{2.1.6}$$

For the OR rule, the cost function $C(g^1, g^2, F)$ is given by

$$C(g^1, g^2; F) = c + \int_{\substack{g^1(\mathbf{y}_1)\le 0 \\ g^2(\mathbf{y}_2)\le 0}} (ap(\mathbf{y}_1, \mathbf{y}_2|\mathbf{H}_1) - bp(\mathbf{y}_1, \mathbf{y}_2|\mathbf{H}_0))\, d\mathbf{y}_1 d\mathbf{y}_2, \tag{2.1.7}$$

and the cost function for the AND rule is given by

$$\begin{aligned} C(g^1, g^2; F) &= c + \int_{g^1(\mathbf{y}_1)\le 0 \text{ or } g^2(\mathbf{y}_2)\le 0} (ap(\mathbf{y}_1, \mathbf{y}_2|\mathbf{H}_1) - bp(\mathbf{y}_1, \mathbf{y}_2|\mathbf{H}_0))\, d\mathbf{y}_1 d\mathbf{y}_2 \\ &= c + \int_{\Re^{n_1+n_2}} (ap(\mathbf{y}_1, \mathbf{y}_2|\mathbf{H}_1) - bp(\mathbf{y}_1, \mathbf{y}_2|\mathbf{H}_0))\, d\mathbf{y}_1 d\mathbf{y}_2 \\ &\quad - \int_{\substack{g^1(\mathbf{y}_1)> 0 \\ g^2(\mathbf{y}_2)> 0}} (ap(\mathbf{y}_1, \mathbf{y}_2|\mathbf{H}_1) - bp(\mathbf{y}_1, \mathbf{y}_2|\mathbf{H}_0))\, d\mathbf{y}_1 d\mathbf{y}_2 \\ &= c' - \int_{\substack{g^1(\mathbf{y}_1)> 0 \\ g^2(\mathbf{y}_2)> 0}} (ap(\mathbf{y}_1, \mathbf{y}_2|\mathbf{H}_1) - bp(\mathbf{y}_1, \mathbf{y}_2|\mathbf{H}_0))\, d\mathbf{y}_1 d\mathbf{y}_2, \end{aligned} \tag{2.1.8}$$

where

$$c' = c + \int_{\Re^{n_1+n_2}} (ap(\mathbf{y}_1, \mathbf{y}_2|\mathbf{H}_1) - bp(\mathbf{y}_1, \mathbf{y}_2|\mathbf{H}_0))\, d\mathbf{y}_1 d\mathbf{y}_2 = c + a - b.$$

Note that in the case of the AND rule, the final "1" decision region $\mathcal{H}_1$ is a product region $\mathcal{H}_1^1 \times \mathcal{H}_1^2$, and for the OR rule, the final "0" decision region $\mathcal{H}_0$ is a product region $\mathcal{H}_0^1 \times \mathcal{H}_0^2$.

Throughout this section, we will say that a pair of sensor rules (g^1, g^2) is equivalent to another pair $(\bar{g}^1, \bar{g}^2)$, and denote it by $(g^1, g^2) \sim (\bar{g}^1, \bar{g}^2)$, if the two have the same decision regions; that is,

$$\mathcal{H}_0^i = \{\mathbf{y}_i \in \Re^{n_i} : g^i(\mathbf{y}_i) \leq 0\} = \{\mathbf{y}_i \in \Re^{n_i} : \bar{g}^i(\mathbf{y}_i) \leq 0\}, \quad i = 1, 2.$$

Clearly, two equivalent pairs of sensor rules will generate identical binary messages d_1, d_2 for all observation vectors $(\mathbf{y}_1, \mathbf{y}_2)$. Since the fusion center's decision F depends solely on the binary messages d_1, d_2, the value of the cost functional $C(g^1, g^2; F)$ (cf. (2.1.4)) remains unchanged if the pair of sensor rules (g^1, g^2) is replaced with an equivalent pair $(\bar{g}^1, \bar{g}^2)$. By (2.1.2), $(g^1, g^2) \sim (\bar{g}^1, \bar{g}^2)$ if and only if

$$g^i(\mathbf{y}_i) = h^i(\mathbf{y}_i)\bar{g}^i(\mathbf{y}_i), \ i = 1, 2, \ \text{for some function } h^i(\mathbf{y}_i) > 0 \text{ a.e.} \quad (2.1.9)$$

Here "a.e." is the quantifier "almost everywhere" that is defined with reference to the conditional distributions $p(\mathbf{y}_1, \mathbf{y}_2|\mathbf{H}_0)$, $p(\mathbf{y}_1, \mathbf{y}_2|\mathbf{H}_1)$.

2.1.2 THE RELATIONSHIP OF DISTRIBUTED AND CENTRALIZED DECISIONS

The above model of distributed multisensor decision has many features in common with the classical Bayes model of hypothesis testing for the single sensor (centralized) case. Specifically, let $Y = (\mathbf{y}_1, \mathbf{y}_2)$ denote the joint observation data vector and suppose that Y has been made available to the fusion center. The decision F of the fusion center is made using a decision rule $g : \Re^{n_1+n_2} \mapsto \Re$ with the interpretation

$$F = \begin{cases} 0, & \text{if } g(Y) \leq 0, \\ 1, & \text{if } g(Y) > 0. \end{cases}$$

The optimal decision function $g(Y)$ is the one that will minimize the cost function

$$\begin{aligned} C(g) &= c + \int_{g(Y)\leq 0} (ap(Y|\mathbf{H}_1) - bp(Y|\mathbf{H}_0))\, dY \\ &= c + \int_{g(\mathbf{y}_1,\mathbf{y}_2)\leq 0} (ap(\mathbf{y}_1, \mathbf{y}_2|\mathbf{H}_1) - bp(\mathbf{y}_1, \mathbf{y}_2|\mathbf{H}_0))\, d\mathbf{y}_1 d\mathbf{y}_2. \end{aligned} \quad (2.1.10)$$

Thus, the difference between the distributed and centralized decisions lies in the "form" of the decision regions: the distributed decision, namely the OR rule case, requires its decision region to be of the "product" form $\mathcal{H}_0 = \mathcal{H}_0^1 \times \mathcal{H}_0^2$

while the centralized decision allows a "coupled" decision region

$$\mathcal{H}_0 = \{(\mathbf{y}_1, \mathbf{y}_2) : g(\mathbf{y}_1, \mathbf{y}_2) \leq 0\}.$$

The reason for this difference is of course that only one bit transmission from the local sensor to the fusion center is allowed for distributed decision. In light of this requirement on its decision region, the performance (e.g., the error probabilities of false alarm or the value of the cost functional) of distributed multisensor decision cannot be better than that of the centralized one.

In the classical decision theory (i.e., the single sensor) case, it is well known [83] that the optimal decision rule is given by the likelihood ratio test. Consequently, the optimal decision region for the centralized multisensor decision has the form

$$\mathcal{H}_0 = \left\{ (\mathbf{y}_1, \mathbf{y}_2) \in \Re^{n_1+n_2} : \frac{p(\mathbf{y}_1, \mathbf{y}_2|\mathbf{H}_1)}{p(\mathbf{y}_1, \mathbf{y}_2|\mathbf{H}_0)} \leq \lambda \right\},$$

where λ is a threshold. In general, for the centralized case $\mathcal{H}_0$ cannot be written in the product form $\mathcal{H}_0^1 \times \mathcal{H}_0^2$. Hence, for the decentralized case there will be a degradation in performance from the centralized multisensor decision scheme. From the geometric standpoint, the problem of finding the optimal distributed decision region becomes that of finding a region of the form $\mathcal{H}_0^1 \times \mathcal{H}_0^2$ to "best" approximate the optimal centralized decision region $\mathcal{H}_0$. Of course, the term "best" here is in the sense of minimizing the cost functionals of (2.1.7) and (2.1.8). Figures 2.5.2 and 2.5.3 in Section 2.5 represent, respectively, the optimal decision regions of distributed and centralized detection schemes for the problem of detecting a real valued Gaussian signal in Gaussian noises with two sensors. It can be seen that the optimal distributed decision region under the OR fusion rule is a rectangle and the optimal distributed decision region under the AND fusion rule is a "cross" region which closely approximate the centralized decision region given by the global likelihood ratio test.

Summing up, from the form of the decision function $g(Y)$ and the decision region $\mathcal{H}_0$, the distributed decision fusion can be viewed as constrained optimization problem while the centralized decision is an unconstrained optimization problem.

2.2 OPTIMAL SENSOR RULE OF BAYES DECISION

In this section, we shall develop a necessary condition for joint local sensor rules (g^1, g^2) to be optimal under a given fusion rule. As will seen later, this condition is represented in terms of a fixed-point of a pair integral equations. We shall also discuss the issues of existence and uniqueness of solutions to these fixed-point problem. In the next section, we shall use these integral equations to design an iterative procedure for computing optimal local sensor rules.

2.2.1 FIXED POINT TYPE NECESSARY CONDITION

We begin with a brief review of the derivation of the optimal centralized decision rule, as this will provide the basis for the derivation of the optimal decentralized decision rules to follow. From the equation (2.1.10) in the previous section, an optimal centralized decision rule g must minimize the following cost functional

$$\int_{g(\mathbf{y}_1,\mathbf{y}_2)\le 0} (ap(\mathbf{y}_1,\mathbf{y}_2|\mathbf{H}_1) - bp(\mathbf{y}_1,\mathbf{y}_2|\mathbf{H}_0))\, d\mathbf{y}_1 d\mathbf{y}_2.$$

Note that the decision rule $g(\mathbf{y}_1,\mathbf{y}_2)$ affects only the integration region but not the function that is integrated. Therefore, to achieve the minimum value of the integral, we should choose the integration region $\{(\mathbf{y}_1,\mathbf{y}_2) : g(\mathbf{y}_1,\mathbf{y}_2) \le 0\}$ to be the region where the integrand is negative. This can be easily accomplished by letting

$$g(\mathbf{y}_1,\mathbf{y}_2) = ap(\mathbf{y}_1,\mathbf{y}_2|\mathbf{H}_1) - bp(\mathbf{y}_1,\mathbf{y}_2|\mathbf{H}_0).$$

Consequently, the optimal centralized decision rule is given by

$$ap(\mathbf{y}_1,\mathbf{y}_2|\mathbf{H}_1) - bp(\mathbf{y}_1,\mathbf{y}_2|\mathbf{H}_0) \le 0,$$

which is equivalent to the likelihood ratio test

$$\frac{p(\mathbf{y}_1,\mathbf{y}_2|\mathbf{H}_0)}{p(\mathbf{y}_1,\mathbf{y}_2|\mathbf{H}_1)} \underset{F=0}{\overset{F=1}{\gtrless}} \frac{b}{a}$$

with b/a acting as the threshold.

For the case of decentralized decision where the fusion center uses the OR rule, the optimal sensor rules (g^1, g^2) must minimize the cost functional (cf. (2.1.7))

$$\int_{\substack{g^1(\mathbf{y}_1)\le 0\\ g^2(\mathbf{y}_2)\le 0}} (ap(\mathbf{y}_1,\mathbf{y}_2|\mathbf{H}_1) - bp(\mathbf{y}_1,\mathbf{y}_2|\mathbf{H}_0))\, d\mathbf{y}_1 d\mathbf{y}_2. \tag{2.2.1}$$

Let us first rewrite this cost functional by integrating over the variable $\mathbf{y}_2$ to obtain

$$\begin{aligned}
&\int_{\substack{g^1(\mathbf{y}_1)\le 0\\ g^2(\mathbf{y}_2)\le 0}} (ap(\mathbf{y}_1,\mathbf{y}_2|\mathbf{H}_1) - bp(\mathbf{y}_1,\mathbf{y}_2|\mathbf{H}_0))\, d\mathbf{y}_1 d\mathbf{y}_2 \\
&= \int_{g^1(\mathbf{y}_1)\le 0} \left(\int_{g^2(\mathbf{y}_2)\le 0} (ap(\mathbf{y}_1,\mathbf{y}_2|\mathbf{H}_1) - bp(\mathbf{y}_1,\mathbf{y}_2|\mathbf{H}_0)) d\mathbf{y}_2 \right) d\mathbf{y}_1 \\
&= \int_{g^1(\mathbf{y}_1)\le 0} f(\mathbf{y}_1) d\mathbf{y}_1,
\end{aligned} \tag{2.2.2}$$

where

$$f(\mathbf{y}_1) = \int_{g^2(\mathbf{y}_2)\leq 0} (ap(\mathbf{y}_1, \mathbf{y}_2|\mathbf{H}_1) - bp(\mathbf{y}_1, \mathbf{y}_2|\mathbf{H}_0)d\mathbf{y}_2.$$

Fix a pair of optimal sensor rules (g^1, g^2). Since (g^1, g^2) jointly minimizes the cost functional (2.2.1), the sensor rule g^1 must minimize the functional

$$\int_{g^1(\mathbf{y}_1)\leq 0} f(\mathbf{y}_1)d\mathbf{y}_1.$$

By the same argument as in the centralized case, the decision region

$$\Omega_1 = \{\mathbf{y}_1 \in \Re^{n_1} : g^1(\mathbf{y}_1) \leq 0\}$$

must coincide with the region where the function $f(\mathbf{y}_1)$ is non-positive. In other words, there exists some positive function $h^1(\mathbf{y}_1)$ such that

$$\begin{aligned} h^1(\mathbf{y}_1)g^1(\mathbf{y}_1) &= f(\mathbf{y}_1) \\ &= \int_{g^2(\mathbf{y}_2)\leq 0} (ap(\mathbf{y}_1, \mathbf{y}_2|\mathbf{H}_1) - bp(\mathbf{y}_1, \mathbf{y}_2|\mathbf{H}_0))d\mathbf{y}_2, \text{ a.e.} \end{aligned} \tag{2.2.3}$$

For simplicity, we will drop "a.e." from our discussion in the remaindes of this section. By a symmetrical argument, there must exist some $h^2(\mathbf{y}_2) > 0$ such that

$$h^2(\mathbf{y}_2)g^2(\mathbf{y}_2) = \int_{g^1(\mathbf{y}_1)\leq 0} (ap(\mathbf{y}_1, \mathbf{y}_2|\mathbf{H}_1) - bp(\mathbf{y}_1, \mathbf{y}_2|\mathbf{H}_0)d\mathbf{y}_2. \tag{2.2.4}$$

Summarizing the preceding argument, we obtain the following necessary condition for the optimality of the sensor rules.

Theorem 2.2.1. Suppose that the fusion center uses the OR rule and the corresponding optimal sensor rules exist. Given any pair of optimal sensor rules (g^1, g^2) which minimizes the cost functional (2.1.7), (g^1, g^2) is equivalent to a pair of optimal decision rules $(\bar{g}^1, \bar{g}^2)$ satisfying the following integral equations

$$\begin{aligned} \bar{g}^1(\mathbf{y}_1) &= \int_{\bar{g}^2(\mathbf{y}_2)\leq 0} (ap(\mathbf{y}_1, \mathbf{y}_2|\mathbf{H}_1) - bp(\mathbf{y}_1, \mathbf{y}_2|\mathbf{H}_0))d\mathbf{y}_2, \\ \bar{g}^2(\mathbf{y}_2) &= \int_{\bar{g}^1(\mathbf{y}_1)\leq 0} (ap(\mathbf{y}_1, \mathbf{y}_2|\mathbf{H}_1) - bp(\mathbf{y}_1, \mathbf{y}_2|\mathbf{H}_0))d\mathbf{y}_1. \end{aligned} \tag{2.2.5}$$

Proof. The proof of the theorem follows from equations (2.2.2)–(2.2.4) and the definition of equivalence (2.1.9). In particular, equations (2.2.2)–(2.2.4) imply that

$$
\begin{aligned}
h^1(\mathbf{y}_1)g^1(\mathbf{y}_1) &= \int_{g^2(\mathbf{y}_2)\leq 0}(ap(\mathbf{y}_1,\mathbf{y}_2|\mathbf{H}_1) - bp(\mathbf{y}_1,\mathbf{y}_2|\mathbf{H}_0))d\mathbf{y}_2,\\
h^2(\mathbf{y}_2)g^2(\mathbf{y}_2) &= \int_{g^1(\mathbf{y}_1)\leq 0}(ap(\mathbf{y}_1,\mathbf{y}_2|\mathbf{H}_1) - bp(\mathbf{y}_1,\mathbf{y}_2|\mathbf{H}_0))d\mathbf{y}_1.
\end{aligned}
\tag{2.2.6}
$$

Let us define

$$
\begin{aligned}
\bar{g}^1(\mathbf{y}_1) &= \int_{g^2(\mathbf{y}_2)\leq 0}(ap(\mathbf{y}_1,\mathbf{y}_2|\mathbf{H}_1) - bp(\mathbf{y}_1,\mathbf{y}_2|\mathbf{H}_0))d\mathbf{y}_2,\\
\bar{g}^2(\mathbf{y}_2) &= \int_{g^1(\mathbf{y}_1)\leq 0}(ap(\mathbf{y}_1,\mathbf{y}_2|\mathbf{H}_1) - bp(\mathbf{y}_1,\mathbf{y}_2|\mathbf{H}_0))d\mathbf{y}_1.
\end{aligned}
$$

Then, we have

$$
\bar{g}^1(\mathbf{y}_1) = h^1(\mathbf{y}_1)g^1(\mathbf{y}_1) \quad \text{and} \quad \bar{g}^2(\mathbf{y}_2) = h^2(\mathbf{y}_2)g^2(\mathbf{y}_2). \tag{2.2.7}
$$

Since h^1, h^2 are positive, this implies that (g^1, g^2) is equivalent to $(\bar{g}^1, \bar{g}^2)$. Consequently, we have

$$
\{\mathbf{y}_1 \in \Re^{n_1} : g^1(\mathbf{y}_1) \leq 0\} = \{\mathbf{y}_1 \in \Re^{n_1} : \bar{g}^1(\mathbf{y}_1) \leq 0\} \tag{2.2.8}
$$

and

$$
\{\mathbf{y}_2 \in \Re^{n_2} : g^2(\mathbf{y}_2) \leq 0\} = \{\mathbf{y}_2 \in \Re^{n_2} : \bar{g}^2(\mathbf{y}_2) \leq 0\}. \tag{2.2.9}
$$

By Eq. (2.2.6), we immediately obtain

$$
\begin{aligned}
\bar{g}^1(\mathbf{y}_1) &= \int_{\bar{g}^2(\mathbf{y}_2)\leq 0}(ap(\mathbf{y}_1,\mathbf{y}_2|\mathbf{H}_1) - bp(\mathbf{y}_1,\mathbf{y}_2|\mathbf{H}_0)d\mathbf{y}_2,\\
\bar{g}^2(\mathbf{y}_2) &= \int_{\bar{g}^1(\mathbf{y}_1)\leq 0}(ap(\mathbf{y}_1,\mathbf{y}_2|\mathbf{H}_1) - bp(\mathbf{y}_1,\mathbf{y}_2|\mathbf{H}_0)d\mathbf{y}_1.
\end{aligned}
$$

This completes the proof of the theorem. ■

Theorem 2.2.1 gives a necessary condition for optimality for the case where the fusion center employs the OR rule. In a manner analogous to the derivation of Theorem 2.2.1, we can establish the following necessary condition for optimality condition for the AND case. The proof of this theorem is omitted.

Theorem 2.2.2. Suppose that the fusion center uses the AND rule and the corresponding optimal sensor rules exist. Given any pair of optimal sensor rules (g^1, g^2) which minimize the cost functional (2.1.8), then (g^1, g^2) is equivalent to a pair of optimal decision rules $(\bar{g}^1, \bar{g}^2)$ satisfying the following integral equations

$$\bar{g}^1(\mathbf{y}_1) = -\int_{\bar{g}^2(\mathbf{y}_2)\le 0} (ap(\mathbf{y}_1,\mathbf{y}_2|\mathbf{H}_1) - bp(\mathbf{y}_1,\mathbf{y}_2|\mathbf{H}_0)d\mathbf{y}_2,$$
$$\bar{g}^2(\mathbf{y}_2) = -\int_{\bar{g}^1(\mathbf{y}_1)\le 0} (ap(\mathbf{y}_1,\mathbf{y}_2|\mathbf{H}_1) - bp(\mathbf{y}_1,\mathbf{y}_2|\mathbf{H}_0)d\mathbf{y}_1. \tag{2.2.10}$$

The integral equations (2.2.5), (2.2.10) are of the fixed-point type. In particular, if we define the mappings

$$\Gamma_{OR}(g^1,g^2) = \begin{bmatrix} \int_{g^2(\mathbf{y}_2)\le 0} (ap(\mathbf{y}_1,\mathbf{y}_2|\mathbf{H}_1) - bp(\mathbf{y}_1,\mathbf{y}_2|\mathbf{H}_0))d\mathbf{y}_2 \\ \int_{g^1(\mathbf{y}_1)\le 0} (ap(\mathbf{y}_1,\mathbf{y}_2|\mathbf{H}_1) - bp(\mathbf{y}_1,\mathbf{y}_2|\mathbf{H}_0))d\mathbf{y}_1 \end{bmatrix} \tag{2.2.11}$$

and

$$\Gamma_{AND}(g^1,g^2) = \begin{bmatrix} -\int_{g^2(\mathbf{y}_2)\le 0} (ap(\mathbf{y}_1,\mathbf{y}_2|\mathbf{H}_1) - bp(\mathbf{y}_1,\mathbf{y}_2|\mathbf{H}_0))d\mathbf{y}_2 \\ -\int_{g^1(\mathbf{y}_1)\le 0} (ap(\mathbf{y}_1,\mathbf{y}_2|\mathbf{H}_1) - bp(\mathbf{y}_1,\mathbf{y}_2|\mathbf{H}_0))d\mathbf{y}_1 \end{bmatrix}, \tag{2.2.12}$$

then the solutions of the integral equations (2.2.5), (2.2.10) are precisely the fixed points of Γ_{OR}, Γ_{AND} respectively.

Theorems 2.2.1 and 2.2.2 show that every pair of optimal sensor rules must be equivalent to some solution of the integral equations (2.2.5) or (2.2.10). Thus, for all practical purposes, it suffices to find an optimal pair (g^1, g^2) that satisfies (2.2.5) or (2.2.10) in the case of the OR or the AND rule, respectively, being used at the fusion center. This reduces the problem of finding the optimal sensor rules for distributed decision to that of solving the integral equations (2.2.5) or (2.2.10). In the next section, we shall describe an iterative procedure for solving these equations.

Although finding the optimal sensor rules here is an infinite dimensional optimization problem, it in fact uses an elementary idea in the finite dimensional optimization problem to reduce the two sensor rule optimization to the single sensor rule optimization problem. It is well known [57] that necessary (and sufficient) conditions for optimality are powerful computational vehicles for the solution of finite dimensional optimization problems. For example, for any bi-variate optimization problem $\min_{x_1,x_2} f(x_1,x_2)$, a necessary condition for optimality is that the two minimizers x_1^* and x_2^* must satisfy the following equations:

$$x_1^* = \min_{x_1} f(x_1, x_2^*), \quad x_2^* = \min_{x_2} f(x_1^*, x_2). \tag{2.2.13}$$

Unfortunately, the above equations may admit local minimizers of $f(x_1,x_2)$, unless $f(x_1,x_2)$ has a unique pair of minimizers. In essence, the integral equations (2.2.10) and (2.2.5) are infinite dimensional analogs of the necessary condition (2.2.13) in finite dimensions. Therefore, the integral equations (2.2.10), (2.2.5) might also contain "spurious" optimal solutions which correspond to the local minimizers of the distributed decision problem. We illustrate this point with the following example.

Example 2.2.1. Suppose that each of the two conditional probability densities are uniform. Hence, it is easy to yield a negative constant valued integrand in the Bayes cost functional on the "cross" region, which consists of two unequal rectangles as shown in the following Fig.2.2.1.

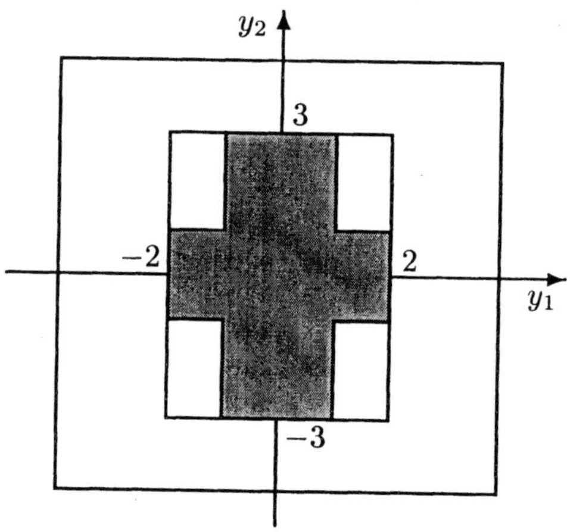

Figure 2.2.1. Example of a local minimum for a cost functional: the grey area is the $\mathcal{H}_0$ region of the centralized decision.

When the OR fusion rule is used, using the idea of Subsection 2.1.2 and noting the negative constant valued integrand on the "cross" region, the optimal sensor rules should generate an $\mathcal{H}_0$ decision region to be the same rectangle as the bigger one $\{(y_1, y_2) : -1 \leq y_1 \leq 1, -3 \leq y_2 \leq 3\}$. However, it is easy to see that the smaller one $\{(y_1, y_2) : -2 \leq y_1 \leq 2, -1 \leq y_2 \leq 1\}$ is a local minimizer as well.

Since Theorems 2.2.1 and 2.2.2 give necessary conditions for the optimal sensor rules for general conditional probability densities, the results in [77], where the two sensor observations are mutually conditionally independent given each hypothesis, are special cases. We have the following corollary:

Corollary 2.2.1. When the sensor observations $\{\mathbf{y}_1, \mathbf{y}_2\}$ are mutually conditionally independent given each hypothesis, for the OR and AND fusion rules, the optimal sensor rules $\{g^1(\mathbf{y}_1), g^2(\mathbf{y}_2)\}$ are given by likelihood ratios, provided that they exist.

Proof. Without loss of generality, we consider the OR fusion rule only. By the conditional independence of the observations, we know that

$$p(\mathbf{y}_1, \mathbf{y}_2|\mathbf{H}_i) = p(\mathbf{y}_1|\mathbf{H}_i)p(\mathbf{y}_2|\mathbf{H}_i), \quad i = 1, 2. \tag{2.2.14}$$

Substituting (2.2.14) into (2.2.5) yields

$$\begin{aligned} g^1(\mathbf{y}_1) &= \int_{g^2(\mathbf{y}_2)\leq 0} (ap(\mathbf{y}_1, \mathbf{y}_2|\mathbf{H}_1) - bp(\mathbf{y}_1, \mathbf{y}_2|\mathbf{H}_0))d\mathbf{y}_2 \\ &= [\int_{g^2(\mathbf{y}_2)\leq 0} ap(\mathbf{y}_2|\mathbf{H}_1)d\mathbf{y}_2]p(\mathbf{y}_1|\mathbf{H}_1) \\ &\quad -[\int_{g^2(\mathbf{y}_2)\leq 0} bp(\mathbf{y}_2|\mathbf{H}_0)d\mathbf{y}_2]p(\mathbf{y}_1|\mathbf{H}_0), \end{aligned} \tag{2.2.15}$$

$$\begin{aligned} g^2(\mathbf{y}_2) &= \int_{g^1(\mathbf{y}_1)\leq 0} (ap(\mathbf{y}_1, \mathbf{y}_2|\mathbf{H}_1) - bp(\mathbf{y}_1, \mathbf{y}_2|\mathbf{H}_0))d\mathbf{y}_1 \\ &= [\int_{g^1(\mathbf{y}_1)\leq 0} ap(\mathbf{y}_1|\mathbf{H}_1)d\mathbf{y}_1]p(\mathbf{y}_2|\mathbf{H}_1) \\ &\quad -[\int_{g^1(\mathbf{y}_1)\leq 0} bp(\mathbf{y}_1|\mathbf{H}_0)d\mathbf{y}_1]p(\mathbf{y}_2|\mathbf{H}_0). \end{aligned} \tag{2.2.16}$$

As the integrals in the above Eqs. (2.2.15) and (2.2.16) are all constants, the corollary holds. ■

2.2.2 EXISTENCE OF THE OPTIMAL SENSOR RULE

Not only might there exist multiple local minimizers for the optimal sensor rules in the distributed decision problem, but the existence of the optimal sensor rules is also questionable in general. Even if $p(\mathbf{y}_1, \mathbf{y}_2|\mathbf{H}_0)$ and $p(\mathbf{y}_1, \mathbf{y}_2|\mathbf{H}_1)$ are both continuous, the optimal sensor rules may still not exist. For example, when $a = b = 1$ and

$$p(\mathbf{y}_1, \mathbf{y}_2|\mathbf{H}_1) \sim N\Big(\begin{pmatrix} 1 \\ 10 \end{pmatrix}, \begin{bmatrix} 1 & -1 \\ -1 & 1 \end{bmatrix}\Big),$$

$$p(\mathbf{y}_1, \mathbf{y}_2|\mathbf{H}_0) \sim N\Big(\begin{pmatrix} 0 \\ 0 \end{pmatrix}, \begin{bmatrix} 1 & -1 \\ -1 & 1 \end{bmatrix}\Big),$$

the two densities actually degenerate to two Gaussian densities over two straight lines. Then, the positive region and the negative region of $ap(\mathbf{y}_1, \mathbf{y}_2|\mathbf{H}_0) - bp(\mathbf{y}_1, \mathbf{y}_2|\mathbf{H}_0)$ degenerate to be two parallel straight lines which are not parallel to the $\mathbf{y}_2$ coordinate axis, at the other points, the values of the integrand is zero. If we use the OR fusion rule, the optimal sensor rules should make the final $\mathcal{H}_0$ decision region a rectangle that covers the negative value straight line as long as possible and the covered interval is symmetrical with respect to the maximizer of the Gaussian density $p(\mathbf{y}_1, \mathbf{y}_2|\mathbf{H}_0)$. Since the maximum of the Gaussian density $p(\mathbf{y}_1, \mathbf{y}_2|\mathbf{H}_1)$ is the same as that of $p(\mathbf{y}_1, \mathbf{y}_2|\mathbf{H}_0)$ and much more larger than the value of $p(\mathbf{y}_1, \mathbf{y}_2|\mathbf{H}_0)$ at the points that are far away from the maximizer, the aforementioned rectangle does not touch the positive value straight line. Obviously, there exists no optimal solution for the two sensor rules in this case. This situation is similar to the case where there is no minimizer of a continuous function on an open set if the real minimizer is on the boundary. On the other hand, from the view point of practical applications, the existence of the optimal sensor rules is not critical. Although the minimizer may not exist, it is easy to see that the infimum of the cost functional must exist and be finite. Therefore, we only care about developing a numerical method to get a discrete optimal approximation to the infimum of the cost functional. We discuss this in the next section.

2.3 AN ALGORITHM FOR COMPUTING THE OPTIMAL SENSOR RULE

In the previous section, we derived a set of integral equations (2.2.5), (2.2.10) whose solutions contain the optimal sensor rules for the distributed decision problem. In what follows, we shall develop an iterative algorithm for computing a discrete approximation of the solutions to the integral equations. We shall also show the finite convergence of this iterative algorithm by using a potential function argument.

2.3.1 GAUSS-SEIDEL ITERATIVE ALGORITHM

An iterative algorithm for the optimal two sensor rules under the OR fusion rule can be defined naturally using the mapping Γ_{OR} given in (2.2.11). Let us use (g_i^1, g_i^2) to denote the decision rules at the ith iteration. We assume that the initial value (g_0^1, g_0^2) is given. In particular, consider the Gauss-Seidel type iteration, for $i = 0, 1, 2, \ldots$,

$$\begin{cases} g_{i+1}^1(\mathbf{y}_1) = \displaystyle\int_{g_i^2(\mathbf{y}_2)\le 0} (ap(\mathbf{y}_1, \mathbf{y}_2|\mathbf{H}_1) - bp(\mathbf{y}_1, \mathbf{y}_2|\mathbf{H}_0))d\mathbf{y}_2, \\ g_{i+1}^2(\mathbf{y}_2) = \displaystyle\int_{g_{i+1}^1(\mathbf{y}_1)\le 0} (ap(\mathbf{y}_1, \mathbf{y}_2|\mathbf{H}_1) - bp(\mathbf{y}_1, \mathbf{y}_2|\mathbf{H}_0))d\mathbf{y}_1. \end{cases} \tag{2.3.1}$$

Similarly, using the mapping Γ_{AND} given in (2.2.12), an iterative algorithm for the two optimal sensor rules under the AND fusion rule is given by the following Gauss-Seidel type iteration

$$\begin{cases} g_{i+1}^1(\mathbf{y}_1) = \int_{g_i^2(\mathbf{y}_2)>0} (ap(\mathbf{y}_1, \mathbf{y}_2|\mathbf{H}_1) - bp(\mathbf{y}_1, \mathbf{y}_2|\mathbf{H}_0))d\mathbf{y}_2, \\ g_{i+1}^2(\mathbf{y}_2) = \int_{g_{i+1}^1(\mathbf{y}_1)>0} (ap(\mathbf{y}_1, \mathbf{y}_2|\mathbf{H}_1) - bp(\mathbf{y}_1, \mathbf{y}_2|\mathbf{H}_0))d\mathbf{y}_1. \end{cases} \tag{2.3.2}$$

2.3.2 THE FINITE CONVERGENCE OF THE DISCRETIZED ALGORITHM

The iterative processes defined by (2.3.1) and (2.3.2) are in terms of the decision rules (g_i^1, g_i^2) which are real-valued functions. Thus, they cannot be implemented exactly in a computer. To facilitate computer simulations, we need to discretize the processes. Specifically, for each $i = 0, 1, 2, ...$, let

$$\{\mathbf{y}_1^1, \mathbf{y}_1^2, \ldots, \mathbf{y}_1^{N_1}\}$$

form an appropriate discretization of $\Re^{n_1}$. Similarly, we choose a suitable discretization

$$\{\mathbf{y}_2^1, \mathbf{y}_2^2, \ldots, \mathbf{y}_2^{N_2}\}$$

of $\Re^{n_2}$. Denote the corresponding real valued discretizations of g_i^1 and g_i^2 by the N_1-vector $\{g_{i,j}^1 : j = 1, 2, \ldots, N_1\}$ and N_2-vector $\{g_{i,j}^2 : j = 1, 2, \ldots, N_2\}$, respectively. For notational simplicity, let the step sizes of the discretizations of $\Re^{n_1}$ and $\Re^{n_2}$ are are both constant. Finally, we let $p(l, m|\mathbf{H}_0)$ and $p(l, m|\mathbf{H}_1)$ denote $p(y_1^l, y_2^m|\mathbf{H}_0)$ and $p(y_1^l, y_2^m|\mathbf{H}_0)$, respectively; that is to say, $\{p(l, m|\mathbf{H}_0) : 1 \le l \le N_1, 1 \le m \le N_2\}$ and $\{p(l, m|\mathbf{H}_1) : 1 \le l \le N_1, 1 \le m \le N_2\}$ are the respective discretizations of $p(\mathbf{y}_1, \mathbf{y}_2|\mathbf{H}_1)$ and $p(\mathbf{y}_1, \mathbf{y}_2|\mathbf{H}_0)$.

With the decision rules (g_i^1, g_i^2) discretized as above, the iterative processes (2.3.1) and (2.3.2) can now be rewritten as

$$\begin{cases} g_{i+1,j}^1 = \sum_{\{m: g_{i,m}^2 \le 0\}} (ap(j,m|\mathbf{H}_1) - bp(j,m|\mathbf{H}_0)), & j = 1, \ldots, N_1, \\ g_{i+1,j}^2 = \sum_{\{l: g^1(i+1,l) \le 0\}} (ap(l,j|\mathbf{H}_1) - bp(l,j|\mathbf{H}_0)), & j = 1, \ldots, N_2, \end{cases} \tag{2.3.3}$$

and

$$\begin{cases} g_{i+1,j}^1 = \sum_{\{m: g_{i,m}^2 > 0\}} (ap(j, m|\mathbf{H}_1) - bp(j, m|\mathbf{H}_0)), & j = 1, \ldots, N_1, \\ g_{i+1,j}^2 = \sum_{\{l: g^1(i+1,l) > 0\}} (ap(l, j|\mathbf{H}_1) - bp(l, j|\mathbf{H}_0)), & j = 1, \ldots, N_2, \end{cases} \tag{2.3.4}$$

respectively, where the step sizes in the summations are omitted since the sensor rules (g^1, g^2) and $(g^1\triangle, g^2\triangle)$, $\triangle > 0$, are equivalent to each other. A simple termination criterion is to stop the iteration as soon as

$$g_{i+1,l}^1 = g_{i,l}^1 \quad \text{and} \quad g_{i+1,m}^2 = g_{i,m}^2, \quad \text{for all } l, m. \tag{2.3.5}$$

An alternative is to stop when

$$\sum_{l,m} \left(|g^1_{i+1,l} - g^1_{i,l}| + |g^2_{i+1,m} - g^2_{i,m}| \right) \le \epsilon,$$

where $\epsilon > 0$ is some prespecified tolerance parameter.

We next show that the Gauss-Seidel iterative algorithm (2.3.3) is actually finitely convergent.

Theorem 2.3.1. For each initial choice of (g^1_0, g^2_0), the Gauss-Seidel type iterative algorithm (2.3.3) terminates with a pair (g^1_i, g^2_i) satisfying (2.3.5) after a finite number of iterations.

Proof. The proof is based on a potential function argument. Let us first define for each $i \ge 0$,

$$\Omega^1_i = \{l : g^1_{i,l} \le 0\} \quad \text{and} \quad \Omega^2_i = \{m : g^2_{i,m} \le 0\}.$$

The iteration formula (2.3.3) can be rewritten as

$$\begin{cases} g^1_{i+1,j} = \sum_{m \in \Omega^2_i} (ap(j,m|\mathbf{H}_1) - bp(j,m|\mathbf{H}_0)), & j = 1, \ldots, N_1, \\ g^2_{i+1,j} = \sum_{l \in \Omega^1_{i+1}} (ap(l,j|\mathbf{H}_1) - bp(l,j|\mathbf{H}_0)), & j = 1, \ldots, N_2. \end{cases}$$

With a slight abuse of notation, we shall still use (g^1_i, g^2_i) to denote the vector $\{(g^1_{i,l}, g^2_{i,m}) : 1 \le l \le N_1, 1 \le m \le N_2\}$, we make three observations.

1 The $(i+1)$th iterate (g^1_{i+1}, g^2_{i+1}) are determined uniquely by the decision region (Ω^1_i, Ω^2_i).

2 Once the condition

$$(\Omega^1_i, \Omega^2_i) = (\Omega^1_{i+1}, \Omega^2_{i+1})$$

is satisfied for some $i = k \ge 0$, then this condition will remain satisfied for all $i \ge k$. Consequently, the termination criterion (2.3.5) will be satisfied at the iteration $i = k+1$.

3 Since there are only a finite number of different choices of (Ω^1_i, Ω^2_i), the vectors (g^1_{i+1}, g^2_{i+1}) can take only finitely many values.

For the purpose of introducing the potential function, we define the indicator variables $(z^1_{i,l}, z^2_{i,m})$ for the decision rules (g^1_i, g^2_i) as follows:

$$z^1_{i,l} = \begin{cases} 1, & \text{if } g^1_{i,l} \le 0 \\ 0, & \text{if } g^1_{i,l} > 0, \end{cases} \qquad \text{for all } 1 \le l \le N_1 \text{ and } i = 0, 1, \ldots \tag{2.3.6}$$

and

$$z_{i,m}^2 = \begin{cases} 1, \text{ if } g_{i,m}^2 \leq 0 \\ 0, \text{ if } g_{i,m}^2 > 0, \end{cases} \quad \text{for all } 1 \leq m \leq N_2 \text{ and } i = 0, 1, ... \qquad (2.3.7)$$

We let z_i^1, z_i^2 denote the indicator vectors with components $\{z_{i,l}^1\}$, $\{z_{i,m}^2\}$ respectively. Since the components of (z_i^1, z_i^2) are either 0 or 1, there are at most $2^{N_1+N_2}$ different values for the indicator vectors. Consider the following potential function

$$\phi(z_i^1, z_i^2) = \sum_{l=1}^{N_1} \sum_{m=1}^{N_2} (ap(l, m|\mathbf{H}_1) - bp(l, m|\mathbf{H}_0)) z_{i,l}^1 z_{i,m}^2, \quad \text{for all } i.$$

Since $z_{i,m}^2 = 1$ if and only if $g_{i,m}^2 \leq 0$, it follows from the iteration formula (2.3.3) that

$$\begin{aligned} \sum_{m=1}^{N_2} (ap(l, m|\mathbf{H}_1) - bp(l, m|\mathbf{H}_0)) z_{i,m}^2 &= \sum_{\substack{m: \\ g_{i,m}^2 \leq 0}} [ap(l, m|\mathbf{H}_1) - bp(l, m|\mathbf{H}_0)] \\ &= g_{i+1,l}^1, \quad \text{for all } l. \end{aligned}$$

Thus, we have

$$\begin{aligned} \phi(z_{i+1}^1, z_i^2) &= \sum_{l=1}^{N_1} \sum_{m=1}^{N_2} (ap(l, m|\mathbf{H}_1) - bp(l, m|\mathbf{H}_0)) z_{i+1,l}^1 z_{i,m}^2 \\ &= \sum_{l=1}^{N_1} z_{i+1,l}^1 g_{i+1,l}^1 \\ &= \phi(z_i^1, z_i^2) + \sum_{l=1}^{N_1} (z_{i+1,l}^1 - z_{i,l}^1) g_{i+1,l}^1 \\ &\leq \phi(z_i^1, z_i^2), \end{aligned}$$

where the last step follows from the facts that $z_{i+1,l}^1 = 1$ if and only if $g_{i+1,l}^1 \leq 0$, and $0 \leq z_{i,l}^1 \leq 1$. Therefore, after a finite number of steps, the potential function $\phi(z_i^1, z_i^2)$ must have converged to a stationary point. Furthermore, the last inequality holds with equality only when

$$(z_{i+1,l}^1 - z_{i,l}^1) g_{i+1,l}^1 = 0, \quad \text{for all } l,$$

which is equivalent to that for all l at least one of $z_{i+1,l}^1 = z_{i,l}^1$ and $g_{i+1,l}^1 = 0$ holds. From the definitions in (2.3.6) and (2.3.7), $g_{i+1,l}^1 = 0$ implies that

$z^1_{i+1,l} = 1$. I.e., there are only two cases: for all l, either

$$z^1_{i+1,l} = z^1_{i,l},$$

or

$$z^1_{i,l} = 0 \quad \text{and} \quad z^1_{i+1,l} = 1.$$

By the finiteness of $z^1_{i+1,l}$, after a finite number of steps, we have

$$z^1_{i+1,l} = z^1_{i,l} \quad \text{for all } l. \tag{2.3.8}$$

Similarly, we have

$$z^2_{i+1,l} = z^2_{i,l} \quad \text{for all } l. \tag{2.3.9}$$

Consequently, it follows that after a finite number of steps $(z^1_{i+1}, z^2_{i+1}) = (z^1_i, z^2_i)$. ■

So far we have developed and analyzed an iterative algorithm for computing a pair of discretized optimal sensor rules. It should be noted that the result on the finite convergence can be extend easily to the case where the fusion center employs an AND type decision rule and the sensor rules are computed via the algorithm (2.3.4). For the case of AND decision rule being used at the fusion center, the above convergence analysis (Theorem 2.3.1) must be modified. Fortunately, the required modifications are only minor. In fact, the argument used in the proof of Theorem 2.3.1 can be modified to show that the Gauss-Seidel type algorithm for the AND fusion rule is still finitely convergent; we shall not give a proof of this fact except to say that it uses a slightly different potential function. However, it is hard to extend the results on the necessary condition and finite convergence Theorems 2.2.1, 2.2.2, and 2.3.1 to a general fusion rule, such as the "XOR" fusion rule:

$$\mathcal{H}_1 = \{(\mathbf{y}_1, \mathbf{y}_2) : (g^1(\mathbf{y}_1) > 0, g^2(\mathbf{y}_2) \le 0), (g^1(\mathbf{y}_1) > 0, g^2(\mathbf{y}_2) \le 0)\}, \tag{2.3.10}$$

$$\mathcal{H}_0 = \{(\mathbf{y}_1, \mathbf{y}_2) : (g^1(\mathbf{y}_1) > 0, g^2(\mathbf{y}_2) > 0), (g^1(\mathbf{y}_1) \le 0, g^2(\mathbf{y}_2) \le 0)\}. \tag{2.3.11}$$

Denote the cost functional for the XOR fusion rule by

$$C(g^1, g^2; XOR) = c + \int_{\mathcal{H}_0} (ap(\mathbf{y}_1, \mathbf{y}_2|\mathbf{H}_1) - bp(\mathbf{y}_1, \mathbf{y}_2|\mathbf{H}_0))\, d\mathbf{y}_1 d\mathbf{y}_2. \tag{2.3.12}$$

By the definitions of $\mathcal{H}_0$ and $\mathcal{H}_0$ in (2.3.10) and (2.3.11), the integral in Eq. (2.3.12) cannot reduce to an integral over single region defined by a pair inequalities of $(g^1(\mathbf{y}_1), g^2(\mathbf{y}_2))$ as done for AND rules in Eq. (2.1.8). Hence, all arguments and methods used in Theorems 2.2.1, 2.2.2, and 2.3.1 are not able to be used for the XOR rule. In the next chapter, we will extend the above algorithms and the convergence results to any general fusion rules by using indicator functions to describe sensor rules and a general fusion rule. In fact, we can also prove that the discrete optimal cost determined by the optimal discretized sensor rules converges to the infimum of the original continuous cost as the discretized step-size goes to zero. In addition, those results can be extended to the corresponding two sensor Neyman-Pearson and sequential binary decision problems. To avoid repetition in the rather similar proofs, we will complete these extensions only for the more general multisensor decision fusion model given in the next chapter.

2.4 RELATIONSHIPS WITH LIKELIHOOD RATIO SENSOR RULES

Tenney and Sandell [77] pointed out that the optimal sensor rules are of the form of likelihood ratios when the sensor observations are conditionally independent of each other. But when this condition is not satisfied, the likelihood ratio test may not be the optimal sensor rule (see Example 2.5.2 below in Section 2.5). An interesting problem is to find conditions on the correlated observations so that the optimal sensor rule remains of the form of a likelihood ratio. In [17], such a condition was given for the following example:

$$p(\mathbf{y}_1, \mathbf{y}_2|\mathbf{H}_1) \sim N\left(\begin{pmatrix} \mu \\ \eta \end{pmatrix}, \begin{bmatrix} \sigma_1^2 & \sigma_{12} \\ \sigma_{21} & \sigma_2^2 \end{bmatrix}\right),$$

$$p(\mathbf{y}_1, \mathbf{y}_2|\mathbf{H}_0) \sim N\left(\begin{pmatrix} 0 \\ 0 \end{pmatrix}, \begin{bmatrix} \sigma_1^2 & \sigma_{12} \\ \sigma_{21} & \sigma_2^2 \end{bmatrix}\right).$$

When $\mu = \eta$, this is a model for the detector of a constant signal μ with dependent Gaussian sensor noises. A sufficient condition under which the optimal sensor rules are of the form of a likelihood ratio is given in [17] as follows:

$$\sigma_{12}(\eta\sigma_1^2 - \mu\sigma_{12})(\mu\sigma_2^2 - \eta\sigma_{12}) \geq 0.$$

We will now give a random signal detection example whose optimal sensor rules are also of the form of likelihood ratios.

The observations consist of a random signal s and noises ν_1, ν_2 such that

$$\begin{aligned} \mathbf{H}_1 &: \mathbf{y}_1 = s + \nu_1, \; \mathbf{y}_2 = s + \nu_2, \\ \mathbf{H}_0 &: \mathbf{y}_1 = \nu_1, \qquad \mathbf{y}_2 = \nu_2, \end{aligned}$$

where s, ν_1 and ν_2 are all mutually independent, and

$$s \sim N(\mu, \sigma^2), \;\; \nu_1 \sim N(0, \sigma_1^2), \;\; \nu_2 \sim N(0, \sigma_2^2).$$

Thus, the two conditional probability density functions are as follows:

$$p(\mathbf{y}_1, \mathbf{y}_2|\mathbf{H}_1) \sim N\left(\begin{pmatrix}\mu \\ \mu\end{pmatrix}, \begin{bmatrix}\sigma^2+\sigma_1^2 & \sigma^2 \\ \sigma^2 & \sigma^2+\sigma_2^2\end{bmatrix}\right),$$

$$p(\mathbf{y}_1, \mathbf{y}_2|\mathbf{H}_0) \sim N\left(\begin{pmatrix}0 \\ 0\end{pmatrix}, \begin{bmatrix}\sigma_1^2 & 0 \\ 0 & \sigma_2^2\end{bmatrix}\right).$$

It is easy to see that the two single sensor likelihood ratio functions $LR(\mathbf{y}_1)$ and $LR(\mathbf{y}_2)$ are both quadratic. Their axes of symmetry are

$$\{\mathbf{y}_1 : \mathbf{y}_1 = \frac{\sigma_1^2\mu}{\sigma^2}\}, \quad \{\mathbf{y}_2 : \mathbf{y}_2 = \frac{\sigma_2^2\mu}{\sigma^2}\}, \tag{2.4.1}$$

respectively. Therefore, changing the thresholds of likelihood ratio test can only change the constant terms of the quadratic functions. Their axes of symmetry are invariant. This property of the likelihood ratio test indicates that the "0" compression region and the "1" compression region for each sensor consist of intervals of the axes of symmetry given in (2.4.1). Obviously, the choice of such sensor rules is too restrictive, and, generally, likelihood ratio is not an optimal sensor rule (for example, see Example 2.5.2 in Section 2.5). However, if $\mu = 0$, then the two conditional probability density functions are centrosymmetric, i.e., the integrand in the cost functional and its negative and positive value regions are both centrosymmetric (see the contours between two parallel straight lines and outside the region between two parallel straight lines in Fig. 2.4.1). Hence, it is expected that the optimal distributed $\mathcal{H}_0$ decision region would

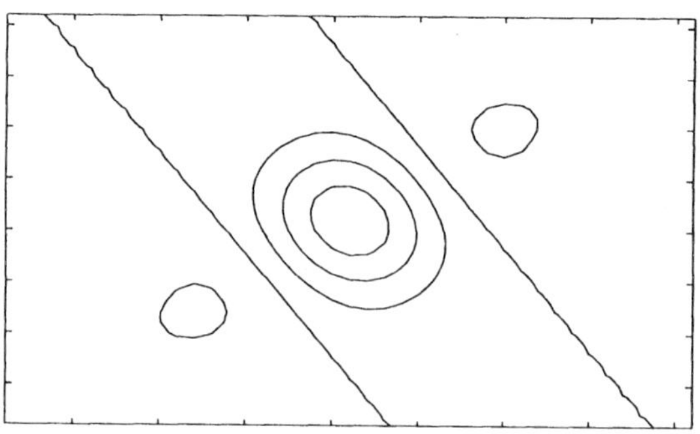

Figure 2.4.1. The contour map of the integrand.

be centrosymmetric as well. On the other hand, by (2.4.1), the two axes of symmetry of $LR(\mathbf{y}_1)$ and $LR(\mathbf{y}_2)$ become $\mathbf{y}_1 = 0$ and $\mathbf{y}_2 = 0$, respectively. It follows that the centrosymmetric optimal distributed $\mathcal{H}_0$ decision region may be determined also by the optimal likelihood ratio sensor rules.

More generally, we conjecture that the same thing may happen for the following case:

$$p(\mathbf{y}_1, \mathbf{y}_2|\mathbf{H}_1) \sim N\left(\begin{pmatrix}\mu\\ \mu\end{pmatrix}, \begin{bmatrix}\sigma^2+\sigma_1^2 & \sigma^2\\ \sigma^2 & \sigma^2+\sigma_2^2\end{bmatrix}\right),$$

$$p(\mathbf{y}_1, \mathbf{y}_2|\mathbf{H}_0) \sim N\left(\begin{pmatrix}\mu\\ \mu\end{pmatrix}, \begin{bmatrix}\sigma_1^2 & 0\\ 0 & \sigma_2^2\end{bmatrix}\right).$$

Many numerical examples support the conjecture, but the rigorous proof for the conjecture have not derived.

2.5 NUMERICAL EXAMPLES

In this section, we present several numerical examples which show the effectiveness of the optimal distributed decision (ODD) algorithms developed in the previous section. The general form of the signal and noise for all the examples is

$$\begin{aligned}\mathbf{H}_1: \quad & \mathbf{y}_1 = \mathbf{s} + \nu_1, \qquad \mathbf{y}_2 = \mathbf{s} + \nu_2,\\ \mathbf{H}_0: \quad & \mathbf{y}_1 = \nu_1, \qquad\quad\;\; \mathbf{y}_2 = \nu_2.\end{aligned} \tag{2.5.1}$$

The signal and noise models will yield the joint PDF of $\mathbf{y}_1$ and $\mathbf{y}_2$ under the two hypotheses. We employ the ratio a/b to control the cost functional where the parameters a and b have been defined in Subsection 2.1.1. The optimal sensor rules $g^1(\mathbf{y}_1)$ and $g^2(\mathbf{y}_2)$ for the particular value of a/b are then evaluated using the method described in the previous section. The cost for the decision can then be evaluated using (2.1.7) or (2.1.8), corresponding to the OR and AND fusion rules, respectively. The receiver operating characteristics (ROC) [63, 83] which plots the probability P_d against the probability P_f can be evaluated at selected values of a/b by the following formulae:

$$P_d = 1 - \int_{\substack{g^1(\mathbf{y}_1)\le 0\\ g^2(\mathbf{y}_2)\le 0}} p(\mathbf{y}_1, \mathbf{y}_2|\mathbf{H}_1) d\mathbf{y}_1 d\mathbf{y}_2, \tag{2.5.2a}$$

$$P_f = 1 - \int_{\substack{g^1(\mathbf{y}_1)\le 0\\ g^2(\mathbf{y}_2)\le 0}} p(\mathbf{y}_1, \mathbf{y}_2|\mathbf{H}_0) d\mathbf{y}_1 d\mathbf{y}_2, \tag{2.5.2b}$$

when the OR fusion rule is used, or

$$P_d = \int_{\substack{g^1(\mathbf{y}_1)> 0\\ g^2(\mathbf{y}_2)> 0}} p(\mathbf{y}_1, \mathbf{y}_2|\mathbf{H}_1) d\mathbf{y}_1 d\mathbf{y}_2, \tag{2.5.3a}$$

$$P_f = \int_{\substack{g^1(\mathbf{y}_1)> 0\\ g^2(\mathbf{y}_2)> 0}} p(\mathbf{y}_1, \mathbf{y}_2|\mathbf{H}_0) d\mathbf{y}_1 d\mathbf{y}_2, \tag{2.5.3b}$$

when the AND fusion rule is used.

Example 2.5.1.

In this example, the sensor outputs are real-valued and the signal s is a Gaussian random variable with mean value equal to 2 and variance equal to 1. The signal is observed by the two sensors under independent zero mean Gaussian noises with variances being equal to 2 and 1 respectively. Since the signal is common to both sensors, the outputs $\mathbf{y}_1$ and $\mathbf{y}_2$ are no longer independent, and the joint PDF under the two hypotheses can easily be shown to be

$$p(\mathbf{y}_1, \mathbf{y}_2|\mathbf{H}_1) \sim N(\mu_1, \mathbf{R}_1), \tag{2.5.4a}$$

$$p(\mathbf{y}_1, \mathbf{y}_2|\mathbf{H}_0) \sim N(\mu_0, \mathbf{R}_0), \tag{2.5.4b}$$

where

$$\mu_1 = [2 \ \ 2]', \qquad \mu_0 = [0 \ \ 0]', \tag{2.5.5a}$$

$$\mathbf{R}_1 = \begin{bmatrix} 3 & 1 \\ 1 & 2 \end{bmatrix}, \quad \mathbf{R}_0 = \begin{bmatrix} 2 & 0 \\ 0 & 1 \end{bmatrix}. \tag{2.5.5b}$$

The ROC for the ODD using the OR and AND rules and that for the optimal centralized decision (OCD) are shown in Fig. 2.5.1. For the OCD method, as mentioned in Section 2.1, the optimal decision rule is the joint LRT. Thus, the probability of detection and the probability of a false alarm are, respectively, given by

$$P_d = \int_{LR(\mathbf{y}_1,\mathbf{y}_2)>\theta} p(\mathbf{y}_1, \mathbf{y}_2|\mathbf{H}_1) d\mathbf{y}_1 d\mathbf{y}_2, \tag{2.5.6a}$$

$$P_f = \int_{LR(\mathbf{y}_1,\mathbf{y}_2)>\theta} p(\mathbf{y}_1, \mathbf{y}_2|\mathbf{H}_0) d\mathbf{y}_1 d\mathbf{y}_2, \tag{2.5.6b}$$

where $\theta = b/a$ is the threshold.

Example 2.5.2.

We will now evaluate the ROC for the optimal distributed decision (ODD) and for the optimal local LRT decision (OLD). The local LRTs at the sensors are given by

$$LR(\mathbf{y}_i) \underset{d\,=\,\mathbf{H}_0}{\overset{d\,=\,\mathbf{H}_1}{\gtrless}} \theta_i, \quad i = 1, 2. \tag{2.5.7}$$

Thus, for the OLD, if we use the OR fusion rule,

$$P_{d(LRT(OR))} = 1 - \int_{\substack{LR_1(\mathbf{y}_1)\le\theta_1 \\ LR_2(\mathbf{y}_2)\le\theta_2}} p(\mathbf{y}_1, \mathbf{y}_2|\mathbf{H}_1) d\mathbf{y}_1 d\mathbf{y}_2, \tag{2.5.8a}$$

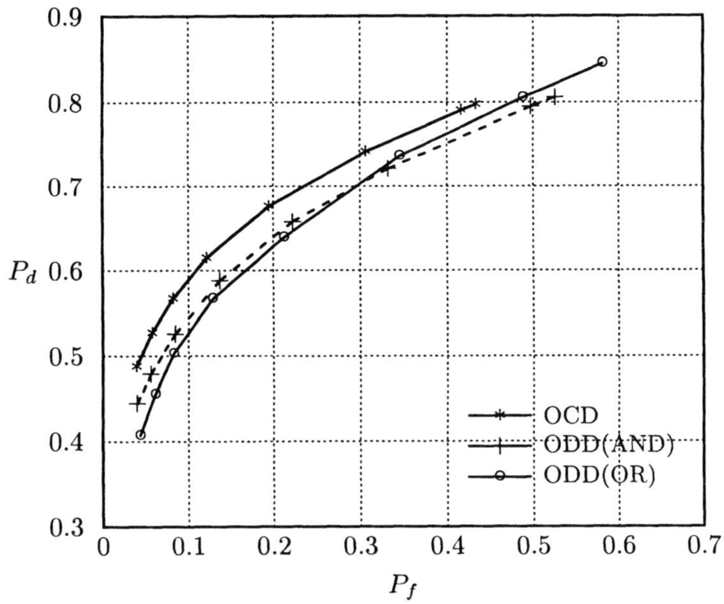

Figure 2.5.1. ROCs of the optimal centralized and optimal AND and OR decision rules.

$$P_{f(LRT(OR))} = 1 - \int_{\substack{LR_1(\mathbf{y}_1)\le\theta_1 \\ LR_2(\mathbf{y}_2)\le\theta_2}} p(\mathbf{y}_1, \mathbf{y}_2|\mathbf{H}_0)d\mathbf{y}_1 d\mathbf{y}_2, \tag{2.5.8b}$$

where θ_1 and θ_2 are, respectively, the thresholds for the first sensor and Sensor 2.

If the AND fusion rule is used, then the OLD will have

$$P_{d(LRT(AND))} = \int_{\substack{LR_1(\mathbf{y}_1)>\theta_1 \\ LR_2(\mathbf{y}_2)>\theta_2}} p(\mathbf{y}_1, \mathbf{y}_2|\mathbf{H}_1)d\mathbf{y}_1 d\mathbf{y}_2, \tag{2.5.9a}$$

$$P_{f(LRT(AND))} = \int_{\substack{LR_1(\mathbf{y}_1)>\theta_1 \\ LR_2(\mathbf{y}_2)>\theta_2}} p(\mathbf{y}_1, \mathbf{y}_2|\mathbf{H}_0)d\mathbf{y}_1 d\mathbf{y}_2. \tag{2.5.9b}$$

In general, the optimal values of θ_1 and θ_2 in Eqs. (2.5.8) and (2.5.9) are unknown [77]. In the following examples, θ_1 and θ_2 are located by globally searching for the minimum of the cost functions in (2.1.7) and (2.1.8)

$$\arg\min_{\theta_1,\theta_2} C_{LRT(OR)} = \arg\min_{\theta_1,\theta_2} \int_{\substack{LR_1(\mathbf{y}_1)\le\theta_1 \\ LR_2(\mathbf{y}_2)\le\theta_2}} L(\mathbf{y}_1, \mathbf{y}_2)d\mathbf{y}_1 d\mathbf{y}_2, \tag{2.5.10a}$$

and

$$\arg\min_{\theta_1,\theta_2} C_{LRT(AND)} = \arg\min_{\theta_1,\theta_2}(- \int_{\substack{LR_1(\mathbf{y}_1)>\theta_1 \\ LR_2(\mathbf{y}_2)>\theta_2}} L(\mathbf{y}_1, \mathbf{y}_2)d\mathbf{y}_1 d\mathbf{y}_2), \tag{2.5.10b}$$

where

$$L(\mathbf{y}_1, \mathbf{y}_2) = ap(\mathbf{y}_1, \mathbf{y}_2|\mathbf{H}_1) - bp(\mathbf{y}_1, \mathbf{y}_2|\mathbf{H}_0).$$

We emphasize that the search for the optimal θ_1 and θ_2 in (2.5.10) is in general computationally inefficient and therefore the method of the local LRT with "optimal" thresholds might not have much practical value.

We now change the means and variances in Example 2.5.1 so that

$$\mu_1 = [-1\ -1]', \qquad \mu_0 = [0\ 0]', \tag{2.5.11a}$$

$$\mathbf{R}_1 = \begin{bmatrix} 1.1 & 1 \\ 1 & 1.1 \end{bmatrix}, \text{ and } \mathbf{R}_0 = \begin{bmatrix} 1 & 0 \\ 0 & 1 \end{bmatrix}. \tag{2.5.11b}$$

In this case, the axes of symmetry of $LR(\mathbf{y}_1)$ and $LR(\mathbf{y}_2)$ in Eq. (2.4.1) become $\mathbf{y}_1 = -0.1$ and $\mathbf{y}_2 = -0.1$. Let $P_0 = P_1 = \frac{1}{2}$, $C_{01} = C_{10} = 1$, $C_{11} = C_{00} = 0$. Now we show that in general, the decision regions of the ODD and the OLD under the OR rule are different (see Fig.2.5.2) therefore the OLD is worse than the ODD.

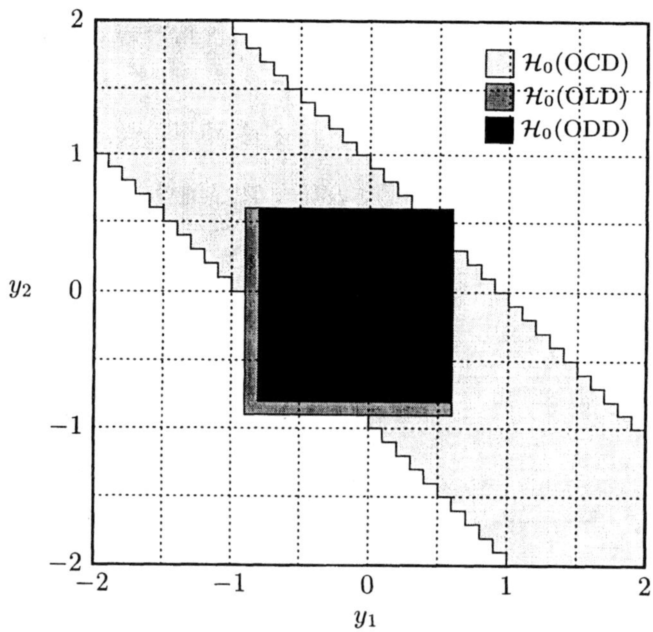

Figure 2.5.2. Decision regions of the centralized, optimal distributed OR and LR sensor rules.

In addition, we present the costs of the OCD, ODD(OR), ODD(AND), OLD(OR), and OLD(AND) decision fusion methods in the following Table 2.5.1.

Table 2.5.1. The costs of five decision fusion methods.

methods	OCD	ODD(OR)	OLD(OR)	ODD(AND)	OLD(AND)
cost	0.1284	0.1480	0.1490	0.1515	0.1590

It shows that the performance of the optimal sensor likelihood ratio rules is in general worse than that of the optimal sensor rules defined in Theorems 2.2.1 and 2.2.2.

Example 2.5.3.

The next example concerns the problem discussed in Section 2.4. In the examples that we have run, the optimal sensor LRT rules were also the optimal distributed sensor rules under both the OR and the AND fusion rules in the model given in Section 2.4.

Now we change the mean of the signal in Example 2.5.2 so that

$$\mu_1 = [0\ 0]', \qquad \mu_0 = [0\ 0]', \tag{2.5.12a}$$

$$\mathbf{R}_1 = \begin{bmatrix} 1.1 & 1 \\ 1 & 1.1 \end{bmatrix}, \quad \text{and } \mathbf{R}_0 = \begin{bmatrix} 1 & 0 \\ 0 & 1 \end{bmatrix}. \tag{2.5.12b}$$

In this case, the axes of symmetry of $LR(\mathbf{y}_1)$ and $LR(\mathbf{y}_2)$ in (2.4.1) become $\mathbf{y}_1 = 0$ and $\mathbf{y}_2 = 0$.

Since the axes of symmetry of the decision regions of the optimal distributed decision fusion problem under both the AND and the OR rules in Fig. 2.5.3 are also $\mathbf{y}_1 = 0$ and $\mathbf{y}_2 = 0$, we can see that the optimal sensor LRT rules in this example are the optimal sensor rules.

Many other examples with different distributions for the signal and noise have been run. It has been observed that, apart from the cases involving Gaussian noises, the ODD methods always outperform the optimal sensor LRT rule methods. On the other hand, the performance of the ODD method using the OR fusion rule differs from that using the AND fusion rule. Depending on the distributions of the signal and noise, the AND rule provides better performance in some cases while the OR rule is superior in others. The ODD method based on the AND rule performs better than that based on the OR rule when $b = P_0(c_{10} - c_{00})$ is large compared to $a = P_1(c_{01} - c_{11})$. This corresponds to the lower left-hand corner of the ROCs in Fig. 2.5.1. However, when the reverse is true, corresponding to the top right-hand corner of the ROCs in Fig. 2.5.1, then the ODD method based on the OR rule is better. This is intuitively clear from the expression for the cost in (2.1.4), and the nature of the two rules:

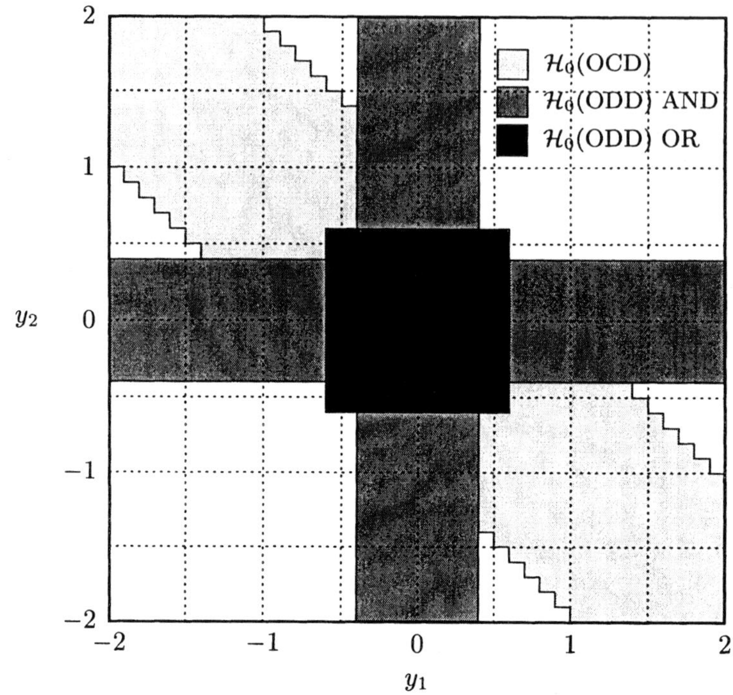

Figure 2.5.3. Decision regions of the optimal distributed decisions under the AND and the OR rules in Example 2.5.3.

the OR rule is more susceptible to a false alarm and the AND rule is more susceptible to a miss. If $b \gg a$, then the term $bp(\mathbf{y}_1, \mathbf{y}_2|\mathbf{H}_0)$ dominates the integrand in the cost functional. This implies that declaring $\mathbf{H}_1$, when $\mathbf{H}_0$ is true, makes a big cost. Under such a condition, the smaller the false alarm rate is, the smaller the cost is. Hence the performance of the AND rule is better than the OR rule in this case. Similarly, the reverse is true when $a \gg b$.

2.6 RANDOMIZED FUSION RULES

In most of the decision fusion part of this book, we restrict ourselves to consider the deterministic decision rules. The Neyman-Pearson decision problem (see Section 1.3.2) was an exception. In the distributed decision problem, to improve the system performance, it is useful to consider randomized fusion rules.

In Example 2.5.1, we found that in the AND and the OR fusion rules there is no one dominant rule for the system performance between (see Fig. 2.5.1). Generally, we do not see a dominant ROC curve among the ROCs of all deterministic fusion rules (see Fig. 2.6.1). Of course, we need not use the same fusion rule on the entire interval [0, 1] of the P_f coordinate axis. Since there is a piece of a dominant ROC between the intersecting points of two ROCs, we can put together all such pieces of dominant ROCs to obtain a dominant ROC. I.e., at each point of P_f, we select a fusion rule with the largest P_d. Although the ROC curve based on this method has the best system performance over all ROCs with single fusion rule, it is not concave. This suggests that it is possible to modify this non-concave curve to be concave by using the linear interpolation method. Let us replace those parts of above derived optimal ROC curve that are not concave by a common tangent as shown in Fig. 2.6.1. The resulting curve

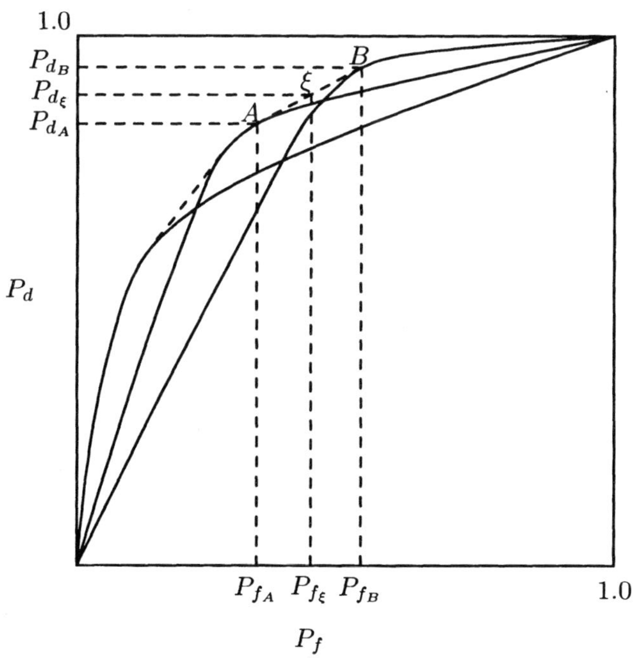

Figure 2.6.1. Constructing better ROC by using randomized fusion rule.

on all of [0, 1] is concave, but it is an artificial curve and not a real ROC curve. If we can give a fusion rule to implement such a curve, it will obviously have better system performance than the original non-concave optimal ROC curve. Since it obviously cannot be implemented by a deterministic fusion rule, we naturally try randomization.

The optimal ROC curve which was constructed piece by piece chose the best fusion rule at each point P_{f_ξ}. Now, we choose fusion rule F_i, $i \leq n$, with probability of $\{P_{P_{f_\xi}}(F_i)$, where the optimal probability is to be determined. Clearly, this is a randomized fusion rule.

The remaining problem is how to implement a randomized rule whose ROC is the tangent interval, namely the straight line interval [A, B] in Fig. 2.6.1. Clearly, the two ends A and B are on the ROC curves of two deterministic fusion rules, namely, fusion rule F_A and fusion rule F_B, respectively. Then the any point ξ on the tangent interval between A and B will be implemented by a proper randomized fusion rule, according to which, we use fusion rule F_A with probability $P_\xi(F_A)$ and use fusion rule F_B with probability $P_\xi(F_B)$, where

$$P_\xi(F_A) + P_\xi(F_B) = 1.$$

Let the two coordinates of A, B, and any point ξ on the tangent interval be (P_{f_A}, P_{d_A}), (P_{f_B}, P_{d_B}) and (P_{f_ξ}, P_{d_ξ}), respectively. The question now is how to get the pair of probabilities $P_\xi(F_A)$ and $P_\xi(F_A)$ that we need for the randomized fusion rule so that under this fusion rule the coordinate of the vertical axis corresponding to the point P_{f_ξ} of the P_f coordinate axis is exactly P_{d_ξ}. By the linear interpolation method, we know that

$$P_{f_\xi} = \frac{P_{f_B} - P_{f_\xi}}{P_{f_B} - P_{f_A}} P_{f_A} + \frac{P_{f_\xi} - P_{f_A}}{P_{f_B} - P_{f_A}} P_{f_B}, \tag{2.6.1}$$

$$P_{d_\xi} = \frac{P_{f_B} - P_{f_\xi}}{P_{f_B} - P_{f_A}} P_{d_A} + \frac{P_{f_\xi} - P_{f_A}}{P_{f_B} - P_{f_A}} P_{d_B}. \tag{2.6.2}$$

By the above two equations and the total probability formula, it is obvious that for the randomized fusion rule with the two assignment probabilities

$$P_\xi(F_A) = \frac{P_{f_B} - P_{f_\xi}}{P_{f_B} - P_{f_A}}, \quad P_\xi(F_B) = \frac{P_{f_\xi} - P_{f_A}}{P_{f_B} - P_{f_A}},$$

the ROC curve over the interval $[P_{f_A}, P_{f_B}]$ of the P_f coordinate axis is the tangent line between points A and B.

It is worth noting that the derivations of the two pairs of probabilities (P_{f_A}, P_{d_A}) and (P_{f_B}, P_{d_B}) in Fig. 2.6.1 depend on not only fusion rules F_A and F_B, but also the associated two pairs of the parameters (a_A, b_A) and (a_B, b_B) of the integrands (2.1.4) in the cost functionals, respectively. In general, (a_A, b_A) and (a_B, b_B) are not the same. For a given decision cost functional with a fixed parameters (a, b), an optimal deterministic fusion rule obviously exists and cannot be improved by the randomized fusion rule. In other words, when we use fusion rule F_A with probability $P_\xi(F_A)$, the parameters (a_A, b_A) in the integrand must be parameters which yield P_{f_A} and P_{d_A} under the fusion rule F_A, and in general is not the same as the parameters (a_B, b_B) when we use fusion rule F_B to yield P_{f_B} and P_{d_B}. Therefore, the randomized fusion rule is suitable to improve the performance of the distributed Neyman-Pearson decision system.

Chapter 3

MULTISENSOR BINARY DECISION

In this chapter, we will not limit the number of sensors to two and not limit sensors transmission to a single binary number as done in the last chapter. In fact, to improve the decision accuracy, if additional communication bandwidth is available, from each sensor more than one binary number could be transmitted out. In the first six sections of this chapter, we will suggest a discretized iterative algorithm to approximate the optimal local (sensor) compression rules under a fixed fusion rule for the distributed multisensor Bayes binary decision problem. First, we will show that any general fusion rule can be formulated as a bi-valued polynomial function of the local compression rules. Then, under a given fusion rule, we will present a fixed point type necessary condition for the optimal local compression rules and propose an efficient discretized iterative algorithm and prove its finite convergence. After this, we will consider the optimal fusion rule problem for a class of the systems with a specific communication pattern. For such a system, namely an l-sensor system, suppose that the total number of bits transmitted by $l-1$ sensors are fixed, we can change the number of bits transmitted by the l sensor to a certain number, which is determined completely by the total number of bits transmitted by other $l-1$ sensors. For those systems with the modified communication pattern, we will present an optimal fusion rule, and prove that this fusion rule is not only superior to all possible fusion rules under the original communication pattern, but also an optimal fusion rule for the modified communication pattern. Furthermore, we will prove that even if the lth sensor now can transmit uncompressed observational data to the fusion center, the performance cannot be better than that achieved by the aforementioned optimal fusion rule. Moreover, this optimal fusion rule does not depend on the statistical properties of the observational data, or even on the decision criteria. It only depend on the total number of bits transmitted by other $l-1$ sensors. Numerical examples support the above results and give some guidance which sensor, to achieve better performance, one should choose as the lth sensor for the system with the modified communication pattern. In the final two sections, the results are extended to the Neyman-Pearson decision and the sequential decision problems.

3.1 THE FORMULATION FOR BAYES BINARY DECISION

We consider the Bayes decision with two hypotheses $\mathbf{H}_0$ and $\mathbf{H}_1$, l sensors, and corresponding observational data $\mathbf{y}_1, \ldots, \mathbf{y}_l$. A set of local compression rules at each of the l sensors, namely the ith sensor: compress the observational data $\mathbf{y}_i$ to r_i bits:

$$I_i^{(1)}(\mathbf{y}_i) : \Re^{n_i} \longmapsto \{0,1\}, \ \cdots, \ I_i^{(r_i)}(\mathbf{y}_i) : \Re^{n_i} \longmapsto \{0,1\}, \quad \forall i \leq l, \tag{3.1.1}$$

which local sensors then transmit to the fusion center. For later usage, we denote this information structure by the expression $(\mathbf{r_1} + \mathbf{r_2} + \cdots + \mathbf{r}_\ell)$. The simplest case is the case of $r_1 = r_2 = \cdots = r_l = 1$, i.e., the information structure $(\mathbf{1} + \mathbf{1} + \cdots + \mathbf{1})$. Unlike we use the real valued function $g^i(\mathbf{y}_i)$ $(i = 1, 2)$ and the inequality

$$g^i(\mathbf{y}_i) \underset{d_1 = 0}{\overset{d_1 = 1}{\gtrless}} 0,$$

to define sensor rules in the last chapter, we now use a bi-valued functions $I_i^{(j)}(\mathbf{y}_i)$, i.e., the indicator function of "1" compression region on $\mathbf{y}_i$ directly as a sensor rule. Although $g^i(\mathbf{y}_i)$ seems more like a straightforward extension of the likelihood ratio functions, it has some drawbacks. First of all, $g^i(\mathbf{y}_i)$ and its "0" and "1" data compression regions are not 1 to 1 corresponding to each other. Infinite different $g^i(\mathbf{y}_i)$ can express the same data compression rule (this was explained around Eqs. (2.2.7)–(2.2.9)). Secondly, it is unconvenient to extend the previous results of Theorems 2.2.1, 2.2.2, and 2.3.1 in the last chapter to general fusion rules, such as the XOR fusion rule.

In terms of the deterministic decision rule, the so-called binary fusion rule of the fusion center is just a partition of the set of $2^{\sum_{i=1}^{l} r_i}$ points of the form of $(I_1^{(1)}, \ldots, I_1^{(r_1)}; I_2^{(1)}, \ldots, I_2^{(r_2)}; \cdots,; I_l^{(1)}, \ldots, I_l^{(r_l)})$ into 2 disjoint subsets. In the sequel of this chapter, for notational simplicity we first consider the $(\mathbf{1} + \mathbf{1} + \cdots + \mathbf{1})$ information structure and its modified version $(\mathbf{1} + \mathbf{1} \cdots + \mathbf{1} + \mathbf{r}_\ell)$, and propose an optimal fusion rule for the second information structure. Then, it is easy to see that the results for the above simple cases can be extended to the above general information structure $(\mathbf{r_1} + \mathbf{r_2} + \cdots + \mathbf{r}_\ell)$ without any difficulty.

Suppose that two joint conditional probability density functions $p(\mathbf{y}_1, \mathbf{y}_2, \ldots, \mathbf{y}_l|\mathbf{H}_1)$ and $p(\mathbf{y}_1, \mathbf{y}_2, \ldots, \mathbf{y}_l|\mathbf{H}_0)$ of arbitrary form are given. The general form of the binary fusion rule will be denoted by a bi-valued function

$$F(I_1(\mathbf{y}_1), \cdots, I_l(\mathbf{y}_l)) : \{0,1\}^l \longmapsto \{0,1\}.$$

The goal of our distributed multisensor decision problem is to minimize the Bayes cost functional

$$\begin{aligned} C(I_1(\mathbf{y}_1), \ldots, I_l(\mathbf{y}_l); F) = {} & C_{00}P_0P(F=0|\mathbf{H}_0) + C_{01}P_1P(F=0|\mathbf{H}_1) \\ & + C_{10}P_0P(F=1|\mathbf{H}_0) + C_{11}P_1P(F=1|\mathbf{H}_1). \end{aligned} \tag{3.1.2}$$

Substituting the definition of F and $I_i(\mathbf{y}_i)$ into (3.1.2) and simplifying, we see that

$$\begin{aligned} & C(I_1(\mathbf{y}_1), \ldots, I_l(\mathbf{y}_l); F) = P_0C_{10} + P_1C_{11} \\ & + \int_{\{(\mathbf{y}_1, \ldots, \mathbf{y}_l):F(I_1, \ldots, I_l)=0\}} \{[P_1(C_{01} - C_{11})p(\mathbf{y}_1, \ldots, \mathbf{y}_l|\mathbf{H}_1)] \\ & \qquad - [P_0(C_{10} - C_{00})p(\mathbf{y}_1, \ldots, \mathbf{y}_l|\mathbf{H}_0)]\} \, d\mathbf{y}_1 \cdots d\mathbf{y}_l \\ & = c + \int_{\{(\mathbf{y}_1, \ldots, \mathbf{y}_l):F(I_1, \ldots, I_l)=0\}} \{ap(\mathbf{y}_1, \ldots, \mathbf{y}_l|\mathbf{H}_1) - bp(\mathbf{y}_1, \ldots, \mathbf{y}_l|\mathbf{H}_0)\} \\ & \qquad d\mathbf{y}_1 \cdots d\mathbf{y}_l, \end{aligned} \tag{3.1.3}$$

where we let

$$c = P_0C_{10} + P_1C_{11}, \quad a = P_1(C_{01} - C_{11}), \quad b = P_0(C_{10} - C_{00}).$$

Our goal is to select a set of optimal sensor rules $(I_1(\mathbf{y}_1), \ldots, I_l(\mathbf{y}_l))$ together with a given fusion rule F which will jointly minimize the integral in (3.1.3). Of course, to compute the cost functional here we only consider those local compression rules under which the integral in (3.1.3) exist for all possible fusion rules.

3.2 FORMULATION OF FUSION RULES VIA POLYNOMIALS OF SENSOR RULES

From the definition of I_i in (3.1.1), the local compression rules $I_i(\mathbf{y}_i)$, $i = 1, 2, \ldots, l$, are all actually indicator functions. Define

$$\Omega_i = \{\mathbf{y}_i \in \Re^{n_i} : I_i(\mathbf{y}_i) = 1\}, \quad i = 1, 2, \ldots, l, \tag{3.2.1}$$

$$\Omega_i^c = \{\mathbf{y}_i \in \Re^{n_i} : I_i(\mathbf{y}_i) = 0\}, \quad i = 1, 2, \ldots, l. \tag{3.2.2}$$

We shall call Ω_i the *1 compression region* for the ith sensor.

The notation (3.2.1) and (3.2.2) define an injective and surjective relationship between a pair of local regions $\{\Omega_i,\ \Omega_i^c\}$ and a pair of simple polynomials $\{I_i,\ 1 - I_i\}$. Furthermore, a local message ($I_1(\mathbf{y}_1) = d_1,\ I_2(\mathbf{y}_2) = d_2, \ldots,$ $I_l(\mathbf{y}_l) = d_l;\ d_j = 0$ or $1,\ j \leq l$) from all sensors corresponds uniquely to a product

$$P_1(\mathbf{y}_1)P_2(\mathbf{y}_2)\cdots P_l(\mathbf{y}_l) \tag{3.2.3}$$

of the above simple polynomials, where for any $j \leq l$,

$$P_j(\mathbf{y}_j) = \begin{cases} 1 - I_j(\mathbf{y}_j) & \text{if } d_j = 0, \\ I_j(\mathbf{y}_j) & \text{if } d_j = 1. \end{cases}$$

We call the polynomial in (3.2.3) the *local message polynomial*. Obviously,

$$\{(\mathbf{y}_1, \mathbf{y}_2,\ \ldots,\ \mathbf{y}_l) : (I_1(\mathbf{y}_1) = i_1, I_2(\mathbf{y}_2) = i_2,\ \ldots,\ I_l(\mathbf{y}_l) = i_l)\}$$

$$= \{(\mathbf{y}_1, \mathbf{y}_2,\ \ldots,\ \mathbf{y}_l) : P_1(\mathbf{y}_1)P_2(\mathbf{y}_2)\cdots P_l(\mathbf{y}_l) = 1\}. \tag{3.2.4}$$

The final decision F of the fusion center is based on the local message ($I_1(\mathbf{y}_1)$, $I_2(\mathbf{y}_1), \ldots,\ I_l(\mathbf{y}_1)$). F will map some of the elements of the set $\{0,1\}^l$ onto the decision "0" and the rest onto the decision "1". Since the set of $\{0,1\}^l$ has 2^l different values therefore has 2^{2^l} possible partitions of two parts, a final $\mathbf{H}_0$ or $\mathbf{H}_1$ decision region associated with any one of the 2^{2^l} possible fusion rules corresponds uniquely to a summation of some of the above local message polynomials. We denote the summation of the local message polynomials associated with $\mathbf{H}_1$ by

$$P_{\mathbf{H}_1}(I_1(\mathbf{y}_1), I_2(\mathbf{y}_2),\ \ldots,\ I_l(\mathbf{y}_l)),$$

which is still a binary-valued function, since for given observational data ($\mathbf{y}_1, \mathbf{y}_2,$ $\cdots, \mathbf{y}_l$) one and only one of the local message polynomials can equals unity. In other words, we have

$$\{(\mathbf{y}_1, \mathbf{y}_2 \cdots \mathbf{y}_l) : F(I_1(\mathbf{y}_1), I_2(\mathbf{y}_2),\ \ldots,\ I_l(\mathbf{y}_l)) = 1\}$$

$$= \{(\mathbf{y}_1, \mathbf{y}_2 \cdots \mathbf{y}_l) : P_{\mathbf{H}_1}(I_1(\mathbf{y}_1), I_2(\mathbf{y}_2),\ \ldots,\ I_l(\mathbf{y}_l)) = 1\} \tag{3.2.5}$$

and

$$\{(\mathbf{y}_1, \mathbf{y}_2 \cdots \mathbf{y}_l) : F(I_1(\mathbf{y}_1), I_2(\mathbf{y}_2),\ \ldots,\ I_l(\mathbf{y}_l)) = 0\}$$

$$= \{(\mathbf{y}_1, \mathbf{y}_2 \cdots \mathbf{y}_l) : P_{\mathbf{H}_1}(I_1(\mathbf{y}_1), I_2(\mathbf{y}_2),\ \ldots,\ I_l(\mathbf{y}_l)) = 0\}. \tag{3.2.6}$$

Thus, the fusion rule $F(I_1, I_2,\ \ldots,\ I_l)$ is finally formulated via the polynomial $P_{\mathbf{H}_1}(I_1(\mathbf{y}_1), I_2(\mathbf{y}_2),\ \ldots,\ I_l(\mathbf{y}_l))$ of the local sensor rules. We call $P_{\mathbf{H}_1}$ the *decision region polynomial*.

Now, let us give a example to show how to write a general fusion rule to be a polynomial of the local sensor rules.

Example 3.2.1. Consider 3-sensor binary decision problem and a fusion rule F which makes the decision $\mathbf{H}_1$ (i.e., $F(I_1(\mathbf{y}_1), I_2(\mathbf{y}_2), I_3(\mathbf{y}_3)) = 1$) if any one of the following 4 local messages

$$\{(1\ 1\ 1), (1\ 1\ 0), (1\ 0\ 1), (0\ 1\ 1)\}$$

is received by the fusion center, where $(1\ 1\ 1) \stackrel{\text{def}}{=} (I_1(\mathbf{y}_1) = 1,\ I_2(\mathbf{y}_2) = 1,\ I_3(\mathbf{y}_3) = 1)$, and so on. This fusion rule is the so-called **two-out-of-three** fusion rule and is neither an AND nor an OR rule. Thus, according to the aforementioned approach given by (3.2.3)–(3.2.6), the decision region polynomial $P_{\mathbf{H}_1}$ is

$$\begin{aligned} P_{\mathbf{H}_1} &= I_1(\mathbf{y}_1)I_2(\mathbf{y}_2)I_3(\mathbf{y}_3) + I_1(\mathbf{y}_1)I_2(\mathbf{y}_2)(1 - I_3(\mathbf{y}_3)) \\ &\quad + I_1(\mathbf{y}_1)(1 - I_2(\mathbf{y}_2))I_3(\mathbf{y}_3) + (1 - I_1(\mathbf{y}_1))I_2(\mathbf{y}_2)I_3(\mathbf{y}_3). \end{aligned}$$

In the sequel, it will be seen that using a polynomial of the indicator functions of the local compression regions to represent a fusion rule is a key technique in deriving a fixed point type necessary condition for the optimal local compression rules and in motivating an iterative algorithm to approximate them.

3.3 FIXED POINT TYPE NECESSARY CONDITION FOR THE OPTIMAL SENSOR RULES GIVEN A FUSION RULE

From (3.1.3), (3.2.5), and (3.2.6), we know that

$$\begin{aligned} &\int_{\substack{\{(\mathbf{y}_1, \ldots, \mathbf{y}_l): \\ F(I_1, \ldots, I_l) = 0\}}} \{ap(\mathbf{y}_1, \ldots, \mathbf{y}_l|\mathbf{H}_1) - bp(\mathbf{y}_1, \ldots, \mathbf{y}_l|\mathbf{H}_0)\} d\mathbf{y}_1 \cdots d\mathbf{y}_l \\ &= \int (1 - P_{\mathbf{H}_1}(I_1(\mathbf{y}_1), I_2(\mathbf{y}_2), \ldots, I_l(\mathbf{y}_l))) \\ &\quad \cdot \{ap(\mathbf{y}_1, \ldots, \mathbf{y}_l|\mathbf{H}_1) - bp(\mathbf{y}_1, \ldots, \mathbf{y}_l|\mathbf{H}_0)\} d\mathbf{y}_1 \cdots d\mathbf{y}_l. \end{aligned} \tag{3.3.1}$$

To show that the solution of the optimization problem in Section 3.2 satisfies a fixed point type necessary condition, we need to further analyze the decision region polynomial $P_{\mathbf{H}_1}$.

Since the polynomial $P_{\mathbf{H}_1}$ is a summation of some of the local message polynomials (3.2.3), we can write $1 - P_{\mathbf{H}_1}$ in the following l different equivalent forms:

$$\begin{aligned}
&1 - P_{\mathbf{H_1}}(I_1, \ldots, I_l) \\
&= (1 - I_1)P_{11}(I_2, I_3, \ldots, I_l) + P_{12}(I_2, I_3, \ldots, I_l) \\
&= (1 - I_2)P_{21}(I_1, I_3, \ldots, I_l) + P_{22}(I_1, I_3, \ldots, I_l) \\
&\qquad \cdots \\
&= (1 - I_l)P_{l1}(I_1, I_2, \ldots, I_{l-1}) + P_{l2}(I_1, I_2, \ldots, I_{l-1}),
\end{aligned} \tag{3.3.2}$$

where P_{i1} and P_{i2}, $i = 1, \ldots, l$, are the polynomials of I_j, $j = 1, \ldots, i-1, i+1, \ldots, l$, respectively, and P_{i1} and P_{i2} are both independent of I_i.

P_{i1} and P_{i2} in the equivalent forms (3.3.2), using the given $P_{\mathbf{H_1}}$, can be determined as follows.

$$\begin{aligned}
P_{i2} &= 1 - P_{\mathbf{H_1}}(I_1(\mathbf{y}_1), \ldots, I_{i-1}(\mathbf{y}_{i-1}), 1, I_{i+1}(\mathbf{y}_{i+1}), \ldots, I_l(\mathbf{y}_l)), \\
P_{i1} &= 1 - P_{\mathbf{H_1}}(I_1(\mathbf{y}_1), \ldots, I_{i-1}(\mathbf{y}_{i-1}), 0, I_{i+1}(\mathbf{y}_{i+1}), \ldots, I_l(\mathbf{y}_l)) - P_{i2}.
\end{aligned} \tag{3.3.3}$$

For notational simplicity, denote

$$L(\mathbf{y}_1, \mathbf{y}_2, \ldots \mathbf{y}_l) = ap(\mathbf{y}_1, \mathbf{y}_2, \ldots, \mathbf{y}_l|\mathbf{H_1}) - bp(\mathbf{y}_1, \mathbf{y}_2, \ldots, \mathbf{y}_l|\mathbf{H_0}). \tag{3.3.4}$$

Now we are prepared to state the following necessary condition for the optimal local compression rules.

Theorem 3.3.1. Suppose that we have a distributed multi-sensor decision system employing a general fusion rule (3.2.5). Given a set of optimal local compression rules $(I_1, I_2, \ldots, I_l)$ which minimizes the cost functional in (3.3.1), then $(I_1, I_2, \ldots, I_l)$ must satisfy the following integral equations:

$$\begin{aligned}
I_1(\mathbf{y}_1) &= I[\int P_{11} L(\mathbf{y}_1, \mathbf{y}_2, \ldots \mathbf{y}_l) d\mathbf{y}_2 d\mathbf{y}_3 \cdots d\mathbf{y}_l], \\
I_2(\mathbf{y}_2) &= I[\int P_{21} L(\mathbf{y}_1, \mathbf{y}_2, \ldots \mathbf{y}_l) d\mathbf{y}_1 d\mathbf{y}_3 \cdots d\mathbf{y}_l], \\
&\cdots \\
I_l(\mathbf{y}_l) &= I[\int P_{l1} L(\mathbf{y}_1, \mathbf{y}_2, \ldots \mathbf{y}_l) d\mathbf{y}_1 d\mathbf{y}_2 \cdots d\mathbf{y}_{l-1}],
\end{aligned} \tag{3.3.5}$$

where $I[\cdot]$ is defined by

$$I[x] = \begin{cases} 1, & \text{if } x > 0, \\ 0, & \text{if } x \leq 0. \end{cases} \tag{3.3.6}$$

Proof. From (3.3.1), (3.3.2) and (3.3.4), we know that the optimal local compression rules $(I_1, I_2, \ldots, I_l)$ must minimize the integral in (3.3.1), which can be written as the following different equivalent forms:

$$\begin{aligned}
&\int [1 - P_{\mathbf{H}_1}(I_1(\mathbf{y}_1), \ldots, I_l(\mathbf{y}_l))] L(\mathbf{y}_1, \ldots, \mathbf{y}_l) d\mathbf{y}_1 \cdots d\mathbf{y}_l \\
=& \int [(1 - I_1(\mathbf{y}_1)) P_{11} + P_{12}] L(\mathbf{y}_1, \ldots, \mathbf{y}_l) d\mathbf{y}_1 \cdots d\mathbf{y}_l \\
=& \int_{I_1(\mathbf{y}_1)=0} \{ \int P_{11} L(\mathbf{y}_1, \ldots, \mathbf{y}_l) d\mathbf{y}_2 \mathbf{y}_3 \cdots d\mathbf{y}_l \} \mathbf{y}_1 + C_{I_1} \\
=& \int [(1 - I_2(\mathbf{y}_2)) P_{21} + P_{22}] L(\mathbf{y}_1, \ldots, \mathbf{y}_l) d\mathbf{y}_1 \cdots d\mathbf{y}_l \\
=& \int_{I_2(\mathbf{y}_2)=0} \{ \int P_{21} L(\mathbf{y}_1, \ldots, \mathbf{y}_l) d\mathbf{y}_1 \mathbf{y}_3 \cdots d\mathbf{y}_l \} \mathbf{y}_2 + C_{I_2} \\
& \cdots \\
=& \int [(1 - I_l(\mathbf{y}_l)) P_{l1} + P_{l2}] L(\mathbf{y}_1, \ldots, \mathbf{y}_l) d\mathbf{y}_1 \cdots d\mathbf{y}_l \\
=& \int_{I_l(\mathbf{y}_l)=0} \{ \int P_{l1} L(\mathbf{y}_1, \ldots, \mathbf{y}_l) d\mathbf{y}_1 \cdots \mathbf{y}_{l-1} \} d\mathbf{y}_l + C_{I_l},
\end{aligned} \tag{3.3.7}$$

where the constants C_{I_i} are all independent of I_i since P_{i2} are all independent of I_i (see the statement after Eq. (3.3.2)). Moreover, to minimize the integrals in Eq. (3.3.7), the integral region $\{\mathbf{y}_i : I_i(\mathbf{y}_i) = 0\}$ should be the non-positive value region of the integrand. Thus, the solutions for the optimal local compressions are given by

$$\begin{aligned}
\{\mathbf{y}_1 : I_1(\mathbf{y}_1) = 0\} &= \{\mathbf{y}_1 : \int P_{11} L(\mathbf{y}_1, \ldots, \mathbf{y}_l) d\mathbf{y}_2 d\mathbf{y}_3 \cdots d\mathbf{y}_l < 0\}, \\
\{\mathbf{y}_2 : I_2(\mathbf{y}_2) = 0\} &= \{\mathbf{y}_2 : \int P_{21} L(\mathbf{y}_1, \ldots, \mathbf{y}_l) d\mathbf{y}_1 d\mathbf{y}_3 \cdots d\mathbf{y}_l < 0\}, \\
& \cdots \\
\{\mathbf{y}_l : I_l(\mathbf{y}_l) = 0\} &= \{\mathbf{y}_l : \int P_{l1} L(\mathbf{y}_1, \ldots, \mathbf{y}_l) d\mathbf{y}_1 d\mathbf{y}_2 \cdots d\mathbf{y}_{l-1} < 0\}.
\end{aligned} \tag{3.3.8}$$

This condition is equivalent to (3.3.5) and (3.3.6). Thus the theorem follows. ■

Obviously, the necessary condition for the optimal sensor compression rules in Theorem 3.3.1 is a fixed point type condition. In particular, if we define the mapping

$$\Gamma(I_1, I_2, ...I_l) = \begin{pmatrix} I[\int P_{11}L(\mathbf{y}_1, \mathbf{y}_2, \ldots, \mathbf{y}_l)d\mathbf{y}_2 d\mathbf{y}_3 \cdots d\mathbf{y}_l] \\ I[\int P_{21}L(\mathbf{y}_1, \mathbf{y}_2, \ldots, \mathbf{y}_l)d\mathbf{y}_1 d\mathbf{y}_3 \cdots d\mathbf{y}_l] \\ \cdots \\ I[\int P_{l1}L(\mathbf{y}_1, \mathbf{y}_2, \ldots, \mathbf{y}_l)d\mathbf{y}_1 d\mathbf{y}_2 \cdots d\mathbf{y}_{l-1}] \end{pmatrix}, \quad (3.3.9)$$

then the solution of (3.3.5) is precisely the fixed point of Γ. Theorem 3.3.1 shows that every set of optimal local compression rules must be a solution of the integral equation (3.3.5). Thus, given a fusion rule, the optimal distributed decision problem is effectively reduced to finding an optimal set $(I_1, I_2, \ldots, I_l)$ that satisfies the integral equation. In general, there is no guarantee of existence of the optimal local compression rules. When a solution exists, it is not necessarily unique, or observable in closed form. However, we know that the infimum of the cost functional exists because it is bounded from below. In the next section, we will prove that the global minimum of the discretized cost functional converges to this infimum as the discretization size tends to zero, under some mild assumptions. Hence, the discretized optimal solution can approximate the solution of the original optimization problem to an arbitrary accuracy, provided that the step size of discretion is small enough.

In particular, from Theorem 3.3.1, it is easy to see that if observational data $\mathbf{y}_1, \mathbf{y}_2, \ldots, \mathbf{y}_l$ are all conditionally independent of each other, then the optimal local compression rules $(I_1, I_2, \ldots, I_l)$ are all of the form of likelihood ratios, a well-known result of Tenney and Sandell [77].

Corollary 3.3.1. Suppose that the observations $\mathbf{y}_1, \mathbf{y}_2, \ldots, \mathbf{y}_l$ are statistically conditionally independent. Then, the optimal sensor compression rules $(I_1, I_2, \ldots, I_l)$ are all of the form of likelihood ratios.

Noting that

$$p(\mathbf{y}_1, \ldots, \mathbf{y}_l|\mathbf{H}_j) = p(\mathbf{y}_1|\mathbf{H}_j) \cdots p(\mathbf{y}_l|\mathbf{H}_j), \quad j = 0, 1,$$

and using (3.3.5), (3.3.6), and the notation (3.3.4), as well as the independence between P_{i1} and I_i, the proof of this corollary is completely similar to that of Corollary 2.2.1.

3.4 THE FINITE CONVERGENCE OF THE DISCRETIZED ALGORITHM

In this section, a computationally efficient iterative algorithm for computing a discrete approximation of the solution of Eq. (3.3.5) is developed. The finite convergence of this iterative algorithm and an approximation of the original continuous problem by its discrete version are also shown.

Noticing that the set of optimal sensor rules is precisely the fixed point of Γ, naturally, we use the mapping Γ given in (3.3.9) to motivate an iterative algorithm to approximate the optimal solution of (3.3.5). Let the local compression rules at the ith stage of the iteration be denoted by $(I_1^{(i)}, I_2^{(i)}, \ldots, I_l^{(i)})$ with the initial values $(I_1^{(0)}, I_2^{(0)}, \ldots, I_l^{(0)})$ being given. Suppose that the general fusion rule described in (3.2.5) is employed. We can now define an iterative algorithm using the mapping Γ by considering the following Gauss-Seidel iterative process.

$$\begin{aligned}
&I_1^{(i+1)}(\mathbf{y}_1)\\
&= I[\int P_{11}(I_2^{(i)}(\mathbf{y}_2), I_3^{(i)}(\mathbf{y}_3), \ldots, I_l^{(i)}(\mathbf{y}_l))L(\mathbf{y}_1, \ldots, \mathbf{y}_l)d\mathbf{y}_2 d\mathbf{y}_3 \cdots d\mathbf{y}_l],\\
&I_2^{(i+1)}(\mathbf{y}_2)\\
&= I[\int P_{21}(I_1^{(i+1)}(\mathbf{y}_1), I_3^{(i)}(\mathbf{y}_3), \ldots, I_l^{(i)}(\mathbf{y}_l))L(\mathbf{y}_1, \ldots, \mathbf{y}_l)d\mathbf{y}_1 d\mathbf{y}_3 \cdots d\mathbf{y}_l],\\
&\qquad\qquad\cdots\\
&I_l^{(i+1)}(\mathbf{y}_l)\\
&= I[\int P_{l1}(I_1^{(i+1)}(\mathbf{y}_1), \ldots, I_{l-1}^{(i+1)}(\mathbf{y}_{l-1}))L(\mathbf{y}_1, \ldots, \mathbf{y}_l)d\mathbf{y}_1 \cdots d\mathbf{y}_{l-1}].
\end{aligned} \tag{3.4.1}$$

To facilitate computer implementation of the above process, we need to discretize the variables. Let the discretization of $\Re^{n_1}, \Re^{n_2}, \ldots, \Re^{n_l}$ be given by selecting a finite numbers of points

$$\{\mathbf{y}_{11}, \mathbf{y}_{12}, \ldots, \mathbf{y}_{1N_1}\}, \{\mathbf{y}_{21}, \mathbf{y}_{22}, \ldots, \mathbf{y}_{2N_2}\}, \cdots; \{\mathbf{y}_{l1}, \mathbf{y}_{l2}, \ldots, \mathbf{y}_{lN_l}\},$$

respectively. For each iteration $i = 0, 1, \ldots$, and $\forall k \leq l$ define

$$I_{kj}^{(i)} = I_k^{(i)}(\mathbf{y}_{kj}).$$

Thus, the iterative processes of Eq. (3.4.1) can be approximated as

$$
\begin{aligned}
I_{1m_1}^{(i+1)} &= I[\sum_{m_2=1}^{N_2}\cdots\sum_{m_l=1}^{N_l} P_{11}(I_2^{(i)}(\mathbf{y}_{2m_2}), I_3^{(i)}(\mathbf{y}_{3m_3}), \ldots, I_l^{(i)}(\mathbf{y}_{lm_l}))\\
&\cdot L(\mathbf{y}_{1m_1}, \ldots, \mathbf{y}_{lm_l})\Delta\mathbf{y}_2\Delta\mathbf{y}_3\cdots\Delta\mathbf{y}_l], \quad m_1 = 1, \ldots, N_1,\\
I_{2m_2}^{(i+1)} &= I[\sum_{m_1=1}^{N_1}\sum_{m_3=1}^{N_3}\cdots\sum_{m_l=1}^{N_l} P_{21}(I_1^{(i+1)}(\mathbf{y}_{1m_1}), I_3^{(i)}(\mathbf{y}_{3m_3}), \ldots,\\
&I_l^{(i)}(\mathbf{y}_{lm_l}))L(\mathbf{y}_{1m_1}, \ldots, \mathbf{y}_{lm_l})\Delta\mathbf{y}_1\Delta\mathbf{y}_3\cdots\Delta\mathbf{y}_l],\ m_2 = 1, \ldots, N_2,\\
&\cdots\\
I_{lm_l}^{(i+1)} &= I[\sum_{m_1=1}^{N_1}\cdots\sum_{m_{l-1}=1}^{N_{l-1}} P_{l1}(I_1^{(i+1)}(\mathbf{y}_{1m_1}), \ldots, I_{l-1}^{(i+1)}(\mathbf{y}_{(l-1)m_{l-1}}))\\
&\cdot L(\mathbf{y}_{1m_1}, \ldots, \mathbf{y}_{lm_l})\Delta\mathbf{y}_1\Delta\mathbf{y}_2\cdots\Delta\mathbf{y}_{l-1}], \quad m_l = 1, \ldots, N_l,
\end{aligned}
\tag{3.4.2}
$$

where $\Delta\mathbf{y}_i$ is the step-size for the discretization of the vectors $\mathbf{y}_i$ in the integral sense. In the sequel, for notational simplicity we will normalize them to unity.

The iterations (3.4.2) are the discrete versions of the iterations (3.4.1). As such, they are readily implementable on a computer. A simple termination criterion is to stop as soon as

$$
I_{1m_1}^{(i+1)} = I_{1m_1}^{(i)}, I_{2m_2}^{i+1} = I_{2m_2}^{(i)}, \ldots, I_{lm_l}^{(i+1)} = I_{lm_l}^{(i)} \quad \text{for all } m_1, m_2, \ldots, m_l. \tag{3.4.3}
$$

An alternative is to stop when

$$
\sum_{m_1, m_2, \ldots, m_l} \left(|I_{1m_1}^{(i+1)} - I_{1m_1}^{(i)}| + \cdots + |I_{lm_l}^{(i+1)} - I_{lm_l}^{(i)}|\right) \leq \epsilon, \tag{3.4.4}
$$

where $\epsilon > 0$ is a prespecified tolerance parameter.

We now examine the convergence of the Gauss-Seidel algorithm defined by (3.4.2). Unfortunately, since the mapping Γ, in general, is not a contraction, we cannot use fixed point theorems to directly claim the convergence.

For convenience later, define the following

$$
\begin{aligned}
&\Phi(I_1^{(i+1)}, \ldots, I_j^{(i+1)}, I_{j+1}^{(i)}, \ldots, I_l^{(i)})\\
&= \sum_{m_1, m_2, \ldots, m_l} [1 - P_{\mathbf{H}_1}(I_{1m_1}^{(i+1)}, \ldots, I_{jm_j}^{(i+1)}, I_{(j+1)m_{j+1}}^{(i)}, \ldots, I_{lm_l}^{(i)})] \\
&\qquad \cdot L(\mathbf{y}_{1m_1}, \ldots, \mathbf{y}_{lm_l}) \quad \text{for } \forall 1 \leq j \leq l
\end{aligned}
\tag{3.4.5}
$$

and

$$
\begin{aligned}
&G_j^{(i+1)}(\mathbf{y}_{jm_j}) \\
&= \sum_{m_1,\ldots,m_{j-1},m_{j+1},\ldots,m_l} P_{j1}(I_{1m_1}^{(i+1)},\ldots,I_{(j-1)m_{j-1}}^{(i+1)},I_{(j+1)m_{j+1}}^{(i)},\ldots,I_{lm_l}^{(i)}) \\
&\qquad \cdot L(\mathbf{y}_{1m_1},\ldots,\mathbf{y}_{lm_l}) \quad \text{for } \forall 1 \le j \le l.
\end{aligned} \tag{3.4.6}
$$

The sum Φ in (3.4.5) is a discrete version of the cost functional in Eqs. (3.3.1) and (3.3.7). It is dependent on the corresponding local compression rule estimates $I_1^{(i+1)}, \ldots, I_j^{(i+1)}, I_{j+1}^{(i)}, \ldots, I_l^{(i)}$, and can be used as a potential function. The sum $G_j^{(i+1)}$ is a discrete version of the integral

$$\int P_{j1}L(\mathbf{y}_1,\ldots,\mathbf{y}_l)d\mathbf{y}_1\cdots d\mathbf{y}_{j-1}d\mathbf{y}_{j+1}\cdots d\mathbf{y}_l$$

in (3.3.7).

In order to simplify the presentation of the proof of convergence, we start with a sequence of lemmas.

Lemma 3.4.1. If the condition

$$(I_1^{(i+1)},\ldots,I_l^{(i+1)}) = (I_1^{(i)},\ldots,I_l^{(i)})$$

is satisfied for some $i = k \ge 0$, then it will remain satisfied for all $i \ge k$. Consequently, the termination criterion of Eq. (3.4.3) will be satisfied at the iteration $i = k + 1$.

The lemma obviously follows from the form of the iterative algorithm (3.4.2).

Lemma 3.4.2. The potential function $\Phi(I_1^{(i+1)},\ldots,I_j^{(i+1)},I_{j+1}^{(i)},\ldots,I_l^{(i)})$ is nonincreasing as j is increasing.

Proof. Using (3.3.2), (3.4.2), (3.4.5), and (3.4.6), we have

$$
\begin{aligned}
&\Phi(I_1^{(i+1)},\ldots,I_j^{(i+1)},I_{j+1}^{(i)},\ldots,I_l^{(i)}) \\
&= \sum_{m_j=1}^{N_j} (1 - I_j^{(i+1)}(\mathbf{y}_{jm_j}))G_j^{(i+1)}(\mathbf{y}_{jm_j}) + C_j \\
&= \sum_{m_j=1}^{N_j} (1 - I_j^{(i)}(\mathbf{y}_{jm_j}))G_j^{(i+1)}(\mathbf{y}_{jm_j}) + C_j \\
&\quad + \textstyle\sum_{m_j=1}^{N_j}(I_j^{(i)}(\mathbf{y}_{jm_j}) - I_j^{(i+1)}(\mathbf{y}_{jm_j}))G_j^{(i+1)}(\mathbf{y}_{jm_j})
\end{aligned} \tag{3.4.7}
$$

$$= \Phi(I_1^{(i+1)}, \ldots, I_{j-1}^{(i+1)}, I_j^{(i)}, \ldots, I_l^{(i)})$$
$$+ \textstyle\sum_{m_j=1}^{N_j} (I_j^{(i)}(\mathbf{y}_{jm_j}) - I_j^{(i+1)}(\mathbf{y}_{jm_j}))G_j^{(i+1)}(\mathbf{y}_{jm_j})$$
$$\leq \Phi(I_1^{(i+1)}, \ldots, I_{j-1}^{(i+1)}, I_j^{(i)}, \ldots, I_l^{(i)}) \quad \text{for } \forall j \leq l,$$

where C_j is a constant independent of $I_j^{(i)}$ and $I_j^{(i+1)}$ by (3.3.2). The first three equalities follow from (3.3.2), (3.4.5) and (3.4.6), and the last inequality holds due to the fact that by (3.4.2) we know that

$$I_j^{(i+1)}(\mathbf{y}_{jm_j}) = 1 \quad \text{if and only if } G_j^{(i+1)}(\mathbf{y}_{jm_j}) > 0, \quad \text{for } \forall\, m_j \leq N_j.$$

That is to say, the all of terms of the summation

$$\sum_{m_j=1}^{N_j} (I_j^{(i)}(\mathbf{y}_{jm_j}) - I_j^{(i+1)}(\mathbf{y}_{jm_j}))G_j^{(i+1)}(\mathbf{y}_{jm_j})$$

are non-positive. ■

From Lemma 3.4.2, we know that $\Phi(I_1^{(i)}, I_2^{(i)}, \ldots, I_l^{(i)})$ must converge to a stationary point after a finite number of iterations because it is finite-valued. Then we have

Lemma 3.4.3. The local compression rules $I_1^{(i)}, I_2^{(i)}, \ldots, I_l^{(i)}$ are also finitely convergent.

Proof. By Lemma 2, after a finite number of iterations, Φ must reach a stationary point. I.e., we must have

$$\Phi(I_1^{(i+1)}, \ldots, I_j^{(i+1)}, I_{j+1}^{(i)}, \ldots, I_l^{(i)}) = \Phi(I_1^{(i+1)}, \ldots, I_{j-1}^{(i+1)}, I_j^{(i)}, \ldots, I_l^{(i)}).$$

Then by (3.4.7), for all m_j, we have

$$(I_j^{(i)}(\mathbf{y}_{jm_j}) - I_j^{(i+1)}(\mathbf{y}_{jm_j}))G_j^{(i+1)}(\mathbf{y}_{jm_j}) = 0,$$

which implies that either

$$I_j^{(i+1)}(\mathbf{y}_{jm_j}) = I_j^{(i)}(\mathbf{y}_{jm_j}),$$

or

$$G_j^{(i+1)}(\mathbf{y}_{jm_j}) = 0, \quad \text{i.e.,} \quad I_j^{(i+1)}(\mathbf{y}_{jm_j}) = 1, I_j^{(i)}(\mathbf{y}_{jm_j}) = 0.$$

It follows that when Φ attains such a point at the ith step, either $I_j^{(i+1)}(\mathbf{y}_{jm_j})$ is invariant, or $I_j^{(i+1)}(\mathbf{y}_{jm_j})$ can change from 0 to 1 at most a finite number of times since all N_j, $\forall j \leq l$, are finite. Thus, the algorithm cannot oscillate infinitely often. ■

Summarizing the above three lemmas, we can assert the following theorem on the finite convergence of the discretized Gauss-Seidel iteration.

Theorem 3.4.1. For any positive discretization stepsizes $\Delta \mathbf{y}_1, \Delta \mathbf{y}_2, \ldots, \Delta \mathbf{y}_l$ and any initial choice of $(I_1^{(0)}, I_2^{(0)}), \ldots, I_l^{(0)})$, the algorithm of Eq. (3.4.2) terminates with a set $(I_1^{(i)}, I_2^{(i)}), \ldots, I_l^{(i)})$ satisfying Eq. (3.4.3) after a finite number of iterations.

Remark 3.4.1. Here we have to emphasize again that Theorem 3.4.1 does not guarantee that the algorithm (3.4.2) converges to globally optimal solution of (3.3.5) for any initial condition. It is possible that the algorithm converges to a local minimum for some initial values. Fortunately, for the probability density functions used in most of the practical applications, the integrand $L(\mathbf{y}_1, \mathbf{y}_2, \ldots \mathbf{y}_l)$ has one minimum or at most only a few minima. The numerical examples in Section 3.5 show that the algorithm converges to a global minimum very often.

Let $\Delta \mathbf{y}_1 = \Delta \mathbf{y}_2 = \cdots = \Delta \mathbf{y}_l = \Delta$ and let C_Δ be the minimum of the discrete version $C_\Delta(I_1, I_2, \ldots, I_l; F)$ of $C(I_1, I_2, \ldots, I_l; F)$. One may question the existence of the limit of C_Δ as Δ goes to zero and its relationship to the infimum of $C(I_1, I_2, \ldots, I_L; F)$. Clearly, it is no guarantee that when two stepsizes Δ_1 and Δ_2 are different, the corresponding two C_{Δ_1} and C_{Δ_2} are both the Riemann sum approximations of the integral in (3.3.1) for the same set of local compression rules $(I_1, I_2, \ldots, I_l)$. Therefore, we have not known the existence of the limit of C_Δ as Δ goes to zero before we complete the proof of the following theorem, which asserts that under mild assumption on integrand $L(\mathbf{y}_1, \mathbf{y}_2, \ldots \mathbf{y}_l)$, not only the limit of C_Δ exists, but also the C_Δ converges to infimum of $C(I_1, I_2, \ldots, I_l; F)$ as Δ tends to zero.

Theorem 3.4.2. Suppose that for a region Ω

$$\Omega = \{(\mathbf{y}_1, \mathbf{y}_2, \ldots \mathbf{y}_l) : F(I_1, I_2, \ldots, I_l) = 0\}$$

defined by any set of local compression rules $(I_1, I_2, \ldots, I_l)$ and any fusion rule $F(I_1, I_2, \ldots, I_l)$, the following inequality

$$\left\| \int_\Omega L(\mathbf{y}_1, \mathbf{y}_2, \ldots \mathbf{y}_l) d\mathbf{y}_1 d\mathbf{y}_2 \cdots d\mathbf{y}_l - S(\Omega, \Delta) \right\| \leq M\Delta \tag{3.4.8}$$

holds, where $S(\Omega, \Delta)$ is a Riemann sum approximation of the integral in (3.4.8) and the constant M does not depend on Ω and Δ. Then we have

$$\lim_{\Delta\to 0} C_\Delta = \inf_{I_1,\ldots,I_l} C(I_1, I_2, \ldots, I_l; F) \triangleq C_{inf}.$$

Proof. By the definition of C_Δ, for arbitrary $\epsilon > 0$, there exists a set of local compression rules $(I_1, I_2, \ldots, I_l)$ such that

$$C(I_1, I_2, \ldots, I_l; F) \le C_{inf} + \frac{1}{2}\epsilon. \tag{3.4.9}$$

Denote the Riemann sum approximation of $C(I_1, I_2, \ldots, I_l; F)$ by $C^\Delta(I_1, I_2, \ldots, I_l; F)$. There exists $\Delta^* > 0$ such that for any $\Delta \le \Delta^*$

$$C^\Delta(I_1, I_2, \ldots, I_l; F) \le C(I_1, I_2, \ldots, I_l; F) + \frac{1}{2}\epsilon. \tag{3.4.10}$$

Thus, combining the above two inequalities yields

$$C^\Delta(I_1, I_2, \ldots, I_l; F) \le C_{inf} + \epsilon. \tag{3.4.11}$$

Furthermore, recalling the definition of C_Δ, we have

$$C_\Delta \le C^\Delta(I_1, I_2, \ldots, I_l; F) \le C_{inf} + \epsilon, \quad \text{for } \forall\Delta \le \Delta^*, \tag{3.4.12}$$

which implies that

$$\limsup_{\Delta\to 0} C_\Delta \le C_{inf} + \epsilon. \tag{3.4.13}$$

Since ϵ is arbitrary, we have

$$\limsup_{\Delta\to 0} C_\Delta \le C_{inf}. \tag{3.4.14}$$

On the other hand, suppose that

$$\liminf_{\Delta\to 0} C_\Delta < C_{inf}. \tag{3.4.15}$$

Then there would be a positive constant $\delta > 0$ and a sequence $\{\Delta_k\}$ such that $\Delta_k \to 0$ and

$$C_{\Delta_k} < C_{inf} - \delta. \tag{3.4.16}$$

For every such C_{Δ_k}, there must be a set of $(I_1^{(k)}, I_2^{(k)}, \ldots, I_l^{(k)})$ such that

$$C_{\Delta_k} = C^{\Delta_k}(I_1^{(k)}, I_2^{(k)}, \ldots, I_l^{(k)}; F). \tag{3.4.17}$$

Using the inequality (3.4.8) and (3.4.16), for large enough K, we have

$$\Delta_K M \le \delta, \quad C(I_1^{(K)}, I_2^{(K)}, \ldots, I_l^{(K)}; F) \le C_{\Delta_K} + \delta < C_{inf}, \tag{3.4.18}$$

which contradicts the definition of C_{inf}. Therefore, the reverse inequality of (3.4.15) should be true and

$$C_{inf} \le \liminf_{\Delta \to 0} C_\Delta \le \limsup_{\Delta \to 0} C_\Delta \le C_{inf}. \tag{3.4.19}$$

The theorem follows. ■

Remark 3.4.2. The assumption in this theorem is not restrictive. If an integral of $L(\mathbf{y}_1, \mathbf{y}_2, ...\mathbf{y}_l)$ over any unbounded region can be approximated by an integral over a bounded region, then we only need to consider the original integral over a bounded region. When the integrand $L(\mathbf{y}_1, \mathbf{y}_2, ...\mathbf{y}_l)$ is locally Lipschitz continuous, one can easily prove that the inequality (3.4.8) holds.

Remark 3.4.3. To improve the decision accuracy, when more communication bandwidth is available, the sensors can compress their own observations to multiple bits and transmit them out. In this case an r_i bit sensor compression can be implemented by a set of indicator functions $I_i^{(1)}(\mathbf{y}_i), \ldots, I_i^{(r_i)}(\mathbf{y}_i)$ at the ith sensor. The local message from all sensors is now a $\sum_{i=1}^{l} r_i$-tuple

$$(I_1^{(1)}(\mathbf{y}_1), \ldots, I_1^{(r_1)}(\mathbf{y}_1); \cdots; I_l^{(1)}(\mathbf{y}_l), \ldots, I_l^{(r_l)}(\mathbf{y}_l)).$$

It is not hard to see that all of the previous analysis, algorithms, and results in Sections 3.2, 3.3, and 3.4 can be extended to the general $(\mathbf{r_1} + \mathbf{r_2} + \cdots + \mathbf{r}_\ell)$ information structure. In particular, for $(\mathbf{1} + \mathbf{1} + \cdots + \mathbf{1} + \mathbf{r}_\ell)$ structure (which we will use in the next section), the algorithm (3.4.1) will be replaced by

$$\begin{aligned}
& I_1^{(i+1)}(\mathbf{y}_1) \\
&= I\Big[\int P_{11}(I_2^{(i)}, I_3^{(i)}, \ldots, I_{l-1}^{(i)}; I_l^{(1)(i)}, \ldots, I_l^{(r_l)(i)}) L(Y) d\mathbf{y}_2 \cdots d\mathbf{y}_l\Big], \\
& I_2^{(i+1)}(\mathbf{y}_2) \\
&= I\Big[\int P_{21}(I_1^{(i+1)}, I_3^{(i)}, \ldots, I_{l-1}^{(i)}; I_l^{(1)(i)}, \ldots, I_l^{(r_l)(i)}) L(Y) d\mathbf{y}_1 d\mathbf{y}_3 \cdots d\mathbf{y}_l\Big], \\
& \qquad \cdots \\
& I_{l-1}^{(i+1)}(\mathbf{y}_{l-1}) \\
&= I\Big[\int P_{(l-1)1}(I_1^{(i+1)}, \ldots, I_{l-2}^{(i+1)}; I_l^{(1)(i)}, \ldots, I_l^{(r_l)(i)}) L(Y) d\mathbf{y}_1 \cdots d\mathbf{y}_{l-2} d\mathbf{y}_l\Big], \\
& I_l^{(1)(i+1)}(\mathbf{y}_l) \\
&= I\Big[\int P_{l^{(1)}1}(I_1^{(i+1)}, \ldots, I_{l-1}^{(i+1)}, I_l^{(2)(i)}, \ldots, I_l^{(r_l)(i)}) L(Y) d\mathbf{y}_1 \cdots d\mathbf{y}_{l-1}\Big],
\end{aligned}$$

$$I_l^{(2)(i+1)}(\mathbf{y}_l)$$
$$= I[\int P_{l^{(2)}1}(I_1^{(i+1)}, \ldots; I_l^{(1)(i+1)}, I_l^{(3)(i)}, \ldots, I_l^{(r_l)(i)})L(Y)d\mathbf{y}_1 \cdots d\mathbf{y}_{l-1}],$$
$$\cdots$$
$$I_l^{(r_l)(i+1)}(\mathbf{y}_l)$$
$$= I[\int P_{l^{(r_l)}1}(I_1^{(i+1)}, \ldots; I_l^{(1)(i+1)}, \ldots, I_l^{(r_l-1)(i+1)})L(Y)d\mathbf{y}_1 \cdots d\mathbf{y}_{l-1}]. \tag{3.4.20}$$

where $L(Y)$ stands for $L(\mathbf{y}_1, \ldots, \mathbf{y}_l)$. Accordingly, there is a discretized version of the above algorithm (3.4.20) to replace the algorithm (3.4.2).

3.5 THE OPTIMAL FUSION RULE AND SOME INTERESTING PROPERTIES

Thus far, we have not considered the optimal fusion rule. Of course, one could use our techniques to find the optimal sensor rules for each of the considered fusion rules. Then one could pick the fusion rule that gives the globally optimal performance. This exhaustive search method could require significant computation if the number of the considered fusion rules is large. For an l sensor binary decision system, we select the optimal fusion rule among 2^{2^l} possible fusion rules. For example, when $l = 4$, the number of all possible rules is 65536. Here we present an alternative for some special cases which generally requires considerably less computation. This alternative can find the best distributed decision system over all possible non-randomized fusion rules. The alternative is based on the following fact. In special cases we can provide a fixed fusion rule that can be used to achieve the optimal performance. These special cases exhibit some interesting properties as we shall discuss in this section. Consider a case with l sensors and assume that $l-1$ of them are required to make binary compressions while one (the last) makes a 2^{l-1} bit compression. This is the $(\mathbf{1} + \mathbf{1} + \cdots + \mathbf{1} + \mathbf{2}^{\ell-1})$ information structure. We will prove that the fusion rule proposed below gives the best performance in the information structure type of $(\mathbf{1} + \mathbf{1} + \cdots + \mathbf{1} + \mathbf{r}_\ell)$ with any $\mathbf{r}_\ell$. More importantly, the performance of the proposed fusion rule cannot be improved even if the lth sensor can transmit the uncompressed data $\mathbf{y}_l$) to the fusion center. When the information structure of a practical system is $(\mathbf{r}_1 + \mathbf{r}_2 + \cdots + \mathbf{r}_\ell)$, without any difficulty we can extend our result to $(\mathbf{r}_1 + \mathbf{r}_2 + \cdots + \mathbf{r}_{\ell-1} + \mathbf{2}^{\sum_{j=1}^{\ell-1} \mathbf{r}_j})$ information structure. Clearly, to adapt this special case and to reduce communications between the lth sensor and the fusion center, one can build the both at the same station. For notational simplicity, here we consider $(\mathbf{1} + \mathbf{1} + \cdots + \mathbf{1} + \mathbf{2}^{\ell-1})$ first.

The fusion rule that we propose is defined by

$$\{I_1,\ldots,I_{l-1};I_l^{(1)},\ \ldots,\ I_l^{(2^{l-1})}:F=0\}$$
$$=\left\{\begin{array}{c} I_1=0,I_2=0,\ \ldots,\ I_{l-1}=0;I_l^{(1)}=0\\ I_1=1,I_2=0,\ \ldots,\ I_{l-1}=0;I_l^{(2)}=0\\ \ldots,\\ I_1=1,I_2=1,\ \ldots,\ I_{l-1}=1;I_l^{(2^{l-1})}=0\end{array}\right\}. \tag{3.5.1}$$

Theorem 3.5.1. By employing the fusion rule in (3.5.1), in the case specified ($l-1$ binary sensors and one 2^{l-1} bit sensor), we obtain optimal performance.

Proof. Obviously, the theorem will be proven if we show that there exists a set of sensor rules which, with this fusion rule, can implement any distributed decision scheme with a non-randomized fusion rule.

Consider a set of sensor rules $I_1,\ldots,I_{l-1};I_l^{(1)},\ldots,I_l^{(2^{l-1})}$ and a general fusion rule

$$\{I_1,\ldots,I_{l-1};I_l^{(1)},\ \ldots,\ I_l^{(2^{l-1})}:F=0\}=$$
$$\left\{\begin{array}{c} I_1=d_1^{(1)},I_2=d_2^{(1)},\ \ldots,\ I_{l-1}=d_{l-1}^{(1)};I_l^{(1)}=d_{l,1}^{(1)},\ldots,I_l^{(2^{l-1})}=d_{l,2^{l-1}}^{(1)}\\ I_1=d_1^{(2)},I_2=d_2^{(2)},\ \ldots,\ I_{l-1}=d_{l-1}^{(2)};I_l^{(1)}=d_{l,1}^{(2)},\ldots,I_l^{(2^{l-1})}=d_{l,2^{l-1}}^{(2)}\\ \ldots,\\ I_1=d_1^{(N)},I_2=d_2^{(N)},\ \ldots,\ I_{l-1}=d_{l-1}^{(N)};I_l^{(1)}=d_{l,1}^{(N)},\ldots,I_l^{(2^{l-1})}=d_{l,2^{l-1}}^{(N)}\end{array}\right\}, \tag{3.5.2}$$

where all $d_i^{(j)}$ and $d_{i,k}^{(j)}$, $j\le N$, equal 0 or 1.

Now divide N rows (local messages) in (3.5.2) into a number of groups, for example, the jth group is of the following form:

$$\left\{\begin{array}{c} I_1=d_1^{(j)},I_2=d_2^{(j)},\ \ldots,\ I_{l-1}=d_{l-1}^{(j)};I_l^{(1)}=d_{l,1}^{(1)},\ldots,I_l^{(2^{l-1})}=d_{l,2^{l-1}}^{(1)}\\ I_1=d_1^{(j)},I_2=d_2^{(j)},\ \ldots,\ I_{l-1}=d_{l-1}^{(j)};I_l^{(1)}=d_{l,1}^{(2)},\ldots,I_l^{(2^{l-1})}=d_{l,2^{l-1}}^{(2)}\\ \ldots,\\ I_1=d_1^{(j)},I_2=d_2^{(j)},\ \ldots,\ I_{l-1}=d_{l-1}^{(j)};I_l^{(1)}=d_{l,1}^{(L)},\ldots,I_l^{(2^{l-1})}=d_{l,2^{l-1}}^{(L)}\end{array}\right\}. \tag{3.5.3}$$

Note that there are at most 2^{l-1} groups, i.e., $j \le 2^{l-1}$. Furthermore, we can define a binary compression rule $\hat{I}_l^{(j)}(\mathbf{y}_l)$ for the lth sensor as follows:

$$\{\mathbf{y}_l : \hat{I}_l^{(j)}(\mathbf{y}_l) = 0\} = \left\{ \mathbf{y}_l : \begin{array}{c} I_l^{(1)} = d_{l,1}^{(1)}, \ldots, I_l^{(2^{l-1})} = d_{l,2^{l-1}}^{(1)} \\ I_l^{(1)} = d_{l,1}^{(2)}, \ldots, I_l^{(2^{l-1})} = d_{l,2^{l-1}}^{(2)} \\ \cdots, \\ I_l^{(1)} = d_{l,1}^{(L)}, \ldots, I_l^{(2^{l-1})} = d_{l,2^{l-1}}^{(L)} \end{array} \right\}. \quad (3.5.4)$$

Noting that every row of the jth group (3.5.3) has the common first $l-1$ sensor decisions ($I_1 = d_1^{(j)}, I_2 = d_2^{(j)}, \ldots, I_{l-1} = d_{l-1}^{(j)}$) and using (3.5.4), the jth group (3.5.3) is equivalent to the local message

$$(I_1 = d_1^{(j)}, I_2 = d_2^{(j)}, \ldots, I_{l-1} = d_{l-1}^{(j)}; \hat{I}_l^{(j)}(\mathbf{y}_l) = 0),$$

which appears in the fusion rule (3.5.1) except the notation difference of $\hat{I}_l^{(j)}(\mathbf{y}_l)$ and $I_l^{(j)}(\mathbf{y}_l)$.

If $j < 2^{l-1}$, i.e., in (3.5.1) a local messages, namely ($I_1 = 0, I_2 = 0, \ldots, I_{l-1} = 0; I_l^{(1)} = 0$), does not appear in the rewritten (3.5.2) via (3.5.4), we can add it into the rewritten (3.5.2) but let the sensor rule $I_l^{(1)}(\mathbf{y}_l) \equiv 1$ for any $\mathbf{y}_l$. Thus, ($I_1 = 0, I_2 = 0, \ldots, I_{l-1} = 0; I_l^{(1)} = 0$) never happens. Obviously, this modification does not change the rewritten (3.5.2) at all. Thus our fusion rule (3.5.1) with a proper set of sensor rules allows us to represent any rule of the form (3.5.2). ∎

Remark 3.5.1. An optimal set of sensor rules and a fusion rule is not necessarily unique. This is clear from Theorem 3.5.1.

The next Theorem shows the special nature of the case we have considered ($l-1$ binary sensors and one 2^{l-1} bit sensor). It says that the performance is not improved if the lth sensor as aforementioned uses more than 2^{l-1} bits. This is true even if this sensor can send uncompressed data (the original observation $\mathbf{y}_l$) to the fusion center.

Theorem 3.5.2. When one sensor, namely the lth sensor, transmits uncompressed observation data and all of the other sensors transmit a total of $l-1$ bits to the fusion center, the optimal performance is equivalent to that of the system in which the lth sensor transmits only 2^{l-1} information bits to the fusion center and uses the fusion rule (3.5.1).

Proof. From the following argument in the proof, without loss of generality assume that the number of the other sensors is $l-1$ and each sensor transmits one bit. It suffices to prove that the general form of fusion rule, $F(I_1(\mathbf{y}_1), \ldots, I_{l-1}(\mathbf{y}_{l-1}), \mathbf{y}_l)$ can be written as the fusion rule (3.5.1). Consider a general scheme with the critical region for the fusion rule of the form

$$\mathcal{H}_0 = \left\{ \begin{array}{l} (\mathbf{y}_1 \cdots \mathbf{y}_l) : \\ F(I_1(\mathbf{y}_1) = d_1^{(1)}, I_2(\mathbf{y}_2) = d_2^{(1)}, \ldots, I_{l-1}(\mathbf{y}_{l-1}) = d_{l-1}^{(1)}; \mathbf{y}_l) = 0 \\ F(I_1(\mathbf{y}_1) = d_1^{(2)}, I_2(\mathbf{y}_2) = d_2^{(2)}, \ldots, I_{l-1}(\mathbf{y}_{l-1}) = d_{l-1}^{(2)}; \mathbf{y}_l) = 0 \\ \qquad \qquad \cdots, \\ F(I_1(\mathbf{y}_1) = d_1^{(N)}, I_2(\mathbf{y}_2) = d_2^{(N)}, \ldots, I_{l-1}(\mathbf{y}_{l-1}) = d_{l-1}^{(N)}; \mathbf{y}_l) = 0 \end{array} \right\}, \tag{3.5.5}$$

where all $d_i^{(j)} = 0$ or 1. Take the same approach as in Theorem 3.5.1, but with $\hat{I}_{l,j}(\mathbf{y}_l) \triangleq F(d_1^{(j)}, \ldots, d_{l-1}^{(j)}, \mathbf{y}_l)$. Using $\hat{I}_{l,j}(\mathbf{y}_l), j = 1, \ldots, 2^{l-1}$ with the other sensor rules $I_i, i = 1, \ldots, l-1$ and (3.5.1) insures that the overall scheme produces the same output as the original scheme using the rule from (3.5.5). ■

It is easy to see that the important point for the proof is that the last sensor has a bit in its sensor decision that can match up with each combination of the other sensor decisions. Thus a total of $l-1$ bits can be used by the first $l-1$ sensors and it is not important that each of the first $l-1$ sensors is restricted to a one bit decision.

Remark 3.5.2. Theorem 3.5.2 is useful in practice. For example, when the performance of a decision system is not satisfactory, we may add a number of extra sensors. The theorem tells us how to construct an optimal fusion decision rule and the number of bits to be used if one sensor is added to binary sensors. We note that increasing the number of bits used in a sensor decision will generally increase performance, except in specific situations. Thus, even in a case with l binary sensors, it is generally advantageous to increase the number of bits used in one of the sensor decisions until the limit of 2^{l-1} is reached.

Remark 3.5.3. Theorem 3.5.1 and Theorem 3.5.2 suggest a way to find optimal distributed signal detection systems which use binary sensor decisions if one can make observations at the fusion center. Start with an $l-1$ sensor system. Allow observations to be made at the fusion center, through an added sensor there, and allow these observations to be used in the fusion. Recall that Theorem 3.5.2 tells us we could quantize these observations to 2^{l-1} bits if desired, without loss of performance and in fact we need to do this to use Theorem 3.5.1.

Then, Theorem 3.5.1 says that (3.5.1) can be used to achieve optimal performance over the class of non-randomized fusion rules. From Theorem 3.5.2, the system would be optimal over all systems that combine binary decisions at the first $l-1$ sensors with an uncompressed observation at the lth sensor. In essence, we have exchanged the complexity of searching over multiple fusion rules for the complexity of designing an extra sensor rule for the added sensor S_l (at the fusion center). The added sensor is more difficult to design than the other sensors since it makes an 2^{l-1} bit sensor decision. Once the optimal sensor rules are found, one can use (3.5.1) to learn exactly how the first $l-1$ single-bit sensor compressions and the observations at the fusion center are used to generate a final decision.

Remark 3.5.4. Note that the added lth sensor could be a dummy which really does not make observations. This might be used as a "trick" to avoid searching for the optimal fusion rule for the $l-1$ sensor problem. In this case we expect that the rules $\hat{I}_{l,j}(\mathbf{y}_l), j = 1, \ldots, 2^{l-1}$ will not depend on $\mathbf{y}_l$ so that each member of $\hat{I}_{l,j}(\mathbf{y}_l), j = 1, \ldots, 2^{l-1}$ will always take on the value 0 or 1 for all $\mathbf{y}_l$. In fact, if we think that the dummy sensor output is fixed at a certain value, maybe 0, then we must acknowledge that this value could be mapped to any of the possible sensor decisions. Since there are 2^{l-1} bits in the sensor S_l's compression then there are 2^{2^l} possible combinations. In this case, trying all possible sensor decision rules at sensor S_l involves trying all of the possible $2^{2^{l-1}}$ bit combinations of the individual bit decisions at the lth sensor. Note that this is exactly the total number of fusion rules for the original $l-1$ sensor problem which makes complete sense. Thus we see there is no magic associated with our results.

Remark 3.5.5. From the proofs of Theorems 3.5.1 and 3.5.2, it is clear that the important property of the case considered, a total of $l-1$ bit decisions at the first $l-1$ sensors and 2^{l-1} bits at the last sensor, is that there is one bit in the last sensor's decision for each of the possible combinations of sensor decisions from the other sensors. Clearly this leads to a generalization of the results in Theorem 3.5.1 and 3.5.2 that will work as long as the last sensor has enough bits in its decision. This leads to fixed fusion rules for other cases that can be used to achieve optimal performance. It also leads to examples where a finite number of bits in a sensors decision will lead to the same performance that can be achieved if s uncompressed data is sent to the fusion center. As one example consider a two sensor case where the first sensor makes a 2-bit decision and the second sensor makes a 4-bit decision. A slight generalization of (3.5.1) gives the fusion rule to use here. A case of this type is considered in the next section.

3.6 NUMERICAL EXAMPLES OF THE ABOVE RESULTS

We will continue to use a discrete version of the Gauss-Seidel algorithm (3.4.20) to compute the optimal local compression rules but based on th fusion rules (3.5.1). Since the polynomials $P_{l^{(1)}1}, P_{l^{(2)}1}, \ldots, P_{l^{(r_l)}1}$ in (3.4.20) are all independent of the local compression rules $I_l^{(1)}, I_l^{(2)}, \ldots, I_l^{(r_l)}$, in order for the iterative algorithm (3.4.20)to start, we only need to give initial compression rules for the first $l-1$ local sensors while no initial rules are needed for the last local sensor S_l, i.e., no initial values for $I_l^{(1)}(\mathbf{y}_l), I_l^{(2)}(\mathbf{y}_l), \ldots, I_l^{(2^{l-1})}(\mathbf{y}_l)$.

In the following numerical investigations, we will consider the problem of detecting a common random Gaussian signal in Gaussian noise with two, three, and four sensors. We will use $\Delta = 0.2$, $\mathbf{y}_i \in [-7, 7]$.

Two sensors

The observations consist of a signal s and noise ν_1, ν_2 so that

$$\mathbf{H_1} : \mathbf{y}_1 = s + \nu_1, \ \mathbf{y}_2 = s + \nu_2,$$
$$\mathbf{H_0} : \mathbf{y}_1 = \nu_1, \qquad \mathbf{y}_2 = \nu_2,$$

where s, ν_1 and ν_2 are all mutually independent, and

$$s \sim N(2, 3), \quad \nu_1 \sim N(0, 3), \quad \nu_2 \sim N(0, 2).$$

Therefore, the two conditional pdfs given $\mathbf{H_0}$ and $\mathbf{H_1}$ are

$$p(\mathbf{y}_1, \mathbf{y}_2 | \mathbf{H_1}) \sim N\left(\begin{pmatrix} 2 \\ 2 \end{pmatrix}, \begin{bmatrix} 6 & 3 \\ 3 & 5 \end{bmatrix}\right),$$

$$p(\mathbf{y}_1, \mathbf{y}_2 | \mathbf{H_0}) \sim N\left(\begin{pmatrix} 0 \\ 0 \end{pmatrix}, \begin{bmatrix} 3 & 0 \\ 0 & 2 \end{bmatrix}\right).$$

First, we consider binary sensor decisions, i.e., the $(1+1)$ information structure (see the definition given in Section 3.1). The ROC (Receiver Operating Characteristics) [63, 83] for the centralized, AND, OR, and XOR (see (2.3.10)) rules are provided in Fig. 3.6.1. We also include the ROC of the optimal scheme using binary decisions at sensor one and two bit decisions at the second sensor in Fig. 3.6.1. This is an example of the type discussed in Theorem 3.5.1, so the fusion rule in (3.5.1) is used here, and it is denoted by "OPT(1+2)". Note that using two bit sensor decisions yields better performance, as we expect.

To show the performances when different sensors are used to transmit the extra bit, in Fig. 3.6.2, we computed the ROC for the optimal schemes "OPT(2+1)" where Sensor 1 transmits two bits and the second sensor transmits one bit, and

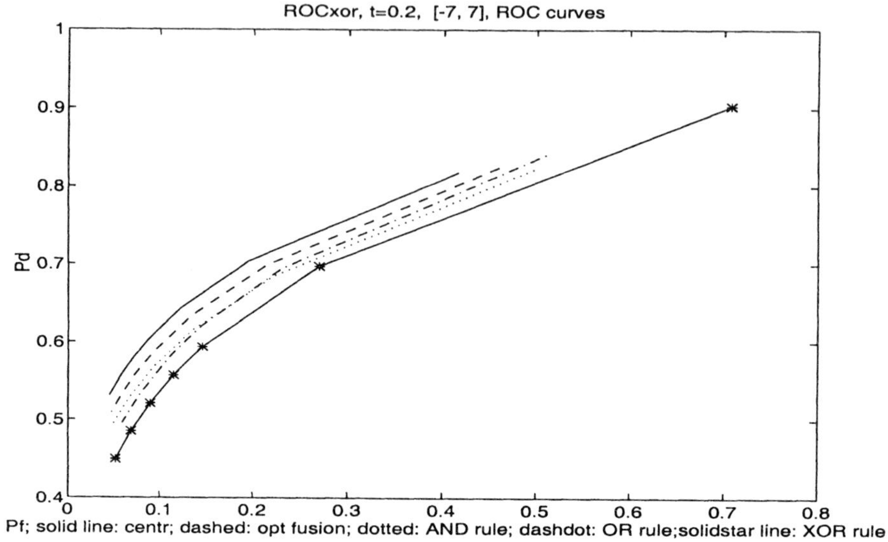

Figure 3.6.1. ROCs for the centralized and distributed OPT(1+2), AND, OR, XOR rules.

also for the optimal schemes "OPT(1+2)" where Sensor 1 transmits 1 bit and the second sensor transmits two bits.

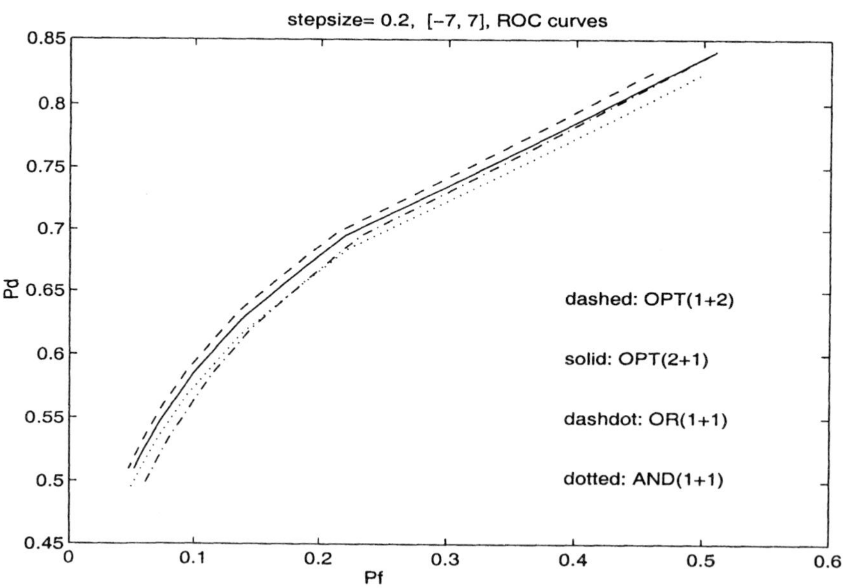

Figure 3.6.2. ROCs for the distributed OPT(2+1), OPT(1+2), AND, OR rules.

Next, we consider cases where one sensor makes two bit decisions and the other sensor makes four bit decisions. This case is covered by the extension to Theorem 3.5.1 which we discussed in Remark 3.5.5. Thus, a fixed fusion rule, which is a slight extension to the one in Theorem 3.5.1, can be used to obtain optimal performance. In Fig. 3.6.3, we have again compared the results labelled "OPT(4+2)" and "OPT(2+4)".

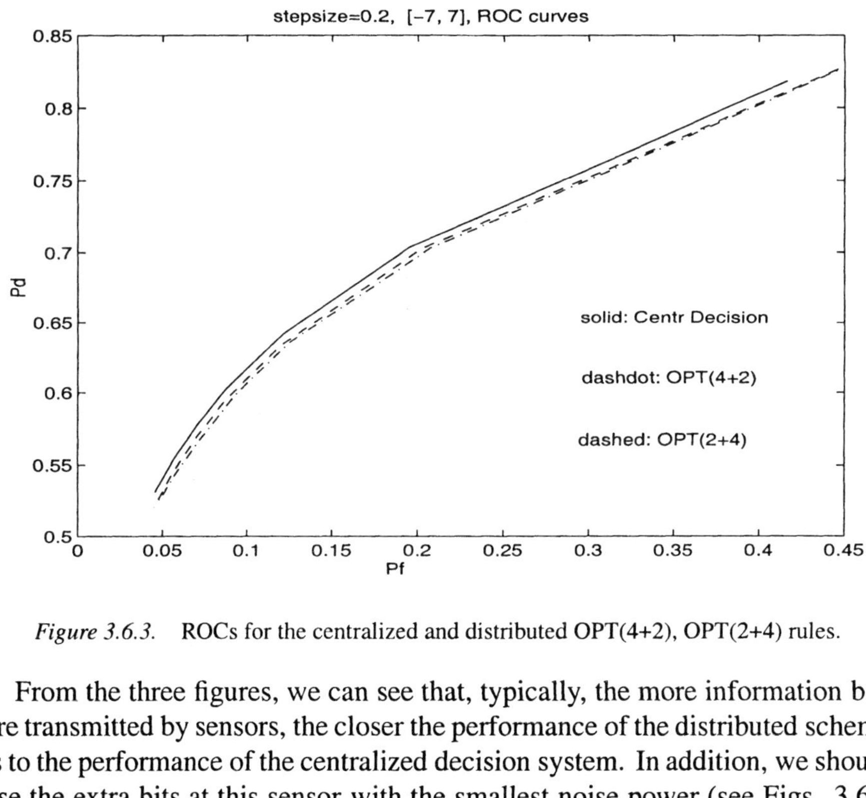

Figure 3.6.3. ROCs for the centralized and distributed OPT(4+2), OPT(2+4) rules.

From the three figures, we can see that, typically, the more information bits are transmitted by sensors, the closer the performance of the distributed scheme is to the performance of the centralized decision system. In addition, we should use the extra bits at this sensor with the smallest noise power (see Figs. 3.6.2 and 3.6.3).

In numerical computations, we tried running the iterative algorithm with many different starting conditions and in each case found only one solution, which is the one shown in Figs. 3.6.1, 3.6.2 and 3.6.3.

Three sensors

Now we add one more sensor with noise $\nu_3 \sim N(0,1)$ to the above system. The resulting two conditional pdfs are given by

$$p(\mathbf{y}_1, \mathbf{y}_2, \mathbf{y}_3 | \mathbf{H}_1) \sim N\left(\begin{pmatrix} 2 \\ 2 \\ 2 \end{pmatrix}, \begin{bmatrix} 6 & 3 & 3 \\ 3 & 5 & 3 \\ 3 & 3 & 4 \end{bmatrix}\right),$$

$$p(\mathbf{y}_1, \mathbf{y}_2, \mathbf{y}_3|\mathbf{H_0}) \sim N\left(\begin{pmatrix}0\\0\\0\end{pmatrix}, \begin{bmatrix}3 & 0 & 0\\0 & 2 & 0\\0 & 0 & 1\end{bmatrix}\right).$$

In Fig. 3.6.4, we show ROC curves for the 2-sensor and 3-sensor centralized decision systems as well as for two distributed detection cases. We consider a two sensor $(1 + 2)$ information structure and a three sensor $(1 + 1 + 4)$ information structure. The latter is another example of the type discussed in Theorem 3.5.1.

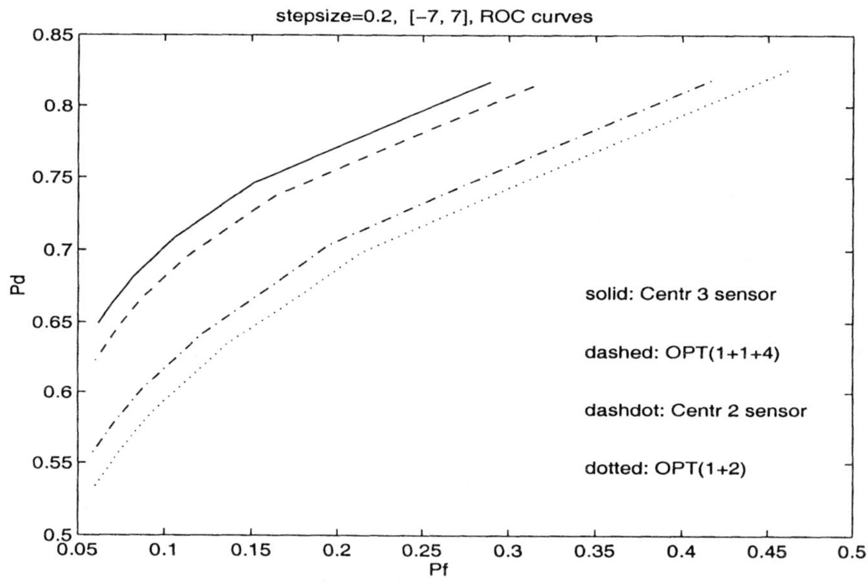

Figure 3.6.4. ROCs for the centralized and distributed OPT(1+2), OPT(1+1+4) rules.

From Fig. 3.6.4 we can see that increasing the number of bits used at one sensor improves the performance significantly in these cases. Also, the three sensor distributed decision system with six bit communication distributed among the sensors can be superior to the two sensor centralized decision system.

Again, in numerical computations, we tried running the iterative algorithm with many different starting conditions and in each case found only one solution that is the one shown in Fig. 3.6.4.

Four sensors

Now we add one more sensor again with noise $\nu_4 \sim N(0, 0.5)$ to the above system. The resulting two conditional pdfs given $\mathbf{H_0}$ and $\mathbf{H_1}$ are

$$p(\mathbf{y}_1, \mathbf{y}_2, \mathbf{y}_3, \mathbf{y}_4|\mathbf{H}_1) \sim N\left(\begin{pmatrix}2\\2\\2\\2\end{pmatrix}, \begin{bmatrix}6 & 3 & 3 & 3\\3 & 5 & 3 & 3\\3 & 3 & 4 & 3\\3 & 3 & 3 & 3.5\end{bmatrix}\right)$$

and

$$p(\mathbf{y}_1, \mathbf{y}_2, \mathbf{y}_3, \mathbf{y}_4|\mathbf{H}_0) \sim N\left(\begin{pmatrix}0\\0\\0\\0\end{pmatrix}, \begin{bmatrix}3 & 0 & 0 & 0\\0 & 2 & 0 & 0\\0 & 0 & 1 & 0\\0 & 0 & 0 & 0.5\end{bmatrix}\right),$$

respectively.

We use the fixed fusion rule from Theorem 3.5.1 for the case considered there with four sensors (single bit decisions at three sensors and eight bit decisions at the other sensor). We know that this fixed fusion rule can achieve optimal performance. We compute the cost functional in (3.1.2) and present the results in Table 3.6.1. The parameters needed to calculate (3.1.2) are $P_1 = \frac{1}{3}$, $P_0 = \frac{2}{3}$, $C_{00} = C_{11} = 0$ and $C_{10} = C_{01} = 1$. Here we tried many different starting conditions and some of these resulted in different solutions as illustrated in Table 3.6.1. This implies that the numerical solution might be a local minimum in this case. In Table 3.6.1, we just show one of the starting conditions that produced a given solution. Actually, we found many starting conditions produced this same solution.

Table 3.6.1. Some centralized and distributed (using our algorithm) designs found and their costs.

Cent/Distr	C (Cost)	initial sensor rules
Cent (4 sen)	0.1055	
Cent (3 sen)	0.1410	
Distr	0.1162	$[I(\mathbf{y}_1), I(\mathbf{y}_2), I(\mathbf{y}_3)]$; $[I(\mathbf{y}_1), I(-\mathbf{y}_2), I(\mathbf{y}_3)]$
Distr	0.1143	$[I(\cos(\mathbf{y}_1)), I(\sin(\mathbf{y}_2)), I(-\sin(\mathbf{y}_3))]$
Distr	0.1144	$[(I(-\sin(\mathbf{y}_1)), I(-\cos(\mathbf{y}_2)), I(\sin(\mathbf{y}_3))]$; $[I(\sin(\mathbf{y}_1)), I(-\cos(\mathbf{y}_2)), I(\cos(\mathbf{y}_3))]$

In Table 3.6.1, $I(\cdot)$ is the indicator function defined as in (3.3.6). Note that we do not really need the initial sensor rules with respect to $\mathbf{y}_4$ in our algorithms since the first three sensor rules define all eight sensor rules $I_4^{(1)}(\mathbf{y}_4), \ldots, I_4^{(8)}(\mathbf{y}_4)$. This can be seen from (3.4.2).

From the given results, we can see that the distributed costs are quite close to the centralized cost. In addition, the initial sensor decision rules in our algorithms do influence the final cost, but the numerical results indicate that the performance may not be too sensitive to the initial rule.

In the first six sections of this chapter, we investigated multisensor Bayes binary decision problems without the assumption that the observations are conditionally independent from sensor to sensor. We have provided necessary conditions for optimal sensor decision rules under a given fusion rule, proposed a discretized Gauss-Seidel iterative algorithm and proved its convergence. Further, we uncovered some highly unexpected results concerning distributed signal detection systems. In certain cases we have shown that a fixed fusion rule can be used to achieve optimal performance. The fusion rule is independent of form of the decision problem (additive noise, known signals, random signals), pdfs, prior probabilities, and all other details, except for the number of bits used in the sensor compressions. This significantly reduces the complexity of finding optimal distributed decision schemes in these cases. The cases for which his fixed fusion rule can be found include those where $l-1$ out of l sensors use a total of $l-1$ bits in their compressions, while the last sensor makes a 2^{l-1} bit compression. Further, we have also shown that the performance is not improved if more than 2^{l-1} bits are used in the last sensor compression. This is true even if fully precise observation of the last sensor is used by the fusion center. All these results have been supported by numerical examples. In the next two sections, the above results will be extended to Neyman-Pearson decision and sequential decision problems.

3.7 OPTIMAL SENSOR RULE OF NEYMAN-PEARSON DECISION

In this section, we consider the distributed Neyman-Pearson decision with general correlated sensor observations and extend the efficient Gauss-Seidel algorithm developed before to search for optimal local compression rules for any fixed fusion rule. Here, the joint probability distributions of the local observations conditional on hypotheses $\mathbf{H}_0$ and $\mathbf{H}_1$ are of general form. No assumptions of either conditional independence or specific probability distribution functions are needed.

To reach this goal, we will use the Lagrange multiplier method (cf. [83]) and the results given in Sections 3.2–3.5, for a given fusion rule. In the first subsection, we present a necessary condition for the optimal sensor rule with a fixed Lagrange multiplier λ. In the next subsection, to facilitate the search for the value of λ whose corresponding P_f best satisfies the constraint, we present an efficient iterative algorithm. Finally, we will give numerical examples in the previous subsection of this section, which provide additional support for the results in the previous two subsections.

3.7.1 NECESSARY CONDITION

In this section, the multisensor decision system consists of l local sensors and a fusion center as defined in Section 3.1. Each local sensor, namely the ith sensor compresses its own observation $\mathbf{y}_i$ to r_i information bits $I_i^{(1)}(\mathbf{y}_i), \ldots, I_i^{(r_i)}(\mathbf{y}_i)$ for the underlying binary hypothesis testing problem and then transmits those bits to the fusion center where the global decision is made. In Subsection 1.3.2, a single sensor Neyman-Pearson decision rule was introduced. Now we consider Neyman-Pearson decision problem for the above l sensor distributed binary decision system.

First, let us recall the desired criterion:

Neyman-Pearson Criterion. Constrain $P_f \leq \alpha$ and design a test to maximizeP_d (or minimize P_m) under this constraint.

In the previous literature on this distributed Neyman-Pearson decision problem, the sensor observational data are assumed to be conditionally independent of each other, and the optimal local compression rules are proved to be of the form of likelihood ratio function (cf. Section 5.2 in [84] and the references therein). In [29, 42], the authors designed an optimal combining scheme at the fusion center when the sensor rules and the correlations between local decisions are all given (therefore, this is a two-level optimization problem). As the same as Bayes test case, there are few results for the general version of the Neyman-Pearson decentralized decision problem with correlated local observations.

In this subsection, we will derive results for a general version of this problem; that is to say, we consider a general correlated local observations and suggest an efficient algorithm to search for the optimal local compression rules for any fixed fusion rule. Here, the joint probabilities of the local observations conditional on the hypotheses $\mathbf{H}_0$ and $\mathbf{H}_1$ are of general form, i.e., they are not assumed to be conditional independent cross sensors or specific probability distribution functions.

A set of sensor compression rules: $(I_1^{(1)}(\mathbf{y}_1),\ldots, I_1^{(r_1)}(\mathbf{y}_1); \cdots; I_l^{(1)}(\mathbf{y}_l),\ldots, I_l^{(r_l)}(\mathbf{y}_l))$ compresses data $\mathbf{y}_i$ to r_i $(1 \leq i \leq l)$ information bits at each sensor, respectively:

$$(I_1^{(1)}(\mathbf{y}_1), \ldots, I_1^{(r_1)}(\mathbf{y}_1)) : (\Re^{n_1})^{r_1} \longmapsto \{0,1\}^{r_1},$$

$$\cdots$$

$$(I_l^{(1)}(\mathbf{y}_l), \ldots, I_l^{(r_l)}(\mathbf{y}_l)) : (\Re^{n_l})^{r_l} \longmapsto \{0,1\}^{r_l}.$$

Then the local sensors transmit their own compressed binary messages to the fusion center. Denote $N = \sum_{i=1}^{l} r_i$. Upon receiving the local message N-tuple $(I_1^{(1)}, \ldots, I_1^{(r_1)}; \cdots; I_l^{(1)}, \ldots, I_l^{(r_l)})$, the fusion center attempts to make a final decision $F \in \{0, 1\}$ with the objective of optimizing a system performance based on the Neyman-Pearson criterion.

In Section 3.2, we showed that any fusion rule can be represented in terms of a polynomial

$$P_{\mathbf{H}_1}(I_1^{(1)}(\mathbf{y}_1), \ldots, I_1^{(r_1)}(\mathbf{y}_1; \cdots; I_l^{(1)}(\mathbf{y}_l), \ldots, I_l^{(r_l)}(\mathbf{y}_l)) \qquad (3.7.1)$$

of the local compression rules in the sense that

$$\begin{aligned} &F(I_1^{(1)}, \ldots, I_1^{(r_1)}; \cdots; I_l^{(1)}, \ldots, I_l^{(r_l)}) = 1 \quad \text{if and only if} \\ &P_{\mathbf{H}_1}(I_1^{(1)}, \ldots, I_1^{(r_1)}; \cdots; I_l^{(1)}, \ldots, I_l^{(r_l)}) = 1, \end{aligned} \qquad (3.7.2)$$

where the polynomial $P_{\mathbf{H}_1}$ in fact is also an indicator function over the final $\mathbf{H}_1$ decision region on $\Re^{n_1} \times \Re^{n_2} \times \cdots \times \Re^{n_l}$.

Thus, as mentioned in Subsection 2.1.2, the difference between the distributed and centralized decisions lies in the "form" of the decision regions: the distributed decision requires its decision region to be of the restricted form given in (3.7.2) while the centralized decision allows an unrestricted decision region

$$\{(\mathbf{y}_1, \ldots, \mathbf{y}_l)) : F = 1\}.$$

With the Neyman-Pearson criterion, we constrain $P_f = \alpha$ (in fact, if P_f cannot be made to be α exactly in the deterministic decision rule, then the randomized decision rule in Subsection 1.3.2 can be used) and design a test to minimize P_m subject to this constraint. The optimal local compression rules ($I_1^{(1)}, \ldots, I_1^{(r_1)}$; ...; $I_l^{(1)}, \ldots, I_l^{(r_l)}$) for a given fusion rule F is derived by using a Lagrange multiplier

$$\begin{aligned} &C(I_1^{(1)}, \ldots, I_1^{(r_1)}; \cdots; I_l^{(1)}, \ldots, I_l^{(r_l)}, F, \lambda) \\ &\stackrel{\text{def}}{=} P_m + \lambda(P_f - \alpha) \\ &= \int_{F=0} p(\mathbf{y}_1, \ldots, \mathbf{y}_l|\mathbf{H}_1) d\mathbf{y}_1 \cdots d\mathbf{y}_l \\ &\qquad + \lambda[\int_{F=1} p(\mathbf{y}_1, \ldots, \mathbf{y}_l|\mathbf{H}_0) d\mathbf{y}_1 \cdots d\mathbf{y}_l - \alpha] \\ &= \lambda(1 - \alpha) \\ &\quad + \int (1 - P_{\mathbf{H}_1})[p(\mathbf{y}_1, \ldots, \mathbf{y}_l|\mathbf{H}_1) - \lambda p(\mathbf{y}_1, \ldots, \mathbf{y}_l|\mathbf{H}_0)] d\mathbf{y}_1 \cdots d\mathbf{y}_l. \end{aligned} \qquad (3.7.3)$$

For notational simplicity, denote

$$L_\lambda(\mathbf{y}_1, \ldots, \mathbf{y}_l) = p(\mathbf{y}_1, \ldots, \mathbf{y}_l|\mathbf{H}_1) - \lambda p(\mathbf{y}_1, \ldots, \mathbf{y}_l|\mathbf{H}_0). \qquad (3.7.4)$$

For a fixed λ, minimizing $C(I_1^{(1)}, ..., I_1^{(r_1)}; \cdots; I_l^{(1)}, \ldots, I_l^{(r_l)}, F, \lambda)$ is equivalent to minimizing

$$C^*(I_1^{(1)}, \ldots, I_l^{(r_l)}, F, \lambda) \stackrel{\text{def}}{=} \int (1 - P_{\mathbf{H}_1}) L_\lambda(\mathbf{y}_1, \ldots, \mathbf{y}_l) d\mathbf{y}_1 \cdots d\mathbf{y}_l. \tag{3.7.5}$$

Now the result on the necessary condition for the optimal sensor compression rules for the distributed Bayes decision in Theorem 3.4.1 can be extended immediately as follows.

Theorem 3.7.1. For a given fusion rule F and a fixed Lagrange multiplier λ, the optimal sensor compression rules $(I_1^{(1)}, ..., I_1^{(r_1)}; ...; I_l^{(1)}, \ldots, I_l^{(r_l)})$, which minimize the cost functional of Eq. (3.7.5) must satisfy the following integral equations

$$\begin{aligned}
I_1^{(1)}(\mathbf{y}_1) &= I[\int P_{11}^{(1)} L_\lambda(\mathbf{y}_1, \ldots, \mathbf{y}_l) d\mathbf{y}_2 \cdots d\mathbf{y}_l], \\
&\cdots, \\
I_1^{(r_1)}(\mathbf{y}_1) &= I[\int P_{11}^{(r_1)} L_\lambda(\mathbf{y}_1, \ldots, \mathbf{y}_l) d\mathbf{y}_2 \cdots d\mathbf{y}_l]; \\
&\cdots; \\
I_l^{(1)}(\mathbf{y}_l) &= I[\int P_{l1}^{(1)} L_\lambda(\mathbf{y}_1, \ldots, \mathbf{y}_l) d\mathbf{y}_1 \cdots d\mathbf{y}_{l-1}], \\
&\cdots, \\
I_l^{(r_l)}(\mathbf{y}_l) &= I[\int P_{l1}^{(r_l)} L_\lambda(\mathbf{y}_1, \ldots, \mathbf{y}_l) d\mathbf{y}_1 \cdots d\mathbf{y}_{l-1}].
\end{aligned} \tag{3.7.6}$$

where $I[\cdot]$ is a indicator function given before (see (3.3.6)).

From the theorem, it is easy to see that this set of optimal sensor compression rules is in general no longer of the form of a likelihood ratio function when the sensor observations are correlated. However, if the sensor observations are uncorrelated, then for this special case the above necessary condition can be easily reduced (as done in Corollary 3.4.1) to a well-known result in [84] that the optimal sensor compression rules are all of likelihood ratio form.

For any positive value of λ, a Gauss-Seidel type iterative algorithm to search for the optimal sensor rules $(I_1^{(1)}, ..., I_1^{(r_1)}; ...; I_l^{(1)}, \ldots, I_l^{(r_l)})$ can be derived, similarly to what was derived in Section 3.5.

Obviously, the above optimal sensor rules depend on λ and the probability of a false alarm also depends on λ, i.e.,

$$P_f = 1 - \int (1 - P_{\mathbf{H}_1}) p(\mathbf{y}_1, \ldots, \mathbf{y}_l | \mathbf{H}_0) d\mathbf{y}_1 \cdots d\mathbf{y}_l = \alpha. \tag{3.7.7}$$

To satisfy the constraint $P_f = \alpha$, we must choose an appropriate λ.

3.7.2 THE ALGORITHM TO SEARCH FOR OPTIMAL SENSOR RULES

The algorithm we propose in this subsection uses Theorems 3.7.1 above and 3.7.2 below and consists of the following two steps:

Step I

For a fixed λ, use the iterative algorithm given in Section 3.5.

$$
\begin{aligned}
I_{1(i+1)}^{(1)}(\mathbf{y}_1) &= I[\int P_{11}^{(1)}(I_{1(i)}^{(2)}, I_{1(i)}^{(3)}, \ldots, I_{l(i)}^{(r_l)})L_\lambda d\mathbf{y}_2 \cdots d\mathbf{y}_l],\\
I_{1(i+1)}^{(2)}(\mathbf{y}_1) &= I[\int P_{11}^{(2)}(I_{1(i+1)}^{(1)}, I_{1(i)}^{(3)}, \ldots, I_{l(i)}^{(r_l)})L_\lambda d\mathbf{y}_2 \cdots d\mathbf{y}_l],\\
&\cdots,\\
I_{l(i+1)}^{(1)}(\mathbf{y}_l) &= I[\int P_{l1}^{(1)}(I_{1(i+1)}^{(1)}, \ldots, I_{(l-1)(i+1)}^{(r_{l-1})}; I_{l(i)}^{(2)}, \ldots, I_{l(i)}^{(r_l)})L_\lambda \\
&\qquad d\mathbf{y}_1 \cdots d\mathbf{y}_{l-1}],\\
&\cdots,\\
I_{l(i+1)}^{(r_l)}(\mathbf{y}_l) &= I[\int P_{l1}^{(1)}(I_{1(i+1)}^{(1)}, \ldots, I_{l(i+1)}^{(r_l-1)})L_\lambda d\mathbf{y}_1 \cdots d\mathbf{y}_{l-1}].
\end{aligned}
\tag{3.7.8}
$$

To facilitate computer implementation of this process, we need to discretize the variables $\mathbf{y}_1, \ldots, \mathbf{y}_l$ and make the discretized version of the above algorithm. In Section 3.5, we have shown this procedure in detail and analyzed its convergence of the algorithm.

Step II

After getting the optimal sensor rules for a fixed λ, choose a suitable value of λ so that the corresponds to P_f satisfies the constraint $P_f = \alpha$.

It is well-known that the probability of a false alarm decreases as λ increases for the centralized Neyman-Pearson decision problem. This can be seen easily from the conventional Neyman-Pearson decision shown in Subsection 1.3.2. This monotonicity is very convenient when searching for the most suitable value of α. Therefore, a new problem has arisen: does the monotonicity hold for the new method presented in the previous subsection? Generally speaking, the answer is no because of the difference between the centralized and distributed decisions. In particular, when the number of local minima of C^* in Eq. (3.7.5) increases as λ increases. In many numerical examples that we have done, however, the monotonicity of P_f with respect to λ still occurs. Although this is not guaranteed for arbitrary distributed observational data, we can use it to determine the searching direction for λ in numerical computations.

Finally, it should be emphasized that all results in this subsection is derived under a given fusion rule which, of course, is the deterministic decision rule. In Section 2.6, we have mentioned that the randomized fusion rule, i.e., using two fusion rules with proper probabilities, can improve P_d under the constraint $P_f \leq \alpha$. However, the implementation of the randomized fusion rule is still based on the result on the optimal sensor rules given a fixed fusion rule in Sections 3.2–3.4 (see Section 2.6).

3.7.3 NUMERICAL EXAMPLES

In the following numerical examples, we consider the same two sensor Gaussian signal detection model as given in Section 3.6. we take the discretization step-size $\Delta = 0.1$.

First, we consider the system information structure with each of two sensors transmitting one bit information to the fusion center. The fusion rules are the AND, OR and XOR rules, respectively. We will prove in Section 5.2 that for the above information structure, an optimal fusion rule among all possible $2^{2^2} = 16$ rules must be one of the AND, OR and XOR rules. For comparison, the numerical results for the centralized Neyman-Pearson decision is presented too.

Then, we consider the system information structure with the first sensor transmitting one bit of information and the second transmitting two bits to the fusion center. The fusion rules are employed here are

$$F = [1 - I_1(\mathbf{y}_1)][1 - I_2^{(1)}(\mathbf{y}_2)] + I_1(\mathbf{y}_1)[1 - I_2^{(2)}(\mathbf{y}_2)],$$

which actually is an optimal rule for the given information structure (for the detailed argument, see Section 3.5). In these examples, the numerical results are not sensitive to the initial values and the algorithms can converge to global minimum. The results are in Table 3.7.1, where OPT(1+2) stands for the above optimal fusion rule.

In Table 3.7.1, we see that the performances for the above five decision rules with constraint $\alpha \leq 0.099$ in order of P_d from big to small are the centralized, OPT(1+2), AND(1+1), OR(1+1), XOR (1+1). The performances with constraint $\alpha \leq 0.2565$ in the same order are the centralized, OPT(1+2), OR(1+1), AND(1+1), XOR (1+1), where only OR(1+1) and AND(1+1) exchange their positions in contrast to the case of $\alpha \leq 0.099$.

In our numerical results, we are not able to keep $P_f \leq 0.1$ and $P_d \geq 0.9$ simultaneously. Therefore, we require more observations to construct a Neyman-Pearson type sequential distributed decision system. The method developed here can be applied to this problem, and will be discussed in the next section.

Table 3.7.1. Performance comparisons with $\alpha \leq 0.099$ and $\alpha \leq 0.2565$.

	P_d	$P_f \leq 0.099$		P_d	$P_f \leq 0.2565$
Centr.	0.5874	0.0987	Centr.	0.7097	0.2499
OPT(1+2)	0.5589	0.0945	OPT(1+2)	0.6945	0.2450
AND(1+1)	0.5418	0.0968	OR(1+1)	0.6789	0.2499
OR(1+1)	0.5327	0.0974	AND(1+1)	0.6749	0.2519
XOR(1+1)	0.4322	0.0976	XOR(1+1)	0.6590	0.2565

3.8 SEQUENTIAL DECISION FUSION GIVEN FUSION RULE

In this section, we combine the Neyman-Pearson type sequential decision method proposed in Subsection 1.3.3 and the multisensor Neyman-Pearson decision proposed in Section 3.7 to develop a Neyman-Pearson type multisensor sequential decision method.

Note that the main feature of the sequential decision method which was proposed in Subsection 1.3.3 is the assignment of two constant decision error probabilities at each stage so that after a finite number of time instants, the two total error probabilities P_f and P_m approximate the given parameters α and β respectively, and at the same time instant the continuation region becomes empty. Thus, the new multisensor sequential method that will be discussed in this section will have two advantages: the stopping time T has a uniform upper bound and takes better advantage of the two constraints $P_f \leq \alpha$ and $P_m \leq \beta$ than Wald's test. In doing so, at each stage, the problem reduces to a multisensor Neyman-Pearson decision problem with two error probability constraints. Indeed, the following method is not new, except the combination of the two decision methods given in Subsection 1.3.3 and the privious section.

3.8.1 ALGORITHM

Note that in Subsection 1.3.3, at each stage, the two decision error probabilities $\triangle\alpha$ and $\triangle\beta$ are both always constant as well as

$$\{Y(k)\} = \{\mathbf{y}_1(k), \mathbf{y}_2(k), \cdots, \mathbf{y}_l(k)\}$$

is a temporarily i.i.d. sequence, but sensor observation $\mathbf{y}_i(k)$ may be dependent from sensor to sensor at each time instant. We propose the distributed sequential decision as follows.

At first stage, using the multisensor Neyman-Pearson decision method given in Section 3.7, for a given $\triangle\alpha$, we compute a set of optimal local compression rules $(I_1^{(1)}(1), \ldots, I_1^{(r_1)}(1); \cdots; I_l^{(1)}(1), \ldots, I_l^{(r_l)}(1))$ which together with the given fusion rule determine a region $\mathcal{H}_1(1)$ in $\Re^{n_1} \times \cdots \times \Re^{n_l}$ so that

$$P(Y(1) \in \mathcal{H}_1(1)|\mathbf{H}_0) = \triangle\alpha. \tag{3.8.1}$$

Then, on the complementary set of $\mathcal{H}_1(1)$ on $\Re^{n_1} \times \cdots \times \Re^{n_l}$, still using the above Neyman-Pearson decision method, for a given $\triangle\beta$, another set of optimal local compression rules $(\bar{I}_1^{(1)}(1), \ldots, \bar{I}_1^{(r_1)}(1); \cdots; \bar{I}_l^{(1)}(1), \ldots, \bar{I}_l^{(r_l)}(1))$ and the corresponding region $\mathcal{H}_0(1) \subset \mathcal{H}_1^c(1)$ can be found similarly, so that

$$P(Y(1) \in \mathcal{H}_0(1)|\mathbf{H}_1) = \triangle\beta. \tag{3.8.2}$$

After obtaining $\mathcal{H}_0(1)$ and $\mathcal{H}_1(1)$, we can calculate two correct-decision probabilities

$$\int_{\mathcal{H}_1(1)} p(Y(1)|\mathbf{H}_1)dY(1) = \beta(1) \tag{3.8.3}$$

and

$$\int_{\mathcal{H}_0(1)} p(Y(1)|\mathbf{H}_0)dY(1) = \alpha(1). \tag{3.8.4}$$

Hence,

$$\int_{\mathcal{C}(1)} p(Y(1)|\mathbf{H}_0)dY(1) = 1 - \triangle\alpha - \alpha(1) \tag{3.8.5}$$

and

$$\int_{\mathcal{C}(1)} p(Y(1)|\mathbf{H}_1)dY(1) = 1 - \triangle\beta - \beta(1). \tag{3.8.6}$$

Obviously, if $P(Y(1) \in \mathcal{C}(1)|\mathbf{H}_1) \leq \triangle\beta$ or $P(Y(1) \in \mathcal{C}(1)|\mathbf{H}_0) \leq \triangle\alpha$, we can choose whole $\mathcal{C}(1) \times \Re^{n_1} \times \cdots \times \Re^{n_l}$ as $\mathcal{H}_0(2)$ or $\mathcal{H}_1(2)$, therefore, $\mathcal{C}(2) = \emptyset$ and the decision is terminated at the second stage w.p.1, i.e., $N = 2$.

At the second stage, using the method similar to those at the first stage, in general, we can partition the space $\Re^{n_1} \times \cdots \times \Re^{n_l}$ into three disjoint regions $\tilde{\mathcal{H}}_1(2)$, $\tilde{\mathcal{H}}_0(2)$ and so that

$$P(Y(2) \in \tilde{\mathcal{H}}_1(2)|\mathbf{H}_0) = \frac{\triangle\alpha}{1 - \triangle\alpha - \alpha(1)} \tag{3.8.7}$$

and

$$P(Y(2) \in \tilde{\mathcal{H}}_0(2)|\mathbf{H}_1) = \frac{\triangle\beta}{1 - \triangle\beta - \beta(1)}. \tag{3.8.8}$$

Now define the second decision regions and continuation region as

$$\mathcal{H}_0(2) = \mathcal{C}(1) \times \tilde{\mathcal{H}}_0(2),$$

$$\mathcal{H}_1(2) = \mathcal{C}(1) \times \tilde{\mathcal{H}}_1(2),$$

$$\mathcal{C}(2) = (\mathcal{H}_0(2) \cup \mathcal{H}_1(2))^c \cap (\mathcal{C}(1) \times (\Re^{n_1} \times \cdots \times \Re^{n_l})).$$

Clearly, we now have

$$P((Y(1), Y(2)) \in \mathcal{H}_1(2)|\mathbf{H}_0) = \triangle\alpha \tag{3.8.9}$$

and

$$P((Y(1), Y(2)) \in \mathcal{H}_0(2)|\mathbf{H}_1) = \triangle\beta. \tag{3.8.10}$$

Similar to (3.8.3)–(3.8.6), we have

$$\int_{\mathcal{H}_1(2)} p(Y(1), Y(2)|\mathbf{H}_1) dY(1) dY(2) = \beta(2), \tag{3.8.11}$$

$$\int_{\mathcal{H}_0(2)} p(Y(1), Y(2)|\mathbf{H}_0) dY(1) dY(2) = \alpha(2), \tag{3.8.12}$$

$$\int_{\mathcal{C}(2)} p(Y(1), Y(2)|\mathbf{H}_0) dY(1) dY(2) = 1 - 2\triangle\alpha - \alpha(1) - \alpha(2) \tag{3.8.13}$$

and

$$\int_{\mathcal{C}(2)} p(Y(1), Y(2)|\mathbf{H}_1) dY(1) dY(2) = 1 - 2\triangle\beta - \beta(1) - \beta(2). \tag{3.8.14}$$

In general, if

$$P((Y(1), \ldots, Y(k)) \in \mathcal{C}(k)|\mathbf{H}_1) \leq \triangle\beta,$$

or

$$P((Y(1), \ldots, Y(k)) \in \mathcal{C}(k)|\mathbf{H}_0) \leq \triangle\alpha$$

at the kth stage, we can choose whole $\mathcal{C}(k) \times \Re^{n_1} \times \cdots \times \Re^{n_l}$ as $\mathcal{H}_0(k+1)$ or $\mathcal{H}_1(k+1)$, and the decision is terminated at the $(k+1)$th stage w.p.1, i.e., $N = k+1$. Otherwise, similar to the preceding stages, we have $\mathcal{H}_0(k+1)$, $\mathcal{H}_1(k+1)$ and $\mathcal{C}(k+1)$ such that

$$P((Y(1), \ldots, Y(k+1)) \in \mathcal{H}_1(k+1)|\mathbf{H}_0) = \triangle\alpha, \tag{3.8.15}$$

$$P((Y(1), \ldots, Y(k+1)) \in \mathcal{H}_0(k+1)|\mathbf{H}_1) = \triangle\beta, \tag{3.8.16}$$

$$\int_{\mathcal{H}_1(k+1)} p(Y(1), \ldots, Y(k+1)|\mathbf{H}_1) dY(1) \cdots dY(k+1) = \beta(k+1), \tag{3.8.17}$$

$$\int_{\mathcal{H}_0(k+1)} p(Y(1), \ldots, Y(k+1)|\mathbf{H}_0)dY(1)\cdots dY(k+1) = \alpha(k+1), \tag{3.8.18}$$

$$\begin{aligned}&\int_{\mathcal{C}(k+1)} p(Y(1), \ldots, Y(k+1)|\mathbf{H}_0)dY(1)\cdots dY(k+1)\\ &= 1-(k+1)\triangle\alpha - \alpha(1) - \cdots - \alpha(k+1),\end{aligned} \tag{3.8.19}$$

and

$$\begin{aligned}&\int_{\mathcal{C}(k+1)} p(Y(1), \ldots, Y(k+1)|\mathbf{H}_1)dY(1)\cdots dY(k+1)\\ &= 1-(k+1)\triangle\beta - \beta(1) - \cdots - \beta(k+1).\end{aligned} \tag{3.8.20}$$

Then, we can use the efficient algorithm proposed in Subsection 1.3.3 to to search for solutions of $\triangle\alpha$ and $\triangle\beta$, uniform upper bounds N w.p.1. of the stopping time T, and the decision region sets $\{\mathcal{H}_0(1), \ldots, \mathcal{H}_0(N)\}$, $\{\mathcal{H}_1(1), \ldots, \mathcal{H}_1(N)\}$, and $\{\mathcal{C}(1), \ldots, \mathcal{C}(N)\}$ so that

$$\begin{aligned}P_f &= \textstyle\sum_{k=1}^N \int_{\mathcal{H}_1(k)} p(Y(1), \ldots, Y(k)|\mathbf{H}_0)dY(1)\cdots dY(k)\\ &= (N-1)\triangle\alpha + \epsilon_0 \le \alpha,\end{aligned} \tag{3.8.21}$$

and

$$\begin{aligned}P_m &= \textstyle\sum_{k=1}^N \int_{\mathcal{H}_0(k)} p(Y(1), \ldots, Y(k)|\mathbf{H}_1)dY(1)\cdots dY(k)\\ &= (N-1)\triangle\beta + \epsilon_1 \le \beta,\end{aligned} \tag{3.8.22}$$

as well as,

$$\mathcal{C}(N) = \emptyset, \tag{3.8.23}$$

where constants $\epsilon_0 \le \triangle\alpha$ and $\epsilon_1 \le \triangle\beta$. In our many numerical example, it was shown that such N can be found.

3.8.2 NUMERICAL EXAMPLE

In the following numerical example, we still consider two sensor Gaussian signal detection model given in Section 3.6. We take $\alpha = \beta = 0.15$. The result of the numerical example shows that it is impossible for one-step-observation Neyman-Pearson decision system to satisfy $P_f \le \alpha$ and $P_m \le \beta$ simultaneously. Hence, we consider the sequential decision system.

First, let us consider the system information structure with each of two sensors transmitting one bit to the fusion center and two fusion rules: the AND and OR at every stage. Then, using the result on the optimal fusion rule in Section 3.5, we consider the system information structure with the first sensor transmitting one bit and the second one transmitting two bits to the fusion center at every stage and use the optimal fusion (OPT) rules under this **(1+2)** information structures.

In the following table, for two centralized decisions, three distributed AND, OR, and OPT decision rules, we show the conditional probabilities

$$P((Y(1), \ldots, Y(k)) \in \mathcal{C}(k)|\mathbf{H}_i), i = 1, 2, \quad k = 1, 2, \ldots,$$

at each stage until the continuation observation region $\mathcal{C}(k)$ becomes empty. Unfortunately, due to intractable (high dimensional numerical integral) computation, we just show $P_W(\mathcal{C}(k)|\mathbf{H}_i)$, $k = 1, 2, 3$.

Table 3.8.1. Performance comparison of the centralized Wald's and new method.

k	$P_W(C\|\mathbf{H}_0)$	$P_W(C\|\mathbf{H}_1)$	$P_N(C\|\mathbf{H}_0)$	$P_N(C\|\mathbf{H}_1)$
1	0.9853	0.6027	0.8264	0.4801
2	0.4534	0.2760	0.4499	0.2024
3	0.2086	0.1267	0.0102	0.0065
4	unclear	unclear	0.0000	0.0000

Table 3.8.2. Numerical results of the distributed AND, OR and OPT(1+2).

k	$P_A(C\|\mathbf{H}_0)$	$P_A(C\|\mathbf{H}_1)$	$P_{OR}(C\|\mathbf{H}_0)$	$P_{OR}(C\|\mathbf{H}_1)$	$P_{OPT}(C\|\mathbf{H}_0)$	$P_{OPT}(C\|\mathbf{H}_1)$
1	0.8475	0.4997	0.9627	0.5616	0.8313	0.4796
2	0.6193	0.2227	0.6007	0.2558	0.5766	0.2016
3	0.2856	0.0705	0.2100	0.0760	0.2132	0.0549
4	0.0000	0.0000	0.0000	0.0000	0.0000	0.0000

Then, using the data in the above tables and the formula

$$E(T|\mathbf{H}_i) = \sum_{k=1}^{N} k(P(\mathcal{C}(k-1)|\mathbf{H}_i) - P(\mathcal{C}(k)|\mathbf{H}_i))$$

$$P(\mathcal{C}(0)|\mathbf{H}_i) = 1, \quad P(\mathcal{C}(N)|\mathbf{H}_i) = 0,$$

it is easy to get

$E_{OR}(T|\mathbf{H}_0) = 2.7734$, $E_{OR}(T|\mathbf{H}_1) = 1.8934$, $E_{AND}(T|\mathbf{H}_0) = 2.7524$, $E_{AND}(T|\mathbf{H}_1) = 1.7929$, $E_{OPT}(T|\mathbf{H}_0) = 2.6211$, $E_{OPT}(T|\mathbf{H}_1) = 1.7359$, $E_N(T|\mathbf{H}_0) = 2.2865$, $E_N(T|\mathbf{H}_1) = 1.5135$.

Since there are no complete data for Wald's sequential decision, no $E_W(T|\mathbf{H}_j)$ can be calculated.

Obviously, the smaller the above two conditional expectations are, the better the performance of the corresponding decision rule should be. The numerical results show that for the above example, the regions $\mathcal{C}(4)$ for all decision systems become empty, except the traditional Wald's sequential decision. Although, due to high dimensional numerical integral computation, we just show $P_W(\mathcal{C}(k)|\mathbf{H}_i)$, $k = 1, 2, 3$, it can be still seen that $P_W(\mathcal{C}(j)|\mathbf{H}_i)$ decreases much more slowly than $P_N(\mathcal{C}(j)|\mathbf{H}_i)$ as j increases. Therefore, $E_N(T|\mathbf{H}_i) < E_W(T|\mathbf{H}_i)$. In addition, the performance of the distributed sequential decision with OPT fusion rule is superior to those of AND and OR rules.

Chapter 4

MULTISENSOR MULTI-HYPOTHESIS NETWORK DECISION

In this chapter, we consider more general multisensor multi-hypothesis decision systems.

Suppose that l sensors observe multiple observational data $(\mathbf{y}_1, \ldots, \mathbf{y}_l)$ and must decide one among m hypotheses $\mathbf{H}_0, \mathbf{H}_1, \ldots, \mathbf{H}_{m-1}$.

At the ith sensor, the r_i binary compression rules: $(I_i^{(1)}(\mathbf{y}_i), \ldots, I_i^{(r_i)}(\mathbf{y}_i))$ compresses the data $\mathbf{y}_i$ to r_i $(i \leq l)$ information bits.

There might be or might not be communication between sensors. We mainly focus on a few fundamental structures. More complicated structures can be generated based on those fundamental structures.

4.1 ELEMENTARY NETWORK STRUCTURES

Three types of elementary distributed multisensor network decision systems are considered and modelled as follows.

4.1.1 PARALLEL NETWORK

We first consider the network without communication between sensors, i.e., the local sensors transmit their own compressed binary messages

$$(I_1^{(1)}(\mathbf{y}_1), \ldots, I_1^{(r_1)}(\mathbf{y}_1)) : \Re^{n_1} \longmapsto \{0, 1\}^{r_1},$$

$$\cdots$$

$$(I_l^{(1)}(\mathbf{y}_l), \ldots, I_l^{(r_l)}(\mathbf{y}_l)) : \Re^{n_l} \longmapsto \{0, 1\}^{r_l}$$

only to the fusion center. Denote

$$N = \sum_{i=1}^{l} r_i.$$

Based on the received N-tuple local message $(I_1^{(1)}, \ldots, I_1^{(r_1)}; \cdots; I_l^{(1)}, \ldots, I_l^{(r_l)})$, under a given fusion rule the fusion center makes a final decision as the following Fig. 4.1.1.

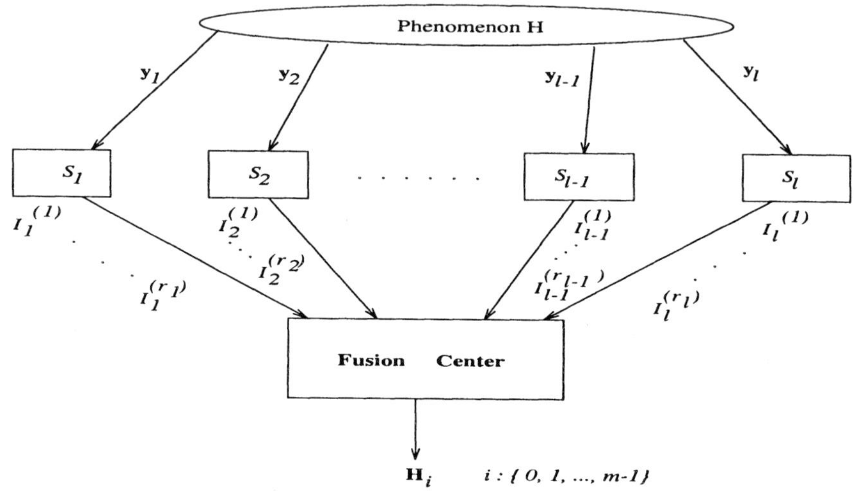

Figure 4.1.1. Parallel network structure.

For notational simplicity, we denote it by

$$(\{\mathbf{y}_1^{(r_1)} // \mathbf{y}_2^{(r_2)} // \cdots // \mathbf{y}_l^{(r_l)}\} \Rightarrow \mathcal{F} \rightarrow \{m\}), \tag{4.1.1}$$

where $\{\cdot // \cdot // \cdots // \cdot\}$ means that all sensors inside "{ }" are in parallel positions without communications between sensors. Besides, "$\mathbf{y}_i^{(r_i)}$" expresses the ith sensor observes own data $\mathbf{y}_i$ and then compresses them to r_i information bits. Moreover, "$\{\mathbf{y}_1^{(r_1)} // \mathbf{y}_2^{(r_2)} // \cdots // \mathbf{y}_l^{(r_l)}\} \Rightarrow \mathcal{F} \rightarrow \{m\}$", implies that all local sensors in "{ }" transmit their compressed data–information bits to the fusion center and then a m-ary decision is made by the fusion center $\mathcal{F}$. In addition, denote m given joint conditional probability density functions by $p(\mathbf{y}_1, \mathbf{y}_2, \cdots, \mathbf{y}_l|\mathbf{H}_0), \ldots, p(\mathbf{y}_1, \mathbf{y}_2, \ldots, \mathbf{y}_l|\mathbf{H}_{m-1})$ which are of arbitrary general forms.

A fusion rule F_{para} of the fusion center for the parallel network is given by an m-valued function

$$\begin{aligned} F_{para}(I_1^{(1)}(\mathbf{y}_1), \ldots, I_1^{(r_1)}(\mathbf{y}_1); \cdots; I_l^{(1)}(\mathbf{y}_l), \ldots, I_l^{(r_l)}(\mathbf{y}_l)) : \\ \{0,1\}^N \longmapsto \{0, 1, \ldots, m-1\}. \end{aligned} \tag{4.1.2}$$

In practice, one could build one of local sensors, namely the lth sensor, and the fusion center at the same station to save communication between this sensor and the fusion center as shown in Fig. 4.1.2. This system is called *the modified*

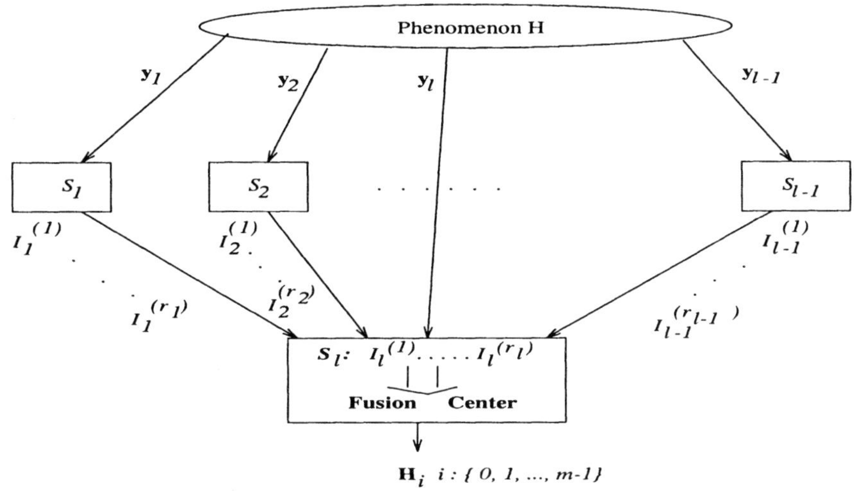

Figure 4.1.2. The modified parallel network structure.

parallel network decision system. Thus, the above information structure can be expressed simply by

$$(\{\mathbf{y}_1^{(r_1)}//\mathbf{y}_2^{(r_2)}//\cdots//\mathbf{y}_{l-1}^{(r_{l-1})}\} \Rightarrow (\mathbf{y}_l^{(r_l)}\cdot \mathcal{F}) \rightarrow \{m\}). \tag{4.1.3}$$

Hence, in this structure, no matter how large the value of r_l is, no communication bandwidth is required by the lth sensor.

Remark 4.1.1. For the above model (4.1.3), when $m = 2^n$ where n is an integer, we can also regard the lth sensor, which compresses its all received messages and observation to n information bits, as an intermediate node in the sensor network, and accordingly, view the fusion rule at the lth sensor as a local compression/fusion rule. Therefore, The parallel network structure is one of basic elements of information structures in the multisensor network decision systems.

4.1.2 TANDEM NETWORK AND TREE NETWORK

In the tandem system, the sensor network is a team of sensors in tandem, i.e., the first sensor compresses its observation data $\mathbf{y}_1$ to r_1 information bits $(I_1^{(1)}(\mathbf{y}_1)$ ..., $I_1^{(r_1)}(\mathbf{y}_1))$ and transmits them to the second sensor. Then due to communication bandwidth limit, the second sensor has to compress its observational data $\mathbf{y}_2$ and the received message $(I_1^{(1)}(\mathbf{y}_1), \ldots, I_1^{(r_1)}(\mathbf{y}_1))$ from the first sensor to s_2 bits $(T_2^{(1)}(I_1^{(1)}, \ldots, I_1^{(r_1)}; \mathbf{y}_2), \cdots, T_2^{(s_2)}(I_1^{(1)}, \ldots, I_1^{(r_1)}; \mathbf{y}_2))$ and then transmit those bits to the next sensor–the third sensor and so on. This procedure is repeated until the $(l-1)$th sensor. The last sensor–the lth sensor also uses its observation data $\mathbf{y}_l$ together with the received message $(T_{l-1}^{(1)}, \ldots, T_{l-1}^{(s_{l-1})})$ from the $(l-1)$th sensor to make a final m-ary decision under a given final fusion rule F_{tan}. So in this model, the lth sensor and the decision center are built in the same station, see the following Fig. 4.1.3.

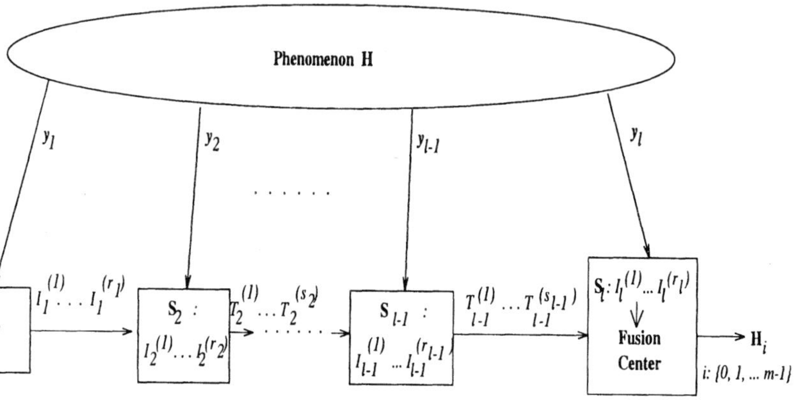

Figure 4.1.3. Tandem network structure.

Noting that $T_i^{(j)}$ $(2 \le i \le l-1$ and $1 \le j \le s_i)$ is also a binary function which we call the *local fusion rule* of the ith sensor. Since all $T_{i-1}^{(1)}, \ldots, T_{i-1}^{(s_{i-1})}$ are binary functions, there are totally $2^{s_{i-1}}$ different s_{i-1}-tuple $(T_{i-1}^{(1)}, \ldots, T_{i-1}^{(s_{i-1})})$. Thus, $T_i^{(j)}(T_{i-1}^{(1)}, \ldots, T_{i-1}^{(s_{i-1})}; \mathbf{y}_i)$ can be equivalently expressed by $2^{s_{i-1}}$ binary functions of $\mathbf{y}_i$, namely $(I_i^{((j-1)2^{s_{i-1}}+1)}(\mathbf{y_i}), \ldots, I_i^{(j2^{s_{i-1}})}(\mathbf{y_i}))$ as follows:

$$\begin{aligned}&\{(\mathbf{y}_1, \ldots, \mathbf{y}_i) : T_i^{(j)}(T_{i-1}^{(1)}, \ldots, T_{i-1}^{(s_{i-1})}; \mathbf{y}_i) = 0 \\ &\qquad\qquad \text{for all } 2^{s_{i-1}}\ (T_{i-1}^{(1)}, \ldots, T_{i-1}^{(s_{i-1})})\} \\ &= \{(\mathbf{y}_1, \ldots, \mathbf{y}_i) : I_i^{((j-1)2^{s_{i-1}}+1)}(\mathbf{y}_i) = 0, \ldots, I_i^{(j2^{s_{i-1}})}(\mathbf{y}_i) = 0\}\end{aligned} \tag{4.1.4}$$

and

$$\begin{aligned}&\{(\mathbf{y}_1, \ldots, \mathbf{y}_i) : T_i^{(j)}(T_{i-1}^{(1)}, \ldots, T_{i-1}^{(s_{i-1})}; \mathbf{y}_i) = 1 \\ &\qquad\qquad \text{for all } 2^{s_{i-1}}\ (T_{i-1}^{(1)}, \ldots, T_{i-1}^{(s_{i-1})})\} \\ &= \{(\mathbf{y}_1, \ldots, \mathbf{y}_i) : I_i^{((j-1)2^{s_{i-1}}+1)}(\mathbf{y}_i) = 1, \ldots, I_i^{(j2^{s_{i-1}})}(\mathbf{y}_i) = 1\}.\end{aligned} \tag{4.1.5}$$

It turns out that the general forms of the s_i local binary fusion rules at the ith sensor

$$T_i^{(j)}(T_{i-1}^{(1)}, \ldots, T_{i-1}^{(s_{i-1})}; \mathbf{y}_i), \quad 1 \leq j \leq s_i,$$

can be rewritten as

$$r_i \triangleq s_i 2^{s_{i-1}} \tag{4.1.6}$$

binary compression rules $I_i^{(1)}(\mathbf{y}_i)$ ..., $I_i^{(s_i 2^{s_k - 1})}(\mathbf{y}_i)$ for $\mathbf{y}_i$ at the ith sensor.

It is worth noting what we have done above is in fact to rewrite a general (whatever local or global) binary fusion rule $F(I_1, \ldots, I_{l-1}; \mathbf{y}_l)$ to 2^{l-1} binary sensor rules $I_l^{(1)}(\mathbf{y}_l), \ldots, I_l^{(2^{l-1})}(\mathbf{y}_l)$ of the the lth sensor. Recall that we have done the same thing in the proof of Theorem 3.5.2. When the general fusion rule F is $m = 2^n$-valued, i.e., the m-ary decision of F consists of n binary decisions, it can be rewritten as $n2^{l-1}$ sensor rules $I_l^{(1)}(\mathbf{y}_l), \ldots, I_l^{(n2^{l-1})}(\mathbf{y}_l)$ of the lth sensor. This technique will play a crucial role in deriving the unified fusion rules for the network decision systems in the next chapter, Chapter 5.

A final fusion rule F_{tan} for the tandem network now is dependent on $(T_{l-1}^{(1)}, \ldots, T_{l-1}^{(r_{l-1})}; \mathbf{y}_l)$. Noting that the local fusion rules $(T_{l-1}^{(1)}, \cdots; T_{l-1}^{(s_{l-1})})$ further depend on all previous local fusion rules $T_i(j)$ $(i \leq l-2, j \leq s_i)$ and the all local compression rules $I_i^{(k)}$ $(1 \leq i \leq l-2$ and $1 \leq k \leq s_{i-1}2^{s_i}$, see Eq. (4.1.6)), F_{tan} finally can be written as an m-valued function of $(I_1^{(1)}(\mathbf{y}_1), \ldots, I_1^{(r_1)}(\mathbf{y}_1); \cdots; I_l^{(1)}(\mathbf{y}_l), \ldots, I_l^{(r_l)}(\mathbf{y}_l))$. Similar to the notations of the information structure in the parallel networks, denote the information structure for the tandem network simply by

$$(\mathbf{y}_1 \overset{(s_1)}{\longmapsto} \mathbf{y}_2 \overset{(s_2)}{\longmapsto} \cdots \overset{(s_{l-2})}{\longmapsto} \mathbf{y}_{l-1} \overset{(s_{l-1})}{\longmapsto} (\mathbf{y}_l \cdot \mathcal{F}) \to \{m\}), \tag{4.1.7}$$

where "$\overset{(s_{k-1})}{\longmapsto}$ $\mathbf{y}_i$ $\overset{(s_i)}{\longmapsto}$" means that the ith sensor compresses its own data $\mathbf{y}_i$ together with the received s_{i-1} bits to s_i bits and finally, transmits those bits to the next sensor. Using (4.1.4) and (4.1.5), we can rewrite (4.1.7) formally as

$$(\mathbf{y}_1^{(r_1)} \overset{(s_1)}{\longmapsto} \mathbf{y}_2^{(r_2)} \overset{(s_2)}{\longmapsto} \cdots \overset{(s_{l-2})}{\longmapsto} \mathbf{y}_{l-1}^{(r_{l-1})} \overset{(s_{l-1})}{\longmapsto} (\mathbf{y}_l^{(r_l)} \cdot \mathcal{F}) \to \{m\}), \tag{4.1.8}$$

where $r_1 = s_1$, and "$\overset{(s_{i-1})}{\longmapsto}$ $\mathbf{y}_i^{(r_i)}$ $\overset{(s_i)}{\longmapsto}$" means that the ith sensor compresses its own data $\mathbf{y}_i$ to $r_i = s_i 2^{s_{i-1}}$ bits first and then compresses them and the received s_{i-1} bits to s_i bits. Finally, the ith sensor transmits s_i bits to the next sensor. Only exception exists at the last sensor, the lth sensor. The number of the sensor compression rules at the lth sensor is

$$r_l = n 2^{s_{l-1}}, \quad \text{as } 2^{n-1} < m \leq 2^n,$$

which will be derived in Section 5.3 below. It must be kept in mind that $(T_{l-1}^{(1)}, \ldots, T_{l-1}^{(s_{l-1})})$ received by the lth sensor, in general, are compressed data from $(I_1^{(1)}(\mathbf{y}_1),\ldots, I_1^{(r_1)}(\mathbf{y}_1);\cdots; I_{l-l}^{(1)}(\mathbf{y}_{l-1}),\ldots, I_{l-l}^{(r_{l-1}2^{r_{l-2}})}(\mathbf{y}_{l-1}))$, i.e., from $\sum_{i=1}^{l-1} s_i 2^{s_{i-1}}$ $(s_0 = 0)$ bits to s_{l-1} bits; hence, the total number of the partitioned sub-sets of the observation space $\Re^{n_1} \times \cdots \times \Re^{n_{l-1}}$ possibly observed by the fusion center (the lth sensor) in the tandem networks is at most $2^{s_{l-1}}$ rather than $2^{\sum_{i=1}^{l-1} s_i}$ although the total amount of communications of the previous $l-1$ sensors is $\sum_{i=1}^{l-1} s_i$ bits. In other words, roughly speaking, the communication in the tandem network gives less information to the fusion center than that in the parallel network. This is a different feature of the tandem network from the parallel network. Clearly, the tandem network is another elementary information structure in the distributed multisensor network decision systems.

4.1.3 HYBRID (TREE) NETWORK

Combining the above two elementary type of networks, a hybrid structure called *tree* network can be constructed, for instance, see the following Fig. 4.1.4 and Example 4.2.3 in Subsection 4.2.2 below.

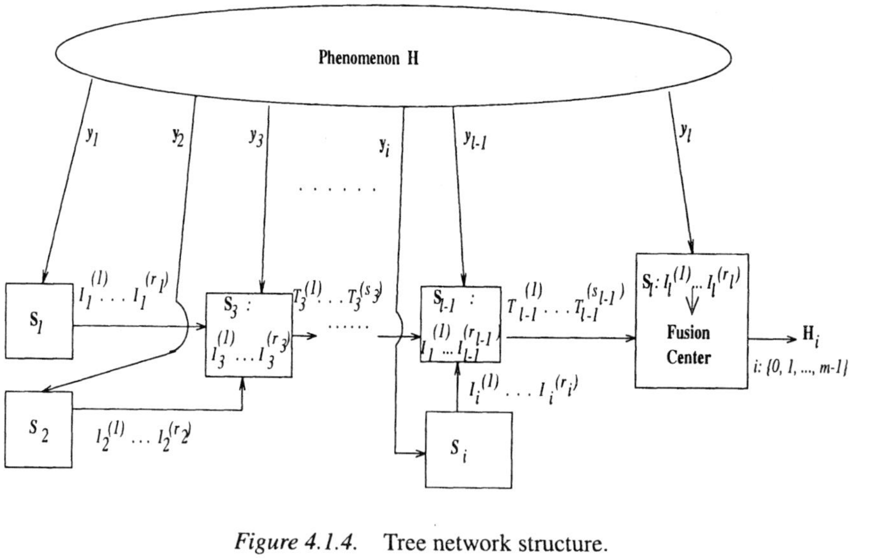

Figure 4.1.4. Tree network structure.

4.2 FORMULATION OF FUSION RULE VIA POLYNOMIALS OF SENSOR RULES

In this section, we extend the formulations of the local messages and fusion rules for the distributed binary decision system in Section 3.2 to the m-ary network decision systems.

As defined in Section 3.2, a local message received by the fusion center, namely a bi-valued element N-tuple ($N = \sum_{i=1}^{l} r_i$) $(I_1^{(1)}, \ldots, I_1^{(r_1)}; \cdots; I_l^{(1)}, \ldots, I_l^{(r_l)})$ in the parallel network (or a bi-valued element $(s_{l-1} + r_l)$-tuple in the tandem network), can be expressed by a local message polynomial of the local sensor rules. A fusion rule for an m-ary decision system divides a set of 2^N different N-tuples $(I_1^{(1)}, \ldots, I_1^{(r_1)}; \cdots; I_l^{(1)}, \ldots, I_l^{(r_l)})$ into m disjoint. The number of all the above different partitions is m^{2^N} in the parallel network and is $m^{2^{s_{l-1}+r_l}}$ in the tandem network. The goal of our distributed decision for those information structures is to find a globally optimal final decision rule consisting of m disjoint decision regions: $\mathcal{H}_0, \mathcal{H}_1, \ldots, \mathcal{H}_{m-1}$, therefore, also consisting of an optimal partition of all possible m^{2^N} partitions of the set of the 2^N different N-tuples $(I_1^{(1)}, \ldots, I_1^{(r_1)}; \cdots; I_l^{(1)}, \ldots, I_l^{(r_l)})$. It turns out from the above analysis that every final $\mathcal{H}_i$ $(0 \leq i \leq m-1)$ uniquely corresponds to a summation polynomial of some local message polynomials with those relevant sensor rules. The summation polynomial is still called *the decision region polynomials* and denoted by

$$P_{\mathbf{H}_i}(I_1^{(1)}, \ldots, I_1^{(r_1)}; \cdots; I_l^{(1)}, \ldots, I_l^{(r_l)}), \quad (0 \leq i \leq m-1). \tag{4.2.1}$$

Of course, we still need only $m-1$ decision region polynomials as the case of $m = 2$ because

$$P_{\mathbf{H}_i} = 1 - \sum_{j \neq i} P_{\mathbf{H}_j}, \quad \forall i : 0 \leq i \leq m-1.$$

Obviously, they are still bi-valued indicator function of a region on $\Re^{n_1} \times \Re^{n_2} \times \cdots \times \Re^{n_l}$ since the regions determined by different local message polynomials are disjoint, for any given observational data $(\mathbf{y}_1, \mathbf{y}_2, \cdots, \mathbf{y}_l)$, one and only one of the values of all local message polynomials equals 1 and others are all equal to 0. In other words, we have for $\forall i : 0 \leq i \leq m-1$

$$\begin{aligned} &\{(\mathbf{y}_1, \mathbf{y}_2 \cdots \mathbf{y}_l) : F(I_1^{(1)}, \ldots, I_1^{(r_1)}; \cdots; I_l^{(1)}, \ldots, I_l^{(r_l)}) = i\} \\ &= \{(\mathbf{y}_1, \mathbf{y}_2 \cdots \mathbf{y}_l) : P_{\mathbf{H}_i}(I_1^{(1)}, \ldots, I_1^{(r_1)}; \cdots; I_l^{(1)}, \ldots, I_l^{(r_l)}) = 1\}. \end{aligned} \tag{4.2.2}$$

Hence, we refer $P_{\mathbf{H}_i}$ as $\mathbf{H}_i$ *decision polynomial.*

Now, let us give two examples to show how to write fusion rules as polynomial versions of the local sensor rules.

Example 4.2.1. Consider a 3-sensor 3-ary decision system of a parallel network and there is only one binary data compress rule at each sensor, i.e.,

$$(\{\mathbf{y}_1^{(1)} // \mathbf{y}_2^{(1)} // \mathbf{y}_3^{(1)}\} \Rightarrow \mathcal{F} \rightarrow \{3\}).$$

A fusion rule F is given by the following equations:

$$\{F(I_1(\mathbf{y}_1), I_2(\mathbf{y}_2), I_3(\mathbf{y}_3)) = 0\} = \{(1\ 1\ 1), (1\ 1\ 0), (1\ 0\ 1), (0\ 1\ 1)\},$$

$$\{F(I_1(\mathbf{y}_1), I_2(\mathbf{y}_2), I_3(\mathbf{y}_3)) = 1\} = \{(0\ 0\ 0)\},$$

where, for example, $(1\ 1\ 1) \triangleq (I_1(\mathbf{y}_1) = 1, I_2(\mathbf{y}_2) = 1, I_3(\mathbf{y}_3) = 1)$. Clearly, The above two equations have also defined $F = 2$. Thus, according to the approach proposed in Section 3.2, the $\mathbf{H}_0$ decision polynomial $P_{\mathbf{H}_0}$ in this case is

$$\begin{aligned} P_{\mathbf{H}_0} = {} & I_1(\mathbf{y}_1)I_2(\mathbf{y}_2)I_3(\mathbf{y}_3) + I_1(\mathbf{y}_1)I_2(\mathbf{y}_2)(1 - I_3(\mathbf{y}_3)) \\ & + I_1(\mathbf{y}_1)(1 - I_2(\mathbf{y}_2))I_3(\mathbf{y}_3) + (1 - I_1(\mathbf{y}_1))I_2(\mathbf{y}_2))I_3(\mathbf{y}_3). \end{aligned}$$

$\mathbf{H}_1$ *decision polynomial* $P_{\mathbf{H}_1}$ and $\mathbf{H}_2$ decision polynomial $P_{\mathbf{H}_2}$ are

$$P_{\mathbf{H}_1} = (1 - I_1(\mathbf{y}_1))(1 - I_2(\mathbf{y}_2))(1 - I_3(\mathbf{y}_3))$$

and

$$P_{\mathbf{H}_2} = 1 - P_{\mathbf{H}_0} - P_{\mathbf{H}_1},$$

respectively.

Example 4.2.2. Consider a tandem network as follows:

$$(\mathbf{y}_1 \overset{(1)}{\longmapsto} \mathbf{y}_2^{(2)} \overset{(1)}{\longmapsto} (\mathbf{y}_3^{(4)} \cdot \mathcal{F}) \to \{3\}),$$

where the local fusion rule $T_2(I_1(\mathbf{y}_1), I_2^{(1)}(\mathbf{y}_2), I_2^{(2)}(\mathbf{y}_2))$ at the second sensor is defined as

$$T_2(I_1(\mathbf{y}_1), I_2^{(1)}(\mathbf{y}_2), I_2^{(2)}(\mathbf{y}_2)) = 0$$

if and only if the local message at the sensor $\mathcal{S}_2$ is either

$$(I_1(\mathbf{y}_1) = 0, I_2^{(1)}(\mathbf{y}_2) = 0) \quad \text{or} \quad (I_1(\mathbf{y}_1) = 1, I_2^{(2)}(\mathbf{y}_2) = 0),$$

otherwise, $T_2(I_1(\mathbf{y}_1), I_2^{(1)}(\mathbf{y}_2), I_2^{(2)}(\mathbf{y}_2)) = 1$. An $\mathbf{H}_0$ decision region $\mathcal{H}_0$ is given by

$$\begin{aligned} & \{(\mathbf{y}_1, \mathbf{y}_2, \mathbf{y}_3) : F(T_2(I_1(\mathbf{y}_1), I_2^{(1)}(\mathbf{y}_2), I_2^{(2)}(\mathbf{y}_2)), I_3^{(1)}(\mathbf{y}_3), I_3^{(2)}(\mathbf{y}_3), I_3^{(3)}(\mathbf{y}_3), \\ & I_3^{(4)}(\mathbf{y}_3)) = 0\} = \{(\mathbf{y}_1, \mathbf{y}_2, \mathbf{y}_3) : (T_2 = 0, I_3^{(1)} = 0), (T_2 = 1, I_3^{(2)} = 0)\}. \end{aligned}$$

Substituting the definition of T_2 into the definition of $\mathcal{H}_0$ above, we have

$$\mathcal{H}_0 = \{(I_1 = 0, I_2^{(1)} = 0, I_3^{(1)} = 0), (I_1 = 1, I_2^{(2)} = 0, I_3^{(1)} = 0),$$

$$(I_1 = 0, I_2^{(1)} = 1, I_3^{(2)} = 0), (I_1 = 1, I_2^{(2)} = 1, I_3^{(2)} = 0)\}.$$

Thus, the $\mathbf{H}_0$ decision polynomial $P_{\mathbf{H}_0}$ now is

$$\begin{aligned}P_{\mathbf{H}_0} = &(1 - I_1(\mathbf{y}_1))(1 - I_2^{(1)}(\mathbf{y}_2))(1 - I_3^{(1)}(\mathbf{y}_3))\\ &+I_1(\mathbf{y}_1)(1 - I_2^{(2)}(\mathbf{y}_2))(1 - I_3^{(1)}(\mathbf{y}_3))\\ &+(1 - I_1(\mathbf{y}_1))I_2^{(1)}(\mathbf{y}_2)(1 - I_3^{(2)}(\mathbf{y}_3)) + I_1(\mathbf{y}_1)I_2^{(2)}(\mathbf{y}_2)(1 - I_3^{(2)}(\mathbf{y}_3)).\end{aligned}$$

An $\mathbf{H}_1$ decision polynomial is given by an analogous form

$$\begin{aligned}P_{\mathbf{H}_1} = &(1 - P_{\mathbf{H}_0})[(1 - I_1(\mathbf{y}_1))(1 - I_2^{(1)}(\mathbf{y}_2))(1 - I_3^{(3)}(\mathbf{y}_3))\\ &+(I_1(\mathbf{y}_1))(1 - I_2^{(2)}(\mathbf{y}_2)(1 - I_3^{(3)}(\mathbf{y}_3))\\ &+(1 - I_1(\mathbf{y}_1))I_2^{(1)}(\mathbf{y}_2)(1 - I_3^{(4)}(\mathbf{y}_3)) + I_1(\mathbf{y}_1)I_2^{(2)}(\mathbf{y}_2)(1 - I_3^{(4)}(\mathbf{y}_3))].\end{aligned}$$

Obviously, two regions

$$\{(\mathbf{y}_1, \mathbf{y}_2, \mathbf{y}_3) : P_{\mathbf{H}_0}(\mathbf{y}_1, \mathbf{y}_2, \mathbf{y}_3) = 0\}$$

and

$$\{(\mathbf{y}_1, \mathbf{y}_2, \mathbf{y}_3) : P_{\mathbf{H}_1}(\mathbf{y}_1, \mathbf{y}_2, \mathbf{y}_3) = 1\}$$

are disjoint. Finally, the $\mathbf{H}_2$ decision polynomial should be simply

$$P_{\mathbf{H}_2} = 1 - P_{\mathbf{H}_0} - P_{\mathbf{H}_1}.$$

From this example, it can be seen that using decision polynomials to present a fusion rule is much more convenient and simpler than using the local messages.

Example 4.2.3. We now use the above two types of information structures to construct a hybrid tree structure as follows:

$$(\{\mathbf{y}_1^{(1)}//\mathbf{y}_2^{(1)}\} \Rightarrow \mathbf{y}_3^{(8)} \overset{(2)}{\longmapsto} (\mathbf{y}_4^{(8)} \cdot \mathcal{F}) \to \{3\}).$$

A fusion rule for this structure is given in versions of the decision polynomials.

Using $T_3^{(1)}$, $T_3^{(2)}$, and $I_4^{(1)}, \ldots,\ I_4^{(8)}$, we construct a fusion rule at the last sensor, the fourth sensor:

$$\begin{aligned}P_{\mathbf{H}_0} = &(1 - T_3^{(1)})(1 - T_3^{(2)})(1 - I_4^{(1)}) + T_3^{(1)}(1 - T_3^{(2)})(1 - I_4^{(2)})\\ &+(1 - T_3^{(1)})T_3^{(2)}(1 - I_4^{(3)}) + T_3^{(1)}T_3^{(2)}(1 - I_4^{(4)}),\\ P_{\mathbf{H}_1} = &(1 - P_{\mathbf{H}_0})[(1 - T_3^{(1)})(1 - T_3^{(2)})(1 - I_4^{(5)}) + T_3^{(1)}(1 - T_3^{(2)})(1 - I_4^{(6)})\\ &+(1 - T_3^{(1)})T_3^{(2)}(1 - I_4^{(7)}) + T_3^{(1)}T_3^{(2)}(1 - I_4^{(8)})],\\ P_{\mathbf{H}_2} = &1 - P_{\mathbf{H}_0} - P_{\mathbf{H}_1}.\end{aligned}$$

Using I_1, I_2, and $I_3^{(1)}, \ldots, I_3^{(8)}$, we construct the four polynomials of the two local fusion rules $T_3^{(1)}$, $T_3^{(2)}$ at the third sensor:

$$T_3^{(1)} = (1 - I_1)(1 - I_2)I_3^{(1)} + I_1(1 - I_2)I_3^{(2)} + (1 - I_1)I_2I_3^{(3)} + I_1I_2I_3^{(4)},$$

$$T_3^{(2)} = (1 - I_1)(1 - I_2)I_3^{(5)} + I_1(1 - I_2)I_3^{(6)} + (1 - I_1)I_2I_3^{(7)} + I_1I_2I_3^{(8)}.$$

Substituting the above two polynomials into $P_{\mathbf{H}_0}$, $P_{\mathbf{H}_1}$ and $P_{\mathbf{H}_2}$, we can get the three final decision polynomials with respect to all local compression rules ($I_1, I_2; I_3^{(1)}, ..., I_3^{(8)}; I_4^{(1)}, \ldots, I_4^{(8)}$). We will use this example again as a numerical example in Section 5.6.

4.3 FIXED POINT TYPE NECESSARY CONDITION FOR OPTIMAL SENSOR RULES GIVEN A FUSION RULE

After the above preliminary work for the sensor rule polynomial expression of a fusion rule, the next problem is to find the corresponding optimal sensor rules under the fusion rule. In this section, we extend the result in Section 3.3 to the network decision systems here, that is, we present a necessary condition of optimal local compression rules for any fixed fusion rule.

For a network, the m-ary Bayes cost is given by

$$\begin{aligned} &C(I_1^{(1)}(\mathbf{y}_1), \ldots, I_1^{(r_1)}(\mathbf{y}_1); \cdots; I_l^{(1)}(\mathbf{y}_l), \ldots, I_l^{(r_l)}(\mathbf{y}_l); F) \\ &= \textstyle\sum_{i=0,j=0}^{m-1} C_{ij}P_jP(F = i|\mathbf{H}_j), \end{aligned} \tag{4.3.1}$$

where each C_{ij} is some suitable cost coefficients; P_j is *a priori* probability for the hypothesis $\mathbf{H}_j$; and each $P(F = i|\mathbf{H}_j)$ denotes the conditional probability of the event that the fusion center's decision F is equal to i when the actual hypothesis is $\mathbf{H}_j$, $i, j = 0, 1, \ldots, m-1$. Here F is any fusion rule chosen by the parallel, tandem, or tree network decision systems.

Substituting the conditional probabilities given $\mathbf{H}_0, \ldots, \mathbf{H}_{m-1}$ into Eq. (4.3.1) and simplifying, we see that

$$\begin{aligned} &C(\mathbf{y}_1, \ldots, \mathbf{y}_l; F) \\ &= \textstyle\sum_{i=0}^{m-1} c_{ii}P_i \\ &+ \textstyle\sum_{i=0}^{m-1} \displaystyle\int_{F=i} \sum_{j=0, j\neq i}^{m-1} P_j(c_{ij} - c_{jj})p(\mathbf{y}_1, \ldots, \mathbf{y}_l|\mathbf{H}_j)d\mathbf{y}_1 \cdots d\mathbf{y}_l. \end{aligned} \tag{4.3.2}$$

Since all $P_{\mathbf{H}_i}$, $i = 0, ..., m-1$ are some sums of the products of $I_1^{(1)}$ (or $1 - I_1^{(1)}$), ..., $I_1^{(r_1)}$ (or $1 - I_1^{(r_1)}$); ...; $I_l^{(1)}$ (or $1 - I_l^{(1)}$), ..., $I_l^{r_l}$ (or $1 - I_l^{(r_l)}$), we can rewrite them as

$$\begin{aligned} P_{\mathbf{H}_i} &= (1 - I_1^{(1)})P_{i(11)}^{(1)} + P_{i(12)}^{(1)} \\ &= (1 - I_1^{(2)})P_{i(11)}^{(2)} + P_{i(12)}^{(2)} \\ &\cdots \\ &= (1 - I_l^{(r_l)})P_{i(l1)}^{(r_l)} + P_{i(l2)}^{(r_l)}, \end{aligned} \tag{4.3.3}$$

where $P_{i(j1)}^{(k)}$ and $P_{i(j2)}^{(k)}$ are both independent of $I_j^{(k)}$ for $j = 1, ..., l$, and $k = 1, ..., r_j$, respectively. Thus, using Eq. (4.3.3), the result in the previous section, Eq. (4.2.2) and the analogous idea of Eq. (3.3.2), we can write the integrand of the right hand side of Eq. (4.3.2) as the following various versions.

$$\begin{aligned} &\textstyle\sum_{i=0}^{m-1} P_{\mathbf{H}_i} \sum_{j=0, j\neq i}^{m-1} P_j(c_{ij} - c_{jj})p(\mathbf{y}_1, \mathbf{y}_2, \ldots, \mathbf{y}_l|\mathbf{H}_j) \\ &= (1 - I_1^{(1)})P_{11}^{(1)}(I_1^{(2)}, ...I_1^{(r_1)}; \cdots; I_l^{(1)}, \ldots, I_l^{r_l}; \mathbf{y}_1, \ldots, \mathbf{y}_l) \\ &\quad + P_{12}^{(1)}(I_1^{(1)}, \ldots, I_1^{(r_1)}; \cdots; I_l^{(1)}, \ldots, I_l^{(r_l)}; \mathbf{y}_1, \ldots, \mathbf{y}_l) \\ &\cdots \\ &= (1 - I_l^{(1)})P_{l1}^{(1)}(I_1^{(1)}, ...I_1^{(r_1)}; \cdots; I_l^{(2)}, \ldots, I_l^{(r_l)}; \mathbf{y}_1, \ldots, \mathbf{y}_l) \\ &\quad + P_{l2}^{(1)}(I_1^{(1)}, \ldots, I_1^{(r_1)}; \cdots; I_l^{(2)}, \ldots, I_l^{(r_l)}; \mathbf{y}_1, \ldots, \mathbf{y}_l) \\ &\cdots \\ &= (1 - I_l^{(r_l)})P_{l1}^{(r_l)}(I_1^{(1)}, ...I_1^{(r_1)}; \cdots; I_l^{(1)}, \ldots, I_l^{(r_l-1)}; \mathbf{y}_1, \ldots, \mathbf{y}_l) \\ &\quad + P_{l2}^{(r_l)}(I_1^{(1)}, ...I_1^{(r_1)}; \cdots; I_l^{(1)}, \ldots, I_l^{(r_l-1)}; \mathbf{y}_1, \ldots, \mathbf{y}_l), \end{aligned} \tag{4.3.4}$$

where $P_{i1}^{(j)}$ and $P_{i2}^{(j)}$ are the functions independent of $I_i^{(j)}$, $i = 1, \ldots, l$ and $j = 1, \ldots, r_i$, respectively.

Now we present the following theorem for a necessary condition of the optimal local compression rules.

Theorem 4.3.1. Suppose that we have a distributed multisensor decision system employing the fusion rule (4.2.2). Given a set of optimal sensor rules $(I_1^{(1)}, ...I_1^{(r_1)}; \cdots; I_l^{(1)}, \ldots, I_l^{r_l})$ which minimizes the cost functional of Eq. (4.3.2), then those sensor rules must satisfy the following integral equations

$$I_1^{(1)}(\mathbf{y}_1) = I[\int P_{11}^{(1)}(I_1^{(2)}, \ldots, I_1^{(r_1)}; \cdots; I_l^{(1)}, \ldots, I_l^{(r_l)}; \mathbf{y}_1, \ldots, \mathbf{y}_l) \\ d\mathbf{y}_2 \cdots d\mathbf{y}_l],$$

$$\cdots;$$

$$I_l^{(1)}(\mathbf{y}_l) = I[\int P_{l1}^{(1)}(I_1^{(1)}, \ldots, I_1^{(r_1)}; \cdots; I_l^{(2)}, \ldots, I_l^{(r_l)}; \mathbf{y}_1, \ldots, \mathbf{y}_l) \\ d\mathbf{y}_1 \cdots d\mathbf{y}_{l-1}],$$

$$\cdots,$$

$$I_l^{(r_l)}(\mathbf{y}_l) = I[\int P_{l1}^{(r_l)}(I_1^{(1)}, ...; I_l^{(1)}, \ldots, I_l^{(r_l-1)}; \mathbf{y}_1, \ldots, \mathbf{y}_l) d\mathbf{y}_1 \cdots d\mathbf{y}_{l-1}], \tag{4.3.5}$$

where I is a indicator function as (3.3.6) in Section 3.3.

Obviously, the necessary condition of the optimal local compression rules in Theorem 4.3.1 is a fixed-point type condition, which is an extension of Theorem 3.3.1. In particular, if we define the mapping

$$\Gamma(I_1^{(1)}, ...I_1^{(r_1)}; \cdots; I_l^{(1)}, \ldots, I_l^{r_l}) =$$

$$\begin{pmatrix} I[\int P_{11}^{(1)}((I_1^{(2)}, \ldots, I_1^{(r_1)}; \cdots; I_l^{(1)}, \ldots, I_l^{(r_l)}; \mathbf{y}_1, \ldots, \mathbf{y}_l) d\mathbf{y}_2 \cdots d\mathbf{y}_l] \\ \cdots \\ I[\int P_{l1}^{(1)}(I_1^{(1)}, \ldots, I_1^{(r_1)}; \ldots; I_l^{(2)}, \ldots, I_l^{(r_l)}; \mathbf{y}_1, \ldots, \mathbf{y}_l) d\mathbf{y}_1 \cdots d\mathbf{y}_{l-1}] \\ \cdots \\ I[\int P_{l1}^{(r_l)}(I_1^{(1)}, \ldots; I_l^{(1)}, \ldots, I_l^{(r_l-1)}; \mathbf{y}_1, \ldots, \mathbf{y}_l) d\mathbf{y}_1 \cdots d\mathbf{y}_{l-1}] \end{pmatrix}, \tag{4.3.6}$$

then Theorem 4.3.1 shows that a set of optimal local compression rules must be a solution of the integral equation (4.3.5) as well as a fixed point of the map Γ in (4.3.6).

4.4 ITERATIVE ALGORITHM AND CONVERGENCE

In this section, the results in Section 3.4 are extended to the more general model under consideration. We present a Gauss-Seidel iterative algorithm and its discrete version, give some convergence results which show the finite convergence of the discretized algorithm. Under mild assumption on the integrand of the cost functional, the global minimum of the discrete cost functional converges to the infimum of the original continuous cost functional as the the discretization step size tends to zero. Since the extensions here is straightforward except notation and formulation, we will just present the relevant results without argument.

Let the local compression rules at the ith stage of iteration be denoted by $(I_1^{(1(i))},..., I_1^{r_1(i)}; ...; I_l^{(1(i))}, \ldots, I_l^{(r_l(i))})$ with the initial rules $(I_1^{(1(0))},..., I_1^{(r_1(0))}; ...; I_l^{(1(0))}, \ldots, I_l^{(r_l(0))})$ being given. Suppose that the given fusion rule described in Eq. (4.2.2) is employed, we can now define an iterative algorithm using the mapping Γ by considering the following Gauss-Seidel iterative process.

$$
\begin{aligned}
I_1^{(1(i+1))}(\mathbf{y}_1) = I[\int P_{11}^{(1)}(I_1^{(2(i))}, \ldots, I_1^{(r_1(i))}; \cdots; I_l^{(1(i))}, \ldots, I_l^{(r_l(i))};\\
\mathbf{y}_1, ...\mathbf{y}_l)d\mathbf{y}_2 \cdots d\mathbf{y}_l],\\
\cdots;\\
I_l^{(1(i+1))}(\mathbf{y}_l) = I[\int P_{l1}^{(1)}(I_1^{(1(i+1))}, \ldots, I_1^{(r_1(i+1))}; ...; I_l^{(2(i))}, \ldots, I_l^{(r_l(i))};\\
\mathbf{y}_1, ...\mathbf{y}_l)d\mathbf{y}_1 \cdots d\mathbf{y}_{l-1}],\\
\cdots,\\
I_l^{(r_l(i+1))}(\mathbf{y}_l) = I[\int P_{l1}^{(r_l)}(I_1^{(1(i+1))}, \ldots, I_1^{(r_1(i+1))}; \cdots; I_l^{(1(i+1))}, \ldots,\\
I_l^{((r_l-1)(i+1))}; \mathbf{y}_1, ...\mathbf{y}_l)d\mathbf{y}_1 \cdots d\mathbf{y}_{l-1}].
\end{aligned}
\tag{4.4.1}
$$

Let the discretization of $\Re^{n_1}, \Re^{n_2}, \ldots, \Re^{n_l}$ be given by

$$\{\mathbf{y}_{11}, \mathbf{y}_{12}, \cdots, \mathbf{y}_{1N_1}\}, \{\mathbf{y}_{21}, \mathbf{y}_{22}, \cdots, \mathbf{y}_{2N_2}\}, \cdots, \{\mathbf{y}_{l1}, \mathbf{y}_{l2}, \cdots, \mathbf{y}_{lN_l}\}$$

respectively. For each iteration $i = 0, 1, \ldots$, let the N_1-vector

$$\{I_{1j}^{(n(i))}; j = 1, \ldots, N_1\}$$

denote the discrete version of the real-valued $I_1^{(n(i))}$, where

$$I_{1j}^{(n(i))} = I_1^{(n(i))}(\mathbf{y}_{1j}), \quad \forall n \le r_1.$$

Similarly, for $\forall k \le l$ let the N_i-vector

$$\{I_{kj}^{(n(i))}; j = 1, \ldots, N_i\}$$

denote the discrete values of $I_k^{n(i))}$ such that $I_{kj}^{(n(i))} = I_k^{(n(i))}(\mathbf{y}_{kj}), \forall n \le r_k$. Thus, the iterative processes of Eq. (4.4.1) can be respectively approximated as

$$I_{1m_1}^{(1(i+1))} = I[\sum_{m_2=1}^{N_2} \cdots \sum_{m_l=1}^{N_l} P_{11}^{(1)}(I_1 m_1^{(2(i))}, \ldots, I_1 m_1^{(r_1(i))}; \ldots; I_{lm_l}^{(1(i))},$$
$$\ldots, I_{lm_l}^{(r_l(i))}; \mathbf{y}_{1m_1}, \ldots, \mathbf{y}_{lm_l}) \triangle \mathbf{y}_2 \triangle \mathbf{y}_3 \cdots \triangle \mathbf{y}_l], \; m_1 \le N_1,$$
$$\cdots;$$
$$I_{lm_l}^{(1(i+1))} = I[\sum_{m_1=1}^{N_1} \cdots \sum_{m_{l-1}=1}^{N_{l-1}} P_{l1}^{(1)}(I_{1m_1}^{(1(i+1))}, \ldots, I_{1m_1}^{(r_1(i+1))}; \ldots,$$
$$I_{(l-1)m_{l-1}}^{(r_{l-1}(i+1))}; I_{lm_l}^{(2(i))}, \ldots, I_{lm_l}^{(r_l(i))}; \mathbf{y}_{1m_1}, \ldots, \mathbf{y}_{lm_l}) \triangle \mathbf{y}_1 \cdots \triangle \mathbf{y}_{l-1}], \; m_l \le N_l,$$
$$\cdots,$$
$$I_{lm_l}^{(r_l(i+1))} = I[\sum_{m_1=1}^{N_1} \cdots \sum_{m_{l-1}=1}^{N_{l-1}} P_{l1}^{(r_l)}(I_{1m_1}^{(1(i+1))}, \ldots, I_{1m_1}^{(r_1(i+1))};$$
$$\cdots; I_{lm_l}^{(1(i+1))}, \ldots, I_{lm_l}^{((r_l-1)(i+1))}; \mathbf{y}_{1m_1}, \ldots, \mathbf{y}_{lm_l}) \triangle \mathbf{y}_1 \cdots \triangle \mathbf{y}_{l-1}], \; m_l \le N_l, \tag{4.4.2}$$

where $\triangle \mathbf{y}_1, \triangle \mathbf{y}_2, \ldots, \triangle \mathbf{y}_l$ are the step-sizes of discretizing of the vectors $\mathbf{y}_1$, $\mathbf{y}_2, \ldots, \mathbf{y}_l$, respectively. The iterations (4.4.2) are the corresponding discretized versions of the continuous iteration processes (4.4.1). Thus, they are readily implementable on computer. A simple termination criterion of these iteration processes is to stop as soon as for all $k \le l$, $m_k \le N_k$ and $n \le r_k$,

$$I_{km_k}^{(n(i+1))} = I_{km_k}^{(n(i))}. \tag{4.4.3}$$

An alternative is to stop when

$$\sum_{m_1, m_2, \ldots, m_l} \left(|I_{1m_1}^{(1(i+1))} - I_{1m_1}^{(1(i))}| + \cdots + |I_{lm_l}^{(r_l(i+1))} - I_{lm_l}^{(r_l(i))}| \right) \le \epsilon, \tag{4.4.4}$$

where $\epsilon > 0$ is some prespecified tolerance parameter.

We now examine the convergence of the Gauss-Seidel iterative algorithm of Eq. (4.4.2), we have the following theorem on the finite convergence of the discrete Gauss-Seidel iteration process.

Theorem 4.4.1. For any positive discretization stepsize of the elements of $\triangle \mathbf{y}_1, \triangle \mathbf{y}_2, \ldots, \triangle \mathbf{y}_r$ and an initial choice of $(I_{1m_1}^{(1(0))}, \ldots, I_{1m_1}^{(r_1(0))}; \ldots; I_{lm_l}^{(1(0))}, \cdots; I_{lm_l}^{(r_l(0))})$ $(j \le l, m_j = 1, \ldots, N_j)$, the algorithm of Eq. (4.4.2) terminates with a set $(I_{1m_1}^{(1(i))}, \ldots, I_{1m_1}^{(r_1(i))}; \ldots; I_{lm_l}^{(1(i))}, \ldots, I_{lm_l}^{(r_l(i))})$ satisfying Eq. (4.4.3) after a finite number of iterations.

The Outline of Proof. The proof can be divided into two steps. The first step is to prove the Bayes cost is decreasing as i is increasing. Hence, it is easy to see the cost reaches a minimum after finite iterations. However, the finite convergence of the cost does not imply the finite convergence of local sensor rules. In addition, one cannot know if the algorithm has attained to the minimal cost while local sensor rules do not converge. Then, we prove that after the cost attains to the minimum, the local sensor rules do not change after at most a finite number of iterations. The detailed proof of this theorem can follow the argument for Theorem 3.4.1 in Section 3.4.

Let $\triangle$ denote the discretization step size of each elements in $\triangle \mathbf{y}_1, \triangle \mathbf{y}_2, ...,$ $\triangle \mathbf{y}_l$ and $C_\triangle$ be the minimum of the discrete version $C_\triangle(I_1^{(1)}, \ldots, I_1^{(r_1)}; \cdots;$ $I_l^{(1)}, \ldots, I_l^{(r_l)}; F)$ of $C(I_1^{(1)}, \ldots, I_1^{(r_1)}; \cdots; I_l^{(1)}, \ldots, I_l^{(r_l)}; F)$. Similarly, we have a convergence result corresponding to Theorem 3.4.2 in Section 3.4, which asserts under mild assumption on the integrand in the above cost functional, not only the limit of $C_\triangle$ exists, but also the $C_\triangle$ converges to the infimum of $C(I_1, I_2, \ldots, I_l; F)$ as $\triangle$ tends to zero. We omit it here since the argument for the proof of this result is the same as the proof of Theorem 3.4.2 except notations.

Chapter 5

OPTIMAL FUSION RULE AND DESIGN OF NETWORK COMMUNICATION STRUCTURES

In chapter 4, we have derived a number of results on the optimal sensor rules with a fixed fusion rule for general multisensor m-ary network decision systems. To achieve globally optimal performance, the previous results are obviously not enough because the number of all possible fusion rules are very large and exhaustive searching for the optimal fusion rule is computationally intractable. Therefore, in Sections 5.3–5.7, we are going to extend the results on the optimal fusion rule and some related interesting properties derived in Section 3.5 to more general network decision systems discussed in Chapter 4. On the other hand, in terms of the two-level optimization strategy in the distributed decision system, a local sensor may first need a locally optimal sensor rule based on its own information for its own local goal, and then send its decision result to the fusion center. Finally, the fusion center fuses all received local decision results to get an optimal final decision. When all sensor rules (not only sensor outputs) are well known, the optimal final fusion can reduce to the conventional decision problem at the second level. We will formulate and solve it thoroughly in Section 5.1. If only the sensor outputs can be known, some results on reducing the number of valuable fusion rules are presented in Section 5.2 in both two-level optimization and global optimization senses. Here, by valuable fusion rules we mean that they must contain an optimal fusion rule.

5.1 OPTIMAL FUSION RULE GIVEN SENSOR RULES

In this section, optimal distributed fusion assuming that sensor decision rules are given is considered. A general and computationally tractable optimal fusion rule is presented, which relies only on the joint conditional probability densities of all sensor observations and all well-known sensor rules. It is valid for general decision systems with any sensor observations and sensor decision rules, regardless of their interdependence, and any network structure. It is also valid for m-ary Bayes decision problems and binary problems under the Neyman-Pearson criterion. Local decision rule of a sensor that is optimal for the sensor itself is also presented, which take the form of a generalized likelihood ratio test. Numerical examples are given, which reveal some interesting phenomena.

5.1.1 PROBLEM FORMULATION

Consider a decision system with a distributed sensor network. Each local sensor observes data and may receive messages from other sensors simultaneously. It locally fuses/compresses all its available information to obtain its own decision result or a communicable message and transmitted them to the fusion center and/or other sensors. The fusion center makes a final decision using all received messages by some fusion rule. Communications are possibly permitted not only between the sensors and the fusion center, but also among sensors themselves.

On the other hand, a sensor decision rule is determined (optimally) in many practical situations based only upon all information available to it including its local objective function, regardless of the whole system's performance. The fusion center only makes a final decision that is optimal subject to the fixed sensor rules. For example, in a decision process of a dynamic system, it is impossible sometimes for a local decision maker to wait and make its own decision until the final decision at the fusion center is known. Another example involves decision systems in a war situation. In order to enhance the survivability of the whole decision system, every local sensor must make a locally optimal decision upon all information available to it and then transmit the decision out. In so doing, even if the fusion center or some local sensors are destroyed, other local sensors can still make decisions. In short, while optimal fusion with already fixed sensor rules does not yield globally optimal performance, it is of strong practical significance and interest.

There are two classes of distributed decision systems in which local (sensor) rules are fixed, and *every* local rule is known to the fusion center and the whole decision system, i.e., either the fusion center or every sensor and the fusion center have complete knowledge of *every* local rules. We will call this class "fusion with given local rules", which is often encountered in practice. The second class is that only the sensor outputs can be known by the fusion center but the sensor rules cannot. A typical example of the second class is a decision system involving partners who do not want to share all intimate details of their own systems. Obviously, it is very difficult to solve the second class decision fusion without other additional information. Such special example is given in this section as well.

Previous work on distributed decision fusion with given sensor outputs and some further knowledge has been reported in [13, 29, 45]. Chair and Varshney [13] presented an optimal fusion rule as a linear combination of local decisions for distributed binary decision with independent local decisions, where the weight for each decision is a ratio of correct-decision probability to incorrect-decision probability. In [29], Drakopoulos and Lee consider cases with dependent local decisions. They in fact used conditional correlation coefficients instead of the joint conditional decision probabilities. Using a similar idea, Kam, Zhu and Gray in [45] normalized local decisions first and then employed the so-called *Bahadur-Lazarsfeld polynomial* and the normalized correlation coefficients to express the optimal fusion of correlated local decisions for distributed binary decision. In fact, for zero-one binary random variables, conditional probabilities can be easily derived from conditional expectations (for details, see the argument at the end of this subsection). Besides, it is hard to extend these results for coupled local decisions to more general cases, e.g., sensors with communications and m-ary decision systems.

The main results of this section lie in the optimal fusion with given sensor rules for general decision systems, in particular those with dependent sensor decisions. There are three cases that lead to dependent sensor decisions:

- sensors with coupled observations but without mutual communications;
- sensors with coupled observations and mutual communications;
- sensor with independent observations but with communications among sensors.

Note that sensor observations of a random signal are coupled even if observation errors are independent.

This section presents a general and computationally tractable optimal decision fusion rule with given sensor rules in terms of the joint conditional probability densities and the sensor rules given. The optimal fusion rule is completely general in that it is valid for all sensor decisions, dependent or not, and all sensor network structures, with or without communications between sensors, provided that the joint conditional probability densities of all sensor observations and the sensor rules are known. It is also valid for both m-ary Bayes decision problems and binary problems under the Neyman-Pearson criterion. Under the same optimality criterion as for the entire system, or under sensor's own criterion, sensor decision rules are also presented that are optimal based on all information available to them individually, including their own observations and the received messages from other sensors. Thus, combining the optimal fusion rule and the locally optimal sensor rules, the optimal performance of a distributed decision system with given locally optimal sensor rules can be obtained. Finally, three numerical examples are given. They not only support the analytic results presented but also demonstrate some interesting properties of a distributed decision system with given sensor rules.

Consider a distributed decision problem of m hypotheses $\mathbf{H}_0, \mathbf{H}_1, \ldots, \mathbf{H}_{m-1}$ and l sensors with multi-dimensional observation data $\mathbf{y}_1, \ldots, \mathbf{y}_l$. Each local sensor i makes a local m-ary decision d_i based upon the information available to it first and then transmits its decision out. If communications between sensors are allowed, the information available to a sensor includes not only its own observation but also messages received of some other sensor decisions. Finally, a fusion center (which may also observe data itself) makes a final m-ary decision F based upon all the received messages of local sensor decisions. Clearly, it is a two level optimization problem.

This is a very general formulation of a distributed decision system. For example, it allows feedback among the fusion and sensors. However, for notational simplicity, we consider Bayes binary decision system first, which consists of only one level of local sensors and a fusion center. Then, we show in Subsections 5.1.3 and 5.1.4 that all results presented for this simpler case can be extended to the more general m-ary decision systems.

At the fusion center, a final decision is made using a non-randomized fusion rule F. Let $(d_1, d_2, \ldots, d_l)$ be the observations of the fusion center. The Bayes cost is

$$\begin{aligned} C(d_1, d_2, \ldots, d_l; F) = {} & P_0C_{10} + P_1C_{11} + P_1(C_{01} - C_{11})P(F = 0|\mathbf{H}_1) \\ & - P_0(C_{10} - C_{00})P(F = 0|\mathbf{H}_0). \end{aligned} \tag{5.1.1}$$

Denote the set for $\mathbf{H}_0$ decision (a finite point set) by

$$\mathcal{H}_0 = \{(d_1, d_2, \ldots, d_l) : F = 0\}. \tag{5.1.2}$$

Hence,

$$\begin{aligned} & P_1(C_{01} - C_{11})P(F = 0|\mathbf{H}_1) - P_0(C_{10} - C_{00})P(F = 0|\mathbf{H}_0) \\ = & \sum_{\mathcal{H}_0}[P_1(C_{01} - C_{11})P(d_1, \ldots, d_l|\mathbf{H}_1) - P_0(C_{10} - C_{00})P(d_1, \ldots, d_l|\mathbf{H}_0)]. \end{aligned} \tag{5.1.3}$$

Using the above three equations, minimizing the cost function is equivalent to defining $\mathcal{H}_0$ as follows:

$$\mathcal{H}_0 = \left\{(d_1, d_2, \ldots, d_l) : \frac{P(d_1, \ldots, d_l|\mathbf{H}_1)}{P(d_1, \ldots, d_l|\mathbf{H}_0)} < \frac{P_0(C_{10} - C_{00})}{P_1(C_{01} - C_{11})}\right\}. \tag{5.1.4}$$

This is the optimal fusion rule mentioned in [13].

If the sensor observations are conditionally independent under the two hypotheses, the above optimal fusion rule easily reduces to the following expression (see [13]):

$$\log \frac{P_1(C_{01} - C_{11})}{P_0(C_{10} - C_{00})} + \sum_{U_1} \log \frac{1 - P_m^i}{P_f^i} + \sum_{U_0} \log \frac{P_m^i}{1 - P_f^i} \underset{F = \mathbf{H}_0}{\overset{F = \mathbf{H}_1}{\gtrless}} 0, \quad (5.1.5)$$

where, U_1 is the set of all subscripts satisfying $d_i = 1$ in a group of sensor observations $(d_1, \ldots, d_l)$, U_0 is the set of all subscripts satisfying $d_i = 0$, i.e.,

$$U_1 = \{i : d_i = 1, d_i \in (d_1, \ldots, d_l)\}, \quad U_0 = \{i : d_i = 0, d_i \in (d_1, \ldots, d_l)\}.$$

In practice, when people do not know the two conditional probabilities $P(d_1, \ldots, d_l|\mathbf{H}_0)$ and $P(d_1, \ldots, d_l|\mathbf{H}_1)$ exactly, but the two error probabilities P_f^i and P_m^i at each sensor may be obtained from previous statistical data, people can employ the optimal fusion rule (5.1.5). However, it is worth noting that the P_f^i and P_m^i here should be obtained from observations of the same object. If not, the optimality of the fusion rule (5.1.5) would not hold. In addition to the restriction of independence of the observations, it is hard to extend this fusion rule to m-ary case.

In [29], Drakopoulos and Lee used the following set of correlation coefficients (see Page 594 in [29]):

$$\mathcal{C} = \{E_j[\prod_{i \in I} d_i] : I \subseteq \{1, \ldots, l\}, I \neq \emptyset, j = 0, 1\}. \quad (5.1.6)$$

Since the set $\{1, \ldots, l\}$ contains $2^l - 1$ different subsets I (note $I \neq \emptyset$) and

$$E_j[\prod_{i \in I} d_i] = E_j[d_{i_1} d_{i_2} \cdots d_{i_k}] = P(d_{i_1} = 1, \ldots, d_{i_k} = 1|\mathbf{H}_j)$$
$$= \sum_{r_1, \ldots, r_{l-k} = 0,1} P(d_{i_1} = 1, \ldots, d_{i_k} = 1; d_{j_1} = r_1, \ldots, d_{j_{l-k}} = r_{l-k}|\mathbf{H}_j), \quad (5.1.7)$$

where

$$\{j_1, \ldots, j_{l-k}\} \subset \{1, \ldots, l\} \text{ and } \{j_1, \ldots, j_{l-k}\} \bigcup \{i_1, \ldots, i_k\} = \{1, \ldots, l\},$$

we can obtain $2^l - 1$ independent [1] linear equations of l-fold joint conditional probabilities given $\mathbf{H}_j$. To get 2^l different l-fold joint conditional probabilities given $\mathbf{H}_j$, we only need to add one more equation

$$\sum_{r_1, \ldots, r_l = 0,1} P(d_1 = r_1, \ldots, d_l = r_l|\mathbf{H}_j) = 1.$$

Thus, we obtain all 2^{l+1} different l-fold joint conditional probabilities given $\mathbf{H}_0$ and $\mathbf{H}_1$ respectively from set $\mathcal{C}$.

---

[1] For a non-degenerate joint probability, it is impossible to define anyone of $E_j[\prod_{i \in I} d_i]$ by linear combination of other $E_j[\prod_{i \in I} d_i]$.

5.1.2 COMPUTATION OF LIKELIHOOD RATIOS

To have optimal fusion performance given the sensor rules, (5.1.4) indicates that all we need to do is computing the required ratio of likelihood or conditional joint sensor decision probabilities. The contribution of (5.1.5) was in essence the simplification of the above likelihood ratio to a product of the ratios of two conditional decision probabilities of every sensor when sensor observations are independent and there are no communications among sensors. The sensor decision probabilities can be calculated easily from the given joint conditional probability densities $p(\mathbf{y}_1, \mathbf{y}_2, \ldots, \mathbf{y}_l|\mathbf{H}_1)$ and $p(\mathbf{y}_1, \mathbf{y}_2, \ldots, \mathbf{y}_l|\mathbf{H}_0)$ (see Eq. (5.1.5)). In practice, if the two conditional decision probability densities of each sensor in (5.1.5) are not known, they may be replaced by their approximate (e.g., empirical) average values obtained from the historical data which, of course, were observed for different objects usually. The fusion rule (5.1.5) thus is still applicable but is, of course, no longer optimal.

Here we point out that when the two conditional probability density functions of the observations, $p(\mathbf{y}_1, \mathbf{y}_2, \ldots, \mathbf{y}_l|\mathbf{H}_1)$ and $p(\mathbf{y}_1, \mathbf{y}_2, \ldots, \mathbf{y}_l|\mathbf{H}_0)$, as well as all sensor decision rules are known to the fusion center, we can compute all conditional joint sensor decision probabilities via the probabilities of subsets in the product space of all sensor observations, no matter how complicated the sensor decision rules are.

We state the above precisely as a proposition.

Proposition 5.1.1. A given set of all local non-randomized decision rules defines a given 1-1 mapping between 2^l subsets $\mathcal{U}_1, \ldots, \mathcal{U}_{2^l}$ of the product space $\Re^{n_1} \times \cdots \times \Re^{n_l}$ of all sensor observations and points of the set of an l-tuple $(d_1, d_2, \ldots, d_l)_i$ $(i \le 2^l)$ of (0,1) binary elements. These 2^l subsets $\mathcal{U}_1, \ldots, \mathcal{U}_{2^l}$ form a partition of the product space $\Re^{n_1} \times \cdots \times \Re^{n_l}$.

Proof. Given a set of all sensor decision rules, no matter how complicated they are, a mapping from the product space $\Re^{n_1} \times \cdots \times \Re^{n_l}$ of all sensor observations onto an l-tuple $(d_1, d_2, \ldots, d_l)$ of (0,1) binary elements is defined. Let $\mathbf{u}_i = (d_1, d_2, \ldots, d_l)_i$ be the ith point in the range of this mapping, that is, the ith possible value of this l-tuple or the ith possible set of sensor decisions. Clearly, there are 2^l distinct $\mathbf{u}_i$s and corresponding to each $\mathbf{u}_i$ there is a unique subset $\mathcal{U}_i$ of the product space $\Re^{n_1} \times \cdots \times \Re^{n_l}$ (if more than one subset is mapped to the same $\mathbf{u}_i$, then $\mathcal{U}_i$ is their union). The proposition thus follows from the fact that these $\mathcal{U}_i$'s are disjoint and exhaustive. ∎

In Subsections 5.1.3 and 5.1.4, we will show this proposition by some quite general examples. Using Proposition 5.1.1, the conditional joint sensor decision probabilities $P(d_1, d_2, \ldots, d_l|\mathbf{H}_0)$ and $P(d_1, d_2, \ldots, d_l|\mathbf{H}_1)$ can be computed easily in principle from the conditional pdfs of the sensor observations as follows:

$$
\begin{aligned}
P(d_1, \ldots, d_l|\mathbf{H}_1) &= \int_{\mathcal{U}_i} p(\mathbf{y}_1, \mathbf{y}_2, \ldots, \mathbf{y}_l|\mathbf{H}_1) d\mathbf{y}_1 d\mathbf{y}_2 \cdots d\mathbf{y}_l, \\
P(d_1, \ldots, d_l|\mathbf{H}_0) &= \int_{\mathcal{U}_i} p(\mathbf{y}_1, \mathbf{y}_2, \ldots, \mathbf{y}_l|\mathbf{H}_0) d\mathbf{y}_1 d\mathbf{y}_2 \cdots d\mathbf{y}_l,
\end{aligned}
\tag{5.1.8}
$$

where

$$
\mathcal{U}_i = \{(\mathbf{y}_1, \ldots, \mathbf{y}_l) : (d_1, d_2, \ldots, d_l)_i\}, \quad i = 0, 1.
$$

To partition the whole observation space $\Re^{n_1} \times \cdots \times \Re^{n_l}$ into two decision regions, we need to compute 2^{l+1} such integrals of $\sum_{i=1}^{l} n_i$ folds because the optimal fusion (5.1.4) requires the computation of probabilities over all possible points $(d_1, d_2, \ldots, d_l)_i$ under both hypotheses. Since each region $\mathcal{U}_i$ can be decomposed into a product of l regions of lower dimensions as

$$
\mathcal{U}_i = \Omega_1 \times \cdots \times \Omega_{l-1} \times \Omega_l, \quad \Omega_i \subset \Re^{n_i},
$$

the above $(\sum_{i=1}^{l} n_i)$-fold integrals can be reduced to an l-stage iterated integrals of n_i $(i \le l)$ folds as follows:

$$
P(d_1, \ldots, d_l|\mathbf{H}_j) = \int_{\Omega_l} \{\cdots\{\int_{\Omega_1} p(\mathbf{y}_1, \ldots, \mathbf{y}_l|\mathbf{H}_j) d\mathbf{y}_1\}\cdots\} d\mathbf{y}_l, \; j = 0, 1. \tag{5.1.9}
$$

If the sensor observations are conditionally independent, the integrals can be further reduced to the product of l integrals of n_i $(i \le l)$ folds

$$
\prod_{i=1}^{l} \int_{\Omega_i} p(\mathbf{y}_i|\mathbf{H}_j) d\mathbf{y}_i, \quad j = 0, 1.
$$

5.1.3 LOCALLY OPTIMAL SENSOR RULES WITH COMMUNICATIONS

All the above results assume that the local decision rules are known. In this subsection, locally optimal sensor decision rules are presented, in particular, while there exist communications among sensors. By a "locally optimal sensor rule" of a local sensor we mean that a sensor rule is optimal using all information it received under the same optimality criterion as the one used for the whole decision system, that is, the Bayes criterion with the same parameters P_0, P_1, and C_{ij}, or under its own criterion, namely parameters C_{ij}^k used by the kth sensor. Such a sensor rule is not be confused with optimal sensor rules under a fixed fusion rule that we discussed in Chapters 2, 3, and 4. There, a fixed fusion rule is given first, then one optimizes the local sensor rules to achieve the best performance for this system (see, for example, [77]). In this case, the function of optimal sensor rules is to optimally compress their observations in terms of final decision performance for whole system under a given fusion rule, is not to make a locally optimal decision by using all possible local information. Hence, the two criteria in the two frameworks are different.

Clearly, in the case of no communications among sensors, regardless the dependence among sensor observations, the decision rule of a sensor relies only on its own observations. Thus, the locally optimal sensor rule is given by the following marginal-likelihood ratio test:

$$\frac{p(\mathbf{y}_i|\mathbf{H}_1)}{p(\mathbf{y}_i|\mathbf{H}_0)} \underset{d_i\,=\,0}{\overset{d_i\,=\,1}{\gtrless}} \frac{P_0(C_{10}-C_{00})}{P_1(C_{01}-C_{11})}, \qquad \forall i \leq l. \tag{5.1.10}$$

If a sensor receives some other sensors' decisions, along with knowledge of their decision rules, its locally optimal decision rule is more complicated due to more information from other sensors but can be obtained according to the definition of joint probability as follows.

Suppose that the ith sensor can receive j local decisions from other sensors along with knowledge of their decision rules. Without loss of generality, denote the received j local decisions by $(d_1, d_2, \ldots, d_j)$ and assume $i > j$ for convenience of presentation. The general decision rule at this sensor is defined by the following mapping:

$$d_i(d_1, \ldots, d_j, \mathbf{y}_i) : \{0,1\}^j \times \Re^{n_i} \longmapsto \{0,1\}. \tag{5.1.11}$$

To define this mapping, we need to determine the values of $d_i(\,)$ for every possible value of the j-tuple $(d_1, d_2, \ldots, d_j)$ and $\mathbf{y}_i$. As these j sensors may also receive local decisions from other sensors, each point of the j-tuple $(d_1, d_2, \ldots, d_j)$ of binary elements is mapped from a subset of $\Re^{n_1} \times \cdots \times \Re^{n_{i-1}} \times \Re^{n_{i+1}} \times \cdots \times \Re^{n_l}$. As we consider non-randomized decisions only, these 2^j subsets are disjoint. Denote them by $\{\mathcal{U}_1, \mathcal{U}_2, \ldots, \mathcal{U}_{2^j}\}$. Since all sensor rules are known, we know exactly what every subset $\mathcal{U}_k$ is. Therefore, according to the definition of a joint probability, we know the two joint probability for $d_i = 0$ and $\mathcal{U}_k$ is

$$P(\mathcal{U}_k, d_i = 0|\mathbf{H}_1) = \int_{(d_i=0)} \int_{\mathcal{U}_k} p(\mathbf{y}_1, \ldots, \mathbf{y}_l|\mathbf{H}_1) d\mathbf{y}_1 d\mathbf{y}_2 \cdots d\mathbf{y}_l, \ \forall k \leq 2^j, \tag{5.1.12}$$

and

$$P(\mathcal{U}_k, d_i = 0|\mathbf{H}_0) = \int_{(d_i=0)} \int_{\mathcal{U}_k} p(\mathbf{y}_1, \ldots, \mathbf{y}_l|\mathbf{H}_0) d\mathbf{y}_1 d\mathbf{y}_2 \cdots d\mathbf{y}_l, \ \forall k \leq 2^j. \tag{5.1.13}$$

Similar to the Bayes cost (5.1.1), we have

$$
\begin{aligned}
&C(\mathcal{U}, d_i, H) - (P_0C_{10} + P_1C_{11}) \\
&= \sum_{k=1}^{2^j}[P_1(C_{01} - C_{11})P(\mathcal{U}_k, d_i = 0|\mathbf{H}_1) - P_0(C_{10} - C_{00})P(\mathcal{U}_k, d_i = 0|\mathbf{H}_0)] \\
&= \sum_{k=1}^{2^j} \int_{d_i=0}\{\int_{\mathcal{U}_k} [P_1(C_{01} - C_{11})p(\mathbf{y}_1, \ldots, \mathbf{y}_l|\mathbf{H}_1) \\
&\quad -P_0(C_{10} - C_{00})p(\mathbf{y}_1, \ldots, \mathbf{y}_l|\mathbf{H}_0)]d\mathbf{y}_1 \cdots d\mathbf{y}_{i-1}d\mathbf{y}_{i+1} \cdots d\mathbf{y}_l\}d\mathbf{y}_i.
\end{aligned}
\tag{5.1.14}
$$

Thus, similarly as for the case without communications, to minimize the above cost, the locally optimal sensor rule at the ith sensor is given by

$$
\frac{\int_{\mathcal{U}_k} p(\mathbf{y}_1, \ldots, \mathbf{y}_l|\mathbf{H}_1)d\mathbf{y}_1 \cdots d\mathbf{y}_{i-1}d\mathbf{y}_{i+1} \cdots d\mathbf{y}_l}{\int_{\mathcal{U}_k} p(\mathbf{y}_1, \ldots, \mathbf{y}_l|\mathbf{H}_0)d\mathbf{y}_1 \cdots d\mathbf{y}_{i-1}d\mathbf{y}_{i+1} \cdots d\mathbf{y}_l} \mathop{\gtrless}_{d_i=0}^{d_i=1} \frac{P_0(C_{10} - C_{00})}{P_1(C_{01} - C_{11})}
\tag{5.1.15}
$$

for $\forall k \leq 2^j$. Note that all the integrals in the above rule are functions of $\mathbf{y}_i$ and this rule consists of 2^j sub-rules corresponding to different values of $(d_1, d_2, \ldots, d_j)$ so that the mapping (5.1.11) is uniquely defined.

When there is no communication between the i sensor and any other sensor, $j = 0$ and thus the only partition $\mathcal{U}_1$ of the product space $\Re^{n_1} \times \cdots \times \Re^{n_{i-1}} \times \Re^{n_{i+1}} \times \cdots \times \Re^{n_l}$ is the product space itself. As such,

$$
p(\mathbf{y}_i|\mathbf{H}_i) = \int_{\mathcal{U}_1} p(\mathbf{y}_1, \ldots, \mathbf{y}_l|\mathbf{H}_i)d\mathbf{y}_1 \cdots d\mathbf{y}_{i-1}d\mathbf{y}_{i+1} \cdots d\mathbf{y}_l.
$$

That is, rule (5.1.15) reduces to rule (5.1.10).

As mentioned above, we can view a locally optimal sensor rule without communication from other sensors as a special case of ignoring the communications from other sensors, i.e., no matter what $(d_1, d_2, \ldots, d_j)$ and $\{\mathcal{U}_1, \mathcal{U}_2, \ldots, \mathcal{U}_{2^j}\}$ are, its local sensor rule is always the rule (5.1.10) independent of $(d_1, d_2, \ldots, d_j)$. Thus, its decision performance obviously is worse than the locally optimal decision rules (5.1.15) because the latter is the optimal solution based upon the received $(d_1, d_2, \ldots, d_j)$. Even if $\{\mathbf{y}_1, \ldots, \mathbf{y}_l\}$ are mutually independent given two hypotheses, the two joint probabilities (5.1.12) and (5.1.13) become simply

$$
P(\mathcal{U}_k|\mathbf{H}_1) \int_{d_i=0} p(\mathbf{y}_i|\mathbf{H}_1)d\mathbf{y}_i \text{ and } P(\mathcal{U}_k|\mathbf{H}_0) \int_{d_i=0} p(\mathbf{y}_i|\mathbf{H}_0)d\mathbf{y}_i,
$$

respectively, the communications among sensors, in general, still improve the local sensor decision performance because the locally optimal sensor rule is now simply

$$\frac{p(\mathbf{y}_i|\mathbf{H}_1)}{p(\mathbf{y}_i|\mathbf{H}_0)} \overset{d_i=1}{\underset{d_i=0}{\gtrless}} \frac{P(\mathcal{U}_k|\mathbf{H}_0)P_0(C_{10}-C_{00})}{P(\mathcal{U}_k|\mathbf{H}_1)P_1(C_{01}-C_{11})} \qquad \forall i \leq l \qquad (5.1.16)$$

still other than the rule (5.1.10). The numerical results in Tables 5.1.2 and 5.1.3 of Examples 5.1.2 and 5.1.3 below support the above analysis.

5.1.4 EXTENSIONS TO MORE GENERAL SYSTEMS

5.1.4.1 EXTENSION TO LOCAL SENSORS WITH FEEDBACK

The above result can be extended to local sensors with feedback from the fusion center.

Two types of feedback

I. Feedback without new observations

Suppose that after the fusion center makes a final decision based upon all the received local decisions at the first stage, the fusion center communicates its decision, namely, the first fusion result $F_1 \subset \Re^{n_1} \times \cdots \times \Re^{n_l}$ $(F_1 = 0, 1)$ to a local sensor. In this case, when the fusion rule can be known by sensors, we can still view the feedback from the fusion center as a communication to the sensor, namely the ith sensor. Then only difference between a locally optimal sensor rule with feedback and without feedback is that the ith sensor receiving feedback can get more detailed information on what regions the observations $\mathbf{y}_j$, $j \neq i$, of other sensors may fall on than the same sensor rule without feedback. Using the same idea that the locally optimal sensor rule with communication outperforms the locally optimal sensor rule without communication, we know that the second locally optimal sensor rule after receiving feedback outperforms the first locally optimal sensor rule without feedback.

II. Feedback with new observations

There is another type of feedback with new observations coming to the local sensors. Suppose that the ith sensor can receive the fusion center's decision at the first stage and j other local decisions, along with knowledge of their decision rules, as well as, a new observation at the second stage. In fact, if we review the new observation as new sensor, this type of feedback can be reduce to the sensor communication problem. That is to say, this is a locally optimal sensor rule problem of the "new" sensor with communications from other sensors. Due to the same reason as aforementioned, the ith sensor's second decision after receiving new observation will improve the local performance of its first decision. The results in Subsection 5.1.3 indicate that the locally optimal sensor rules as well as the optimal fusion rule given these local rules depend only on the joint probability densities in a form well known as the likelihood ratio test. In view of this, the optimal fusion rule s can be extended to a variety of very general distributed decision systems.

5.1.4.2 EXTENSION TO SOPHISTICATED NETWORK STRUCTURES

A multi-level decision system, such as a *tandem* or a *tree* network system, can be viewed as the above two-level decision system with possible communications among sensors and between sensors and the fusion center. Sensors at a higher level in the multi-level system may be treated by fictitious sensors that receive new messages at a new stage in the above system. This should be the case since a two-level system that allows communications between any two sensors and between any sensor and the fusion center is actually a system of a general structure. Note that the optimal fusion rule and the locally optimal sensor rules presented in the above subsections are valid for this general system.

5.1.4.3 EXTENSION TO M-ARY DECISION SYSTEMS

The above results can be easily extended to an m-ary decision system because the optimal decision rule for a centralized m-ary decision problem can be reduced to a set of likelihood ratio tests (see, e.g., [83]).

For an m-ary decision system, the Bayes cost is given as in Eq. (4.3.1)

$$C(d_1, d_2, \ldots, d_l; F) = \sum_{i=0}^{m-1} \sum_{F=i} \sum_{j=0, j\neq i}^{m-1} C_{ij} P_j P(d_1, \ldots, d_l | \mathbf{H}_j).$$

Similarly, to minimize the above cost, the optimal decision regions should be $\mathcal{H}_i$ for $\mathbf{H}_i$ are defined as

$$\begin{aligned} \mathcal{H}_i = \{(d_1, \ldots, d_l) : & \sum_{j=0, j\neq i}^{m-1} C_{ij} P_j P(d_1, \ldots, d_l | \mathbf{H}_j) \\ & \leq \sum_{j=0, j\neq i}^{m-1} C_{kj} P_j P(d_1, \ldots, d_l | \mathbf{H}_j), \forall k \neq i\}, \end{aligned} \tag{5.1.17}$$

where those points $(d_1, \ldots, d_l)$ satisfying multiple decision regions $\mathcal{H}_i$ can be defined to belong to anyone of them.

5.1.4.4 EXTENSION TO NEYMAN-PEARSON DECISION SYSTEMS

For a distributed Neyman-Pearson decision system, the major task for its optimal decision rules is still the computation of the conditional joint sensor decision probabilities $P(d_1, d_2, \ldots, d_l|\mathbf{H}_i)$, $i = 0, 1$. The only thing that differs from the Bayes decision in this case is that $P(d_1, d_2, \ldots, d_l|\mathbf{H}_i)$, $i = 0, 1$ are in general nonzero over the region

$$\left\{(\mathbf{y}_1, \ldots, \mathbf{y}_l) : \frac{\int_{\mathcal{U}_k} p(\mathbf{y}_1, \ldots, \mathbf{y}_l|\mathbf{H}_1) d\mathbf{y}_1 \cdots d\mathbf{y}_{i-1} d\mathbf{y}_{i+1} \cdots d\mathbf{y}_l}{\int_{\mathcal{U}_k} p(\mathbf{y}_1, \ldots, \mathbf{y}_l|\mathbf{H}_0) d\mathbf{y}_1 \cdots d\mathbf{y}_{i-1} d\mathbf{y}_{i+1} \cdots d\mathbf{y}_l} = \lambda\right\} \tag{5.1.18}$$

for $\forall k \leq 2^j$, where λ ($\lambda \geq 0$) is an appropriate parameter for the probability of making $\mathbf{H}_1$ decision while observation falls into the following region

$$\left\{(\mathbf{y}_1, \ldots, \mathbf{y}_l) : \frac{\int_{\mathcal{U}_k} p(\mathbf{y}_1, \ldots, \mathbf{y}_l|\mathbf{H}_1) d\mathbf{y}_1 \cdots d\mathbf{y}_{i-1} d\mathbf{y}_{i+1} \cdots d\mathbf{y}_l}{\int_{\mathcal{U}_k} p(\mathbf{y}_1, \ldots, \mathbf{y}_l|\mathbf{H}_0) d\mathbf{y}_1 \cdots d\mathbf{y}_{i-1} d\mathbf{y}_{i+1} \cdots d\mathbf{y}_l} \geq \lambda\right\} \tag{5.1.19}$$

for $\forall k \leq 2^j$ required in order for the actual type I error (false-alarm) probability P_f to best approximate (but not exceed) its maximum allowable value α ($0 \leq \alpha \leq 1$, see Subsection 1.3.2).

5.1.5 NUMERICAL EXAMPLES

In the following numerical examples, we consider distributed systems of two and three sensors, respectively, for detecting Gaussian signals in Gaussian noise. All computation intervals are $-8 \leq \mathbf{y}_i \leq 8$.

5.1.5.1 TWO-SENSOR NEYMAN-PEARSON DECISION SYSTEM

The two hypotheses are

$$\begin{aligned} \mathbf{H}_0 &: \mathbf{y}_1 = \nu_1, \qquad \mathbf{y}_2 = \nu_2, \\ \mathbf{H}_1 &: \mathbf{y}_1 = s + \nu_1, \ \mathbf{y}_2 = s + \nu_2, \end{aligned}$$

where the signal s and the two sensor-observation noises ν_1, and ν_2 are Gaussian and all mutually independent:

$$s \sim N(2, 2), \quad \nu_1 \sim N(0, 0.3), \quad \nu_2 \sim N(0, 0.2).$$

Thus, the two conditional pdfs under $\mathbf{H}_0$ and $\mathbf{H}_1$, respectively, are

$$\begin{aligned} p(\mathbf{y}_1, \mathbf{y}_2|\mathbf{H}_0) &\sim N\left(\begin{pmatrix} 0 \\ 0 \end{pmatrix}, \begin{bmatrix} 0.3 & 0 \\ 0 & 0.2 \end{bmatrix}\right), \\ p(\mathbf{y}_1, \mathbf{y}_2|\mathbf{H}_1) &\sim N\left(\begin{pmatrix} 2 \\ 2 \end{pmatrix}, \begin{bmatrix} 2.3 & 2 \\ 2 & 2.2 \end{bmatrix}\right). \end{aligned} \tag{5.1.20}$$

Example 5.1.1.

Consider Neyman-Pearson detection with false-alarm probability $P_f \le 0.092$. Table 5.1.1 gives the detection probabilities, false-alarm probabilities and the thresholds of the two-sensor centralized decision, single-sensor decisions, and two-sensor distributed decision with given two sensor decision rules, where the step size used for the discretized algorithm was 0.05.

Table 5.1.1. Performance comparison of Neyman-Pearson systems.

	P_f	P_d	λ
Centralized	0.0913	0.8805	0.375
Sensor 1	0.0919	0.8087	0.656625
Sensor 2	0.0919	0.8437	0.51
Distributed	0.0919	0.8584	0.65

It is observed that the distributed decision system outperforms the single sensor decision systems but of course is worse than the centralized decision system. Sensor 2 with a greater signal-to-noise ratio (SNR) performs better than the first sensor.

5.1.5.2 THREE-SENSOR BAYES DECISION SYSTEM

It was set in all the numerical examples below for the single sensor and three sensor Bayes decision systems that $C_{ij} = 1$ for $i \neq j$, $C_{ii} = 0$, $P_0 = 1/2$, $P_1 = P_2 = 1/4$, the stepsize $\Delta \mathbf{y}_i = 0.025$. In this case, the Bayes cost functional, denoted as P_e, is actually *decision error probability*.

The hypotheses are

$$\begin{aligned}
&\mathbf{H}_0 : \mathbf{y}_1 = \nu_1, \qquad \mathbf{y}_2 = \nu_2, \qquad \mathbf{y}_3 = \nu_3,\\
&\mathbf{H}_1 : \mathbf{y}_1 = s_1 + \nu_1,\ \mathbf{y}_2 = s_1 + \nu_2,\ \mathbf{y}_3 = s_1 + \nu_3,\\
&\mathbf{H}_2 : \mathbf{y}_1 = s_2 + \nu_1,\ \mathbf{y}_2 = s_2 + \nu_2,\ \mathbf{y}_3 = s_2 + \nu_3,
\end{aligned}$$

where the two signals s_1, s_2 and the three sensor observation noises ν_1, ν_2 and ν_3 are all Gaussian and mutually independent:

$$\begin{aligned}
&s_1 \sim N(2,3), \quad s_2 \sim N(-2,3),\\
&\nu_1 \sim N(0,3),\ \nu_2 \sim N(0,2),\ \ \nu_3 \sim N(0,1).
\end{aligned} \tag{5.1.21}$$

Therefore, the three conditional pdfs under $\mathbf{H}_0$, $\mathbf{H}_1$ and $\mathbf{H}_2$, respectively, are

$$
\begin{aligned}
p(\mathbf{y}_1, \mathbf{y}_2, \mathbf{y}_3|\mathbf{H}_0) &\sim N\left(\begin{pmatrix}0\\0\\0\end{pmatrix}, \begin{bmatrix}3&0&0\\0&2&0\\0&0&1\end{bmatrix}\right),\\
p(\mathbf{y}_1, \mathbf{y}_2, \mathbf{y}_3|\mathbf{H}_1) &\sim N\left(\begin{pmatrix}2\\2\\2\end{pmatrix}, \begin{bmatrix}6&3&3\\3&5&3\\3&3&4\end{bmatrix}\right),\\
p(\mathbf{y}_1, \mathbf{y}_2, \mathbf{y}_3|\mathbf{H}_2) &\sim N\left(\begin{pmatrix}-2\\-2\\-2\end{pmatrix}, \begin{bmatrix}6&3&3\\3&5&3\\3&3&4\end{bmatrix}\right).
\end{aligned}
\tag{5.1.22}
$$

Example 5.1.2.

Consider a parallel Bayes decision system with the above ternary hypotheses without communications among sensors. According to (5.1.10), the locally optimal sensor rule at each sensor can be derived. Table 5.1.2 gives the decision error probabilities of the centralized decision, single sensor decisions, and distributed decision with given sensor decision rules.

Again, the distributed decision system outperforms all single sensor decision systems but of course performs slightly worse than the centralized decision system. Among the three single sensor decisions, the greater the SNR of a local sensor is, the better the performance is.

Table 5.1.2. Performance comparison of Bayes decision systems.

	Centr.	Sensor 1	Sensor 2	Sensor 3	Distr.
P_e	0.2157	0.3650	0.3274	0.2644	0.2498

Example 5.1.3.

Consider again the above three-sensor decision system, but with one extra communication channel from sensor i to sensor j, denoted by "Sensor $i \mapsto j$," $i, j = 1, 2, 3, i \neq j$, in addition to transmitting all local decisions to the fusion center.

Sensor decision rules can be obtained by (5.1.17). For example, for "Sensor $1 \mapsto 2$", the three local decision rules (regions) for Sensor 1 are given by

$$
\mathcal{H}_0^{(1)}(\mathbf{y}_1) = \left\{ \begin{array}{l} \frac{1}{4}(p(\mathbf{y}_1|\mathbf{H}_1) + p(\mathbf{y}_1|\mathbf{H}_2)) \leq \frac{1}{2}p(\mathbf{y}_1|\mathbf{H}_0) + \frac{1}{4}p(\mathbf{y}_1|\mathbf{H}_2), \\ \frac{1}{4}(p(\mathbf{y}_1|\mathbf{H}_1) + p(\mathbf{y}_1|\mathbf{H}_2)) \leq \frac{1}{2}p(\mathbf{y}_1|\mathbf{H}_0) + \frac{1}{4}p(\mathbf{y}_1|\mathbf{H}_1) \end{array} \right\},
\tag{5.1.23}
$$

$$\mathcal{H}_1^{(1)}(\mathbf{y}_1) = \left\{ \begin{array}{l} \frac{1}{2}p(\mathbf{y}_1|\mathbf{H}_0) + \frac{1}{4}p(\mathbf{y}_1|\mathbf{H}_2) \leq \frac{1}{2}p(\mathbf{y}_1|\mathbf{H}_0) + \frac{1}{4}p(\mathbf{y}_1|\mathbf{H}_1), \\ \frac{1}{2}p(\mathbf{y}_1|\mathbf{H}_0) + \frac{1}{4}p(\mathbf{y}_1|\mathbf{H}_2) < \frac{1}{4}(p(\mathbf{y}_1|\mathbf{H}_1) + p(\mathbf{y}_1|\mathbf{H}_2)) \end{array} \right\}, \tag{5.1.24}$$

$$\mathcal{H}_2^{(1)}(\mathbf{y}_1) = \left\{ \begin{array}{l} \frac{1}{2}p(\mathbf{y}_1|\mathbf{H}_0) + \frac{1}{4}p(\mathbf{y}_1|\mathbf{H}_1) < \frac{1}{4}(p(\mathbf{y}_1|\mathbf{H}_1) + p(\mathbf{y}_1|\mathbf{H}_2)) \\ \frac{1}{2}p(\mathbf{y}_1|\mathbf{H}_0) + \frac{1}{4}p(\mathbf{y}_1|\mathbf{H}_1) < \frac{1}{2}p(\mathbf{y}_1|\mathbf{H}_0) + \frac{1}{4}p(\mathbf{y}_1|\mathbf{H}_2) \end{array} \right\}. \tag{5.1.25}$$

To let people easily see that the above formulae come from (5.1.17), they are all not simplified by removing redundant terms from both sides of the inequalities. Since we can assign the sets

$$\{\mathbf{y}_1 : \frac{1}{4}(p(\mathbf{y}_1|\mathbf{H}_1) + p(\mathbf{y}_1|\mathbf{H}_2)) = \frac{1}{2}p(\mathbf{y}_1|\mathbf{H}_0) + \frac{1}{4}p(\mathbf{y}_1|\mathbf{H}_2)\},$$

$$\{\mathbf{y}_1 : \frac{1}{4}(p(\mathbf{y}_1|\mathbf{H}_1) + p(\mathbf{y}_1|\mathbf{H}_2)) = \frac{1}{2}p(\mathbf{y}_1|\mathbf{H}_0) + \frac{1}{4}p(\mathbf{y}_1|\mathbf{H}_1)\},$$

$$\{\mathbf{y}_1 : \frac{1}{2}p(\mathbf{y}_1|\mathbf{H}_0) + \frac{1}{4}p(\mathbf{y}_1|\mathbf{H}_2) = \frac{1}{2}p(\mathbf{y}_1|\mathbf{H}_0) + \frac{1}{4}p(\mathbf{y}_1|\mathbf{H}_1)\}$$

only to one decision region, respectively, we use "$<$" in the definitions of $\mathcal{H}_1^{(1)}(\mathbf{y}_1)$ and $\mathcal{H}_2^{(1)}(\mathbf{y}_1)$ other than "$\leq$". By an extension of (5.1.15) (see the likelihood ratio test given in [83] for a ternary decision system), the locally optimal sensor decision rules at Sensor 2 are defined by the following 9 regions of $\mathbf{y}_2$, where $\mathcal{H}_i^{(2)}(\mathcal{H}_j^{(1)}(\mathbf{y}_1))(\mathbf{y}_2)$, $i, j = 0, 1, 2$, denotes the region for Sensor 2 to decide on $\mathbf{H}_i$ while the received Sensor 1's decision is $\mathbf{H}_j$.

$$\mathcal{H}_0^{(2)}(\mathcal{H}_j^{(1)}) = \left\{ \begin{array}{l} \frac{1}{4}(\int_{\mathcal{H}_j^{(1)}} p(\mathbf{y}_1, \mathbf{y}_2|\mathbf{H}_1)d\mathbf{y}_1 + \int_{\mathcal{H}_j^{(1)}} p(\mathbf{y}_1, \mathbf{y}_2|\mathbf{H}_2)d\mathbf{y}_1) \\ \leq \frac{1}{2}\int_{\mathcal{H}_j^{(1)}} p(\mathbf{y}_1, \mathbf{y}_2|\mathbf{H}_0)d\mathbf{y}_1 + \frac{1}{4}\int_{\mathcal{H}_j^{(1)}} p(\mathbf{y}_1, \mathbf{y}_2|\mathbf{H}_2)d\mathbf{y}_1, \\ \frac{1}{4}(\int_{\mathcal{H}_j^{(1)}} p(\mathbf{y}_1, \mathbf{y}_2|\mathbf{H}_1)d\mathbf{y}_1 + \int_{\mathcal{H}_j^{(1)}} p(\mathbf{y}_1, \mathbf{y}_2|\mathbf{H}_2)d\mathbf{y}_1) \\ \leq \frac{1}{2}\int_{\mathcal{H}_j^{(1)}} p(\mathbf{y}_1, \mathbf{y}_2|\mathbf{H}_0)d\mathbf{y}_1 + \frac{1}{4}\int_{\mathcal{H}_j^{(1)}} p(\mathbf{y}_1, \mathbf{y}_2|\mathbf{H}_1)d\mathbf{y}_1 \end{array} \right\}, \tag{5.1.26}$$

$$
\mathcal{H}_1^{(2)}(\mathcal{H}_j^{(1)}) = \left\{ \begin{array}{l} \frac{1}{2}\int_{\mathcal{H}_j^{(1)}} p(\mathbf{y}_1,\mathbf{y}_2|\mathbf{H}_0)d\mathbf{y}_1 + \frac{1}{4}\int_{\mathcal{H}_j^{(1)}} p(\mathbf{y}_1,\mathbf{y}_2|\mathbf{H}_2)d\mathbf{y}_1 \\ \leq \frac{1}{2}\int_{\mathcal{H}_j^{(1)}} p(\mathbf{y}_1,\mathbf{y}_2|\mathbf{H}_0)d\mathbf{y}_1 + \frac{1}{4}\int_{\mathcal{H}_j^{(1)}} p(\mathbf{y}_1,\mathbf{y}_2|\mathbf{H}_1)d\mathbf{y}_1, \\ \frac{1}{2}\int_{\mathcal{H}_j^{(1)}} p(\mathbf{y}_1,\mathbf{y}_2|\mathbf{H}_0)d\mathbf{y}_1 + \frac{1}{4}\int_{\mathcal{H}_j^{(1)}} p(\mathbf{y}_1,\mathbf{y}_2|\mathbf{H}_2)d\mathbf{y}_1 \\ < {}_{\bar{1}}4(\int_{\mathcal{H}_j^{(1)}} p(\mathbf{y}_1,\mathbf{y}_2|\mathbf{H}_1)d\mathbf{y}_1 + \int_{\mathcal{H}_j^{(1)}} p(\mathbf{y}_1,\mathbf{y}_2|\mathbf{H}_2)d\mathbf{y}_1) \end{array} \right\}, \tag{5.1.27}
$$

$$
\mathcal{H}_2^{(2)}(\mathcal{H}_j^{(1)}) = \left\{ \begin{array}{l} \frac{1}{2}\int_{\mathcal{H}_j^{(1)}} p(\mathbf{y}_1,\mathbf{y}_2|\mathbf{H}_0)d\mathbf{y}_1 + \frac{1}{4}\int_{\mathcal{H}_j^{(1)}} p(\mathbf{y}_1,\mathbf{y}_2|\mathbf{H}_1)d\mathbf{y}_1 \\ < {}_{\bar{1}}2\int_{\mathcal{H}_j^{(1)}} p(\mathbf{y}_1,\mathbf{y}_2|\mathbf{H}_0)d\mathbf{y}_1 + \frac{1}{4}\int_{\mathcal{H}_j^{(1)}} p(\mathbf{y}_1,\mathbf{y}_2|\mathbf{H}_2)d\mathbf{y}_1, \\ \frac{1}{2}\int_{\mathcal{H}_j^{(1)}} p(\mathbf{y}_1,\mathbf{y}_2|\mathbf{H}_0)d\mathbf{y}_1 + \frac{1}{4}\int_{\mathcal{H}_j^{(1)}} p(\mathbf{y}_1,\mathbf{y}_2|\mathbf{H}_1)d\mathbf{y}_1 \\ < \frac{1}{4}(\int_{\mathcal{H}_j^{(1)}} p(\mathbf{y}_1,\mathbf{y}_2|\mathbf{H}_1)d\mathbf{y}_1 + \int_{\mathcal{H}_j^{(1)}} p(\mathbf{y}_1,\mathbf{y}_2|\mathbf{H}_2)d\mathbf{y}_1) \end{array} \right\}. \tag{5.1.28}
$$

Table 5.1.3 gives the performances of all locally optimal sensor decision at the jth sensor with communication from the ith sensor denoted by Sensor $i \mapsto j$, respectively. The performances are better than that without communications

Table 5.1.3. Performance comparison of locally optimal sensor decisions with additional communications.

Local Optim.	Sensor 1 $\mapsto$ 2	Sensor 1 $\mapsto$ 3	Sensor 2 $\mapsto$ 3
P_e	0.2957	0.2494	0.2421
Local Optim.	Sensor 3 $\mapsto$ 2	Sensor 3 $\mapsto$ 1	Sensor 2 $\mapsto$ 1
P_e	0.2472	0.2556	0.2986

(see Table 5.1.2 above) but worse than that of the corresponding triple sensor distributed decision (see Table 5.1.4 below).

Table 5.1.4 gives the performances of the three sensor **distributed decision fusion** with given sensor rules for the systems with all possible Sensor $i \mapsto j$, respectively.

Table 5.1.4 carries very interesting information. Comparing it with the results in Example 5.1.2, we have the following **observations**:

- *Fusion does improve performance.* All distributed decision systems still outperform the three single sensor decision systems. This makes sense

Table 5.1.4. Performance comparison of distributed Bayes decision fusion with different sensor communications.

Distr.	Sensor 1 $\mapsto$ 2	Sensor 1 $\mapsto$ 3	Sensor 2 $\mapsto$ 3
P_e	0.2518	0.2433	0.2381
Distr.	Sensor 3 $\mapsto$ 2	Sensor 3 $\mapsto$ 1	Sensor 2 $\mapsto$ 1
P_e	0.2441	0.2509	0.2577

since more information is used in the fused decision than in any single-sensor decision.

- *The direction of communication affects the performance of the local sensor decision significantly.* For two given sensors, communication from the one with a smaller SNR to the one having a greater SNR leads to better local sensor performance than the other way round (see Table 5.1.3). This is understandable from the following perspective. In the example of "Sensor 1$\mapsto$2" (see (5.1.23)–(5.1.28), in Example 5.1.3), it can be seen from a comparison between Sensor 1 (with three decision regions) and Sensor 2 (with nine regions) that in terms of decision rules, a sensor receiving information from another sensor is equivalent to having a more refined partition of its observation space. Therefore, communication to a more reliable sensor will result in better performance than if the communication direction is reversed.

- *Communication does not necessarily improve global performance*: Although communications between sensors do improve the local decision performance at the sensor receiving communications (see the analysis in Subsection 5.1.3 above and Tables 5.1.3 and 5.1.4), not all the distributed decision fusion with communication between sensors outperform the corresponding distributed decision fusion without communication. While communication between sensors with a greater SNR improves fusion performance, communication between sensors with a smaller SNR degrades the fusion performance. This is somewhat counter-intuitive at first glance. Note, however, that what we considered is distributed fusion with given sensor rules. This is a two separate level optimization issue. When a local decision performance at a less reliable sensor was improved, it would be possible to lead the global system performance to be worse. In fact, this kind of phenomenon appears very often in a multi-level technical and economic systems. For example, consuming resource to improve a sub-system with lower efficiency may result in a worse global system performance. Therefore, while design a distributed decision system, one has to choose appropriate communication direction as mentioned in the last observation, and balance the local and global decision performances according to the requirement of a practical application.

These observations give an insight into the problem of distributed decision with sensor-wise communications and provide guidelines for the design of communications between sensors in practice.

Summing up, in this section, it was shown that if the sensor rules are known completely by the fusion center and all sensors, locally optimal sensor rules and the optimal fusion rule can be derived for very general multisensor decision systems. It is valid for any sensor observations and any given local decision rules (whether they are dependent or not) and any network structure (with or without communications between any two sensors). We have also shown that the decision rules of a sensor that are locally optimal – in the sense that all information available to it is used optimally — are of the form of a generalized likelihood ratio test. Thus, we can compute various numerical examples, and reveal some interesting phenomenon, such as the second and third observations above, which might not be expected before the numerical computations. This suggests us that one should get the knowledge of sensor rules as much as possible in order to obtain the optimal fusion. The numerical examples also provide useful information for the design of communications among sensors in practice.

5.2 THE EQUIVALENT CLASSES OF FUSION RULES

The optimal fusion rules in the previous section are all derived in two level optimization sense, that is to say, the sensor rules are given first. Those results are not suitable to the optimal fusion in the global optimization sense. To obtain globally optimal performance of a system, we should find an optimal fusion rule and the corresponding set of optimal local compression rules. However, this is a very difficult task since the optimal fusion rule in general depends on the statistical properties of the observational data as well as the objective of optimizing the system performance. Besides, the number of all possible fusion rules is increasing very quick as the number of sensors becomes large. To reduce the number of fusion rules, one needs to check which fusion rules are worth evaluating and which fusion rules can be given up.

When all sensor outputs are given, sensor observations are conditionally independent mutually, and $P_d^i \geq P_f^i$, where P_d^i and P_f^i are the sensor probabilities of detection and false alarm respectively, a monotonic fusion rule was defined in [78] and Remark 3.3.1, Page 63 of [84], and the optimal fusion rule can be found in the class of the monotonic fusion rules.

When the sensor decisions are conditionally independent mutually, we can rewrite Eq. (5.1.5) of the optimal fusion rule as

$$\log \frac{P_1(C_{01} - C_{11})}{P_0(C_{10} - C_{00})} + \sum_{U_1} \log \frac{P_d^i}{P_f^i} + \sum_{U_0} \log \frac{1 - P_d^i}{1 - P_f^i} \overset{F = \mathbf{H}_1}{\underset{F = \mathbf{H}_0}{\gtrless}} 0. \qquad (5.2.1)$$

Since $P_d^i \geq P_f^i$, on the left hand side of the above inequality, the two summations

$$\sum_{U_1} \log \frac{P_d^i}{P_f^i} \quad \text{and} \quad \sum_{U_0} \log \frac{1 - P_d^i}{1 - P_f^i}$$

are both non-decreasing as the sensors deciding $\mathbf{H}_1$ are increasing. Hence, using this monotonicity in the optimal fusion rule (5.2.1), we have the following necessary condition for the optimal fusion rule (5.2.1):

The necessary condition of the fusion rule (5.2.1). When an optimal fusion rule is defined by (5.2.1), the optimal fusion rule must satisfy the following monotonicity: If a local message $(d_1, \ldots, d_l)^1$, denoting its subscript set of those $d_i = 1$ by

$$U^1 = \{i : d_i = 1, d_i \in (d_1, \ldots, d_l)^1\},$$

belongs to $\mathcal{H}_1$ decision region of the optimal fusion rule, then any other local message $(d_1, \ldots, d_l)^2$, denoting its subscript set of those $d_i = 1$ by

$$U^2 = \{i : d_i = 1, d_i \in (d_1, \ldots, d_l)^2\},$$

must also belong to $\mathbf{H}_1$ decision region of the optimal fusion rule whenever $U^1 \subset U^2$.

From the monotonicity of the two summations in (5.2.1), the above necessary condition is obviously true. Therefore, we need to search for the optimal fusion rule only within the above monotonic fusion rules.

For example, there are 16 possible fusion rules to choose for a distributed binary decision system with two bi-decision sensors, but only the following 6 fusion rules are monotonic fusion rules:

$$\mathcal{H}_1 = \{(1\ \ 1)\}; \mathcal{H}_1 = \{(1\ \ 0); (0\ \ 1), (1\ \ 1)\}; \mathcal{H}_1 = \{(1\ \ 0), (1\ \ 1)\};$$

$$\mathcal{H}_1 = \{(0\ \ 1), (1\ \ 1)\}, ; \mathcal{H}_1 = \{(0\ \ 0), (1\ \ 0); (0\ \ 1), (1\ \ 1)\}; \mathcal{H}_1 = \emptyset. \qquad (5.2.2)$$

For a distributed binary decision system with three bi-decision sensors, there are 21 fusion rules to be monotonic fusion rules (see [78]). However, this result is not suitable to global system optimization because all sensor rules here were assumed to be given already before fusion. Besides, it is not easy to extend this result to more general systems, such as, m-ary decision system, dependent sensor observation system.

In this section, from the point of view of globally optimal performance, we analyze the equivalence of some of the fusion rules for general distributed multisensor multi-hypothesis decision problem. To focus on comparing the performances of different fusion rules, suppose that one can always find out optimal sensor compression rules for any given fusion rule (for specific algorithms, see Chapter 2-4 above). After making such an assumption, we can concentrate only on the fusion rules and show under what condition two different fusion rules are actually equivalent with the same final cost, as well as, one may be superior to another. Furthermore, by partitioning the set of all fusion rules to equivalent classes, comparing performance between some equivalent classes and dropping off those "worse" fusion rule classes, the number of the *valuable* fusion rule classes will be reduced, such as from 16 to 3 for two sensor binary decision system, or from 256 to 32 for three sensor binary decision system. For more general multisensor m-ary decision systems, one can use computer to analyze and determine valuable fusion rule classes according to the results in this section.

5.2.1 PRELIMINARY DEFINITIONS

We still consider a general multisensor m-ary distributed decision system. Clearly, a fixed fusion rule with a variety of local compression rules can produce a set of different final decision regions, which is called the *variety set of final decision regions generated by this fusion rule.* The goal of our distributed multisensor decision is to find a globally optimal final decision region of $(\mathbf{y}_1, \mathbf{y}_2, ..., \mathbf{y}_l)$ determined by an optimal partition of the set of $\sum_{i=1}^{l} r_i$-tuples $(I_1^{(1)}, \ldots, I_1^{(r_1)}; \cdots; I_l^{(1)}, \ldots, I_l^{(r_l)})$ and the corresponding optimal sensor compression rules so that the cost functional is minimized. Suppose that one can always find out optimal sensor rules for any given cost functional and any fixed fusion rule; thus, minimizing cost functional just depends on the variety set of the final decision regions generated by the fixed fusion rule. Using the aforementioned convention, we give the following three definitions straightforwardly.

Definition 5.2.1. If two fusion rules can generate the same variety set of final decision regions, they are called to be *equivalent.* Clearly, the minimum costs for the equivalent fusion rules are identical although they may have different optimal sensor rules to achieve such minimum cost. Collecting all equivalent fusion rules yields a *equivalent class of fusion rules.*

Definition 5.2.2. If the variety set of final decision regions generated by the first fusion rule is a real subset of the variety set of final decision regions generated by the second one, we call the latter to be *superior* to former.

Obviously, Definition 5.2.2 is reasonable because optimization result over larger domain is superior to that over smaller domain.

Definition 5.2.3. We call $1 - I_i^{(j)}(\mathbf{y}_i)$ the *complementary indicator function* of $I_i^{(j)}(\mathbf{y}_i)$.

Obviously, $1 - I_i^{(j)}(\mathbf{y}_i)$ is also a compression rule for data $\mathbf{y}_i$, and its *0 compression region* is just the *1 compression region* of $I_i^{(j)}(\mathbf{y}_i)$, i.e.,

$$\{\mathbf{y}_i : 1 - I_i^{(j)}(\mathbf{y}_i) = 0\} = \{\mathbf{y}_i : I_i^{(j)}(\mathbf{y}_i) = 1\}.$$

5.2.2 PROPOSITIONS

Using the above definitions, we have the following two propositions to classify fusion rules and compare performances between some of two fusion rules.

Proposition 5.2.1. If a fusion rule can become another fusion rule by changing some of the local compression rules in the first fusion rule to their complementary indicator functions, the both fusion rules are equivalent.

Now we give an example to show two fusion rules are equivalent. Suppose that

$$\mathcal{H}_0^{(1)} = \{(\mathbf{y}_1, \mathbf{y}_2) : (I_1(\mathbf{y}_1) = 0, I_2(\mathbf{y}_2) = 0)\}$$

and

$$\mathcal{H}_0^{(2)} = \{(\mathbf{y}_1, \mathbf{y}_2) : (I_1(\mathbf{y}_1) = 1, I_2(\mathbf{y}_2) = 1)\}$$

are two different $\mathbf{H}_0$ decision regions for the two sensor binary decision problem. In the sequel, for notational simplicity denote the above two fusion rules by

$$\{(0^{(1)}, 0^{(2)})\}_{\mathbf{H}_0} \quad \text{and} \quad \{(1^{(1)}, 1^{(2)})\}_{\mathbf{H}_0},$$

respectively. By Proposition 5.2.1, the two fusion rules are equivalent.

From Definition 5.2.2, we have

Proposition 5.2.2. If restricting some of the local compression rules in a fusion rule to be identical to 1 or 0 no matter what those sensors observe can yields another fusion rule, then the former is superior to the latter.

For example, the fusion rule $\{(1^{(1)}, 1^{(2)})\}_{\mathbf{H}_0}$ is superior to the fusion rule $\{(1^{(1)}, 1^{(2)}), (1^{(1)}, 0^{(2)})\}_{\mathbf{H}_0}$ because restricting the second local compression rule $I_2(\mathbf{y}_2) \equiv 1$ in the first fusion rule yields the second fusion rule which just like ignoring the observation data of second sensor.

In the above argument, only the variety set of final decision regions are considered, no any test type and cost functional are involved; therefore, the above results do not depend on statistical properties of observational data, and hold for all distributed decision problems, such as Neyman-Pearson test.

5.2.3 APPLICATIONS OF PROPOSITIONS

We now apply the above two propositions to analyze any two fusion rules whether or not they are equivalent and one may be superior to another. As results of this analysis, the equivalent classes of the valuable fusion rules for two and three sensor distributed multisensor binary decision systems are presented below.

5.2.3.1 TWO SENSOR BINARY DECISION

Suppose that each sensor compresses its observation to one information bit. Therefore, there exist four different sets of local messages as follows:

$$(0^{(1)}, 0^{(2)}), (1^{(1)}, 0^{(2)}), (0^{(1)}, 1^{(2)}), (1^{(1)}, 1^{(2)}).$$

Clearly, they can produce sixteen different fusion rules.

I. 1-local-message fusion rules.

$$\{(0^{(1)}, 0^{(2)})\}_{\mathbf{H}_0}, \{(1^{(1)}, 0^{(2)})\}_{\mathbf{H}_0}, \{(0^{(1)}, 1^{(2)})\}_{\mathbf{H}_0}, \{(1^{(1)}, 1^{(2)})\}_{\mathbf{H}_0},$$

which are all equivalent to each other (OR rule for $\mathbf{H}_1$ decision).

II. 2-local-message fusion rules.

II.1.

$$\{(0^{(1)}, 0^{(2)}), (1^{(1)}, 0^{(2)})\}_{\mathbf{H}_0}, \{(0^{(1)}, 0^{(2)}), (0^{(1)}, 1^{(2)})\}_{\mathbf{H}_0},$$

$$\{(1^{(1)}, 1^{(2)}), (1^{(1)}, 0^{(2)})\}_{\mathbf{H}_0}, \{(1^{(1)}, 1^{(2)}), (0^{(1)}, 1^{(2)})\}_{\mathbf{H}_0}.$$

They are classified to be two equivalent classes (ignoring Sensor 1 or Sensor 2) and are both worse than Class **I**.

II.2.

$$\{(0^{(1)}, 0^{(2)}), (1^{(1)}, 1^{(2)})\}_{\mathbf{H}_0}, \{(0^{(1)}, 1^{(2)}), (1^{(1)}, 0^{(2)})\}_{\mathbf{H}_0}.$$

They are both equivalent to each other (XOR rule).

III. 3-local-message fusion rules.

$$\{(0^{(1)},0^{(2)}),(0^{(1)},1^{(2)}),(1^{(1)},0^{(2)})\}_{\mathbf{H}_0},$$

$$\{(1^{(1)},1^{(2)}),(0^{(1)},1^{(2)}),(1^{(1)},0^{(2)})\}_{\mathbf{H}_0}.$$

They are both equivalent to each other (AND rule).

IV. 4-local-message fusion rule.

$$\{(0^{(1)},0^{(2)}),(0^{(1)},1^{(2)}),(1^{(1)},0^{(2)}),(1^{(1)},1^{(2)})\}_{\mathbf{H}_0}.$$

This fusion rule implies the fusion center always says $\mathbf{H}_0$ occurs no matter what data are observed at the two local sensors. It can be also regarded as a special OR rule with the two particular local compression rules $(I_1(\mathbf{y}_1) \equiv 0, I_2(\mathbf{y}_2) \equiv 0))$; therefore, it is worse than the OR rule as well as Class **II**.

V. 0-local-message fusion rule.

Here 0-local-message fusion rule means that the fusion center always chooses $\mathbf{H}_1$ for any local observation data. Similar to the last case, it is worse than the OR rule and Class **II**.

Summing up, in two sensor binary decision case, 16 fusion rules are classified to be 7 equivalent classes where only 3 classes: AND, OR, and XOR rules are valuable to search for the global optimal fusion rule. In contrast to the monotonic fusion rules in (5.2.2) in the two level optimization sense, the valuable fusion rules in the global optimization sense are significantly different from each other. For example, there is no the XOR fusion rule in (5.2.2) but it does appear in the valuable fusion rules. Besides, the following example shows that the XOR fusion rule even can achieve the optimal performance of a centralized decision system.

Example 5.2.1. Suppose that for a two sensor binary decision system, the negative-valued region of the integrand $ap(\mathbf{y}_1,\mathbf{y}_2|\mathbf{H}_1)-bp(\mathbf{y}_1,\mathbf{y}_2|\mathbf{H}_0)$ is given by

$$\{(\mathbf{y}_1,\mathbf{y}_2) : (\mathbf{y}_1 \le 0, \mathbf{y}_2 > 0), \quad (\mathbf{y}_1 > 0, \mathbf{y}_2 \le 0)\}.$$

That is, the negative value region of the integrand consists of the two rectangles at the diagonal positions. Then, let the two local sensor rules are defined as

$$I_1(\mathbf{y}_1) = \begin{cases} 0 & \text{as } \mathbf{y}_1 \le 0, \\ 1 & \text{as } \mathbf{y}_1 > 0. \end{cases} \qquad I_1(\mathbf{y}_2) = \begin{cases} 0 & \text{as } \mathbf{y}_2 \le 0, \\ 1 & \text{as } \mathbf{y}_2 > 0. \end{cases}$$

Obviously, using XOR rule $\{(0^{(1)},1^{(2)}),(1^{(1)},0^{(2)})\}_{\mathbf{H}_0}$, the distributed decision can have the best performance as the centralized decision.

5.2.3.2 THREE SENSOR BINARY DECISION

Still, suppose that each sensor compresses its observation data to one information bit. Therefore, there exist 8 different local messages:

$$(0^{(1)}, 0^{(2)}, 0^{(3)}), (1^{(1)}, 0^{(2)}, 0^{(3)}), (0^{(1)}, 1^{(2)}, 0^{(3)}),$$

$$(0^{(1)}, 0^{(2)}, 1^{(3)}), (1^{(1)}, 1^{(2)}, 0^{(3)}), (1^{(1)}, 0^{(2)}, 1^{(3)}),$$

$$(0^{(1)}, 1^{(2)}, 1^{(3)}), (1^{(1)}, 1^{(2)}, 1^{(3)}).$$

Clearly, they can produce 256 different fusion rules. Using the above two propositions and the similar method we can show there are only 32 valuable equivalent classes of fusion rules. If we know more information about an integrand in a cost functional, for example, the integrand is a symmetric function with respect to $\mathbf{y}_1, \ldots, \mathbf{y}_l$, i.e., the sensors are all identical, the number of the valuable equivalent classes will be reduced again, for the above three sensor binary decision system, reduced to 16 valuable equivalent classes of fusion rules.

Although the equivalent class method can reduce the number of the valuable fusion rules, the number of fusion rules to need to evaluate by computing its final cost are still quite large when the fusion rule receives more sensor bits. Even for four sensor binary decision system, the number of valuable fusion rules is more than thousand. Therefore, we will develop other method to handle the tough problem of searching optimal fusion rules for various distributed network decision systems.

5.3 UNIFIED FUSION RULE FOR PARALLEL NETWORK

In this section, we present a *unified fusion rule* in the distributed multi-hypothesis multi-sensor parallel network decision systems, where the fusion center can also observe data. For the above decision network, we propose a specific fusion rule, which is in fact of the most general form (i.e. other fusion rules are all its special cases) and independent of the statistical characteristics of observations and decision criteria. It is therefore called the unified fusion rule of all fusion rules. To achieve globally optimal decision performance, people only need to optimize sensor rules under the unified fusion rule for given conditional distributions of observations and decision criterion. Numerical examples support the above results. To globally optimize performance of the system, we usually should find an optimal fusion rule from a huge number of possible fusion rules. An exhaustive method obviously is computationally intractable provided that the decision center receives more information bits.

In previous Chapter 3, we have discussed the optimal local sensor rules under a fixed fusion rule and proposed the optimal fusion rules for some specific distributed parallel binary Bayes decision systems, where one of sensors, namely the lth sensor, compresses its observation to $r_l = 2^L$, $L = \sum_{i=1}^{l-1} r_i$. In fact, if we can combine the lth sensor and the fusion center in the system there, those optimal fusion rule is just the unified fusion rule for the distributed parallel binary Bayes decision system with the fusion center observing data.

In this section, we extend the results there to the aforementioned general distributed multi-hypothesis multisensor parallel network decision system, and then, extend them to the tandem and tree network decision systems in the next section. In comparison with the above exhaustive searching, the unified fusion rule method can reduce much computation and more tractable in practice. For example, for an m-ary decision system with totally k information bits from sensors collected by the fusion center, the number of possible fusion rules are m^{2^k}. Although the number of valuable fusion rules are much less than m^{2^k} (see the analysis in the previous section, Section 5.2), it is still very large as mentioned at the end of the previous section. However, to achieve globally optimal performance, using the unified fusion rules, one needs to calculate $k + n2^k$ optimal sensor (binary compression) bits (integer n defined as $2^{n-1} < m \leq 2^n$), which is much less than that by using the equivalent class method. for example, for a three sensor 4-ary decision system, when the 10 sensor bits are designed as $I_1(\mathbf{y}_1), I_2(\mathbf{y}_2), I_3^{(1)}(\mathbf{y}_3), \ldots, I_3^{(8)}(\mathbf{y}_8)$, using the unified fusion rule, only 10 optimal sensor bits need to be calculated in order to achieve the global optimal performance. However, even using the equivalent class method, at least we need to calculate more than billion optimal sensor bits because the number of all different fusion rules in this case is $4^{2^{10}} > 10^{300}$! Clearly, this unified fusion rule method can save huge computations to obtain globally optimal decision performance.

In the existing literature [60, 79, 86, 88], the above network decision systems were discussed as well, how to find an optimal final fusion rule was a open question in those papers and no aforementioned result was known before. Besides, we also analyze the performances of the parallel and tandem networks with the same communication amount, the result is that theoretically, there is no dominance between the two networks even when the number of sensors goes to infinity, which is partly different from the result in [60, 89]. It is, however, very likely that the performance for the parallel network may be better than the tandem network in many practical applications with given statistically distributed data.

We have done a number of numerical examples, which support the above results and show how to choose the local sensors to assign properly lower compression rates at them by comparing their noise powers. More precisely speaking, the smaller the noise power at a sensor is, the lower the compression rate at the sensor should be. The numerical examples also show that the parallel network performed better than the tandem network while detecting a Gaussian signal in Gaussian noises.

We now consider the unified fusion rule for the parallel network first, then will extend the results in this section to the tandem or tree networks in the next section.

Suppose that the l sensor parallel decision network is given by

$$(\{\mathbf{y}_1^{(r_1)}//\mathbf{y}_2^{(r_2)}//\cdots//\mathbf{y}_{l-1}^{(r_{l-1})}\} \Rightarrow (\mathbf{y}_l \cdot \mathcal{F}) \rightarrow \{m\}). \tag{5.3.1}$$

Let us consider the special case $m = 2^n$ first, i.e., the fusion center compresses all received and observed information to a binary code of length n $(d_1, \dots, d_n)$ $(d_i = 0 \text{ or } 1, \forall i \leq n)$. This case just likes the intermediate node case in a tree network as mentioned around Eqs. (4.1.4) and (4.1.5). Since the lth sensor is also the fusion center, the general fusion rule should be given formally by

$$\begin{aligned} F(I_1^{(1)}(\mathbf{y}_1), \dots, I_1^{(r_1)}(\mathbf{y}_1); \cdots; I_{l-1}^{(1)}(\mathbf{y}_{l-1}), \dots, I_{l-1}^{(r_{l-1})}(\mathbf{y}_{l-1}); \mathbf{y}_l) :& \\ \{0,1\}^L \times \Re^{n_l} \longmapsto \{0, \dots, m-1\} = \{0,1\}^n,& \end{aligned} \tag{5.3.2}$$

where $L = \sum_{i=1}^{l-1} r_i$. Clearly, this fusion rule consists of n binary sub-fusion-rules:

$$\begin{cases} F_1(I_1^{(1)}(\mathbf{y}_1), \dots, I_1^{(r_1)}(\mathbf{y}_1); \cdots; I_{l-1}^{(1)}(\mathbf{y}_{l-1}), \dots, I_{l-1}^{(r_{l-1})}(\mathbf{y}_{l-1}); \mathbf{y}_l) : \\ \qquad \{0,1\}^L \times \Re^{n_l} \longmapsto \{0,1\}, \\ \qquad \cdots \\ F_n(I_1^{(1)}(\mathbf{y}_1), \dots, I_1^{(r_1)}(\mathbf{y}_1); \cdots; I_{l-1}^{(1)}(\mathbf{y}_{l-1}), \dots, I_{l-1}^{(r_{l-1})}(\mathbf{y}_{l-1}); \mathbf{y}_l) : \\ \qquad \{0,1\}^L \times \Re^{n_l} \longmapsto \{0,1\}. \end{cases} \tag{5.3.3}$$

Note that the above every F_j is a general/unified version of the fusion rule for an l-sensor parallel bi-decision system with the lth sensor being the fusion center. In other words, every F_j is a general/unified version of the fusion rule for the following system:

$$(\{\mathbf{y}_1^{(r_1)}//\mathbf{y}_2^{(r_2)}//\cdots//\mathbf{y}_{l-1}^{(r_{l-1})}\} \Rightarrow (\mathbf{y}_l \cdot \mathcal{F}) \rightarrow \{2\}). \tag{5.3.4}$$

Recall the basic idea and technique in Section 3.5. There we assert that when all numbers $r_1, \dots, r_{l-1}$ of the local compression rules at the first $l-1$ sensors are fixed, along with taking all 2^L different L-tuple

$$(I_1^{(1)} = d_1^{(1)}, \dots, I_1^{(r_1)} = d_1^{(r_1)}; \cdots; I_{l-1}^{(1)} = d_l^{(1)}, \dots, I_{l-1}^{(r_{l-1})} = d_{l-1}^{(r_{l-1})}),$$

every F_j in fact defines 2^L sensor compression rules at the last sensor, the lth sensor as follows (also see the proof of Theorem 3.5.2):

$$\begin{cases} I_l^{((j-1)2^L+1)}(\mathbf{y}_l) = F_j(0,0,\ldots,0;\mathbf{y}_l), \\ I_l^{((j-1)2^L+2)}(\mathbf{y}_l) = F_j(1,0,\ldots,0;\mathbf{y}_l), \\ \cdots \\ I_l^{(j2^L)}(\mathbf{y}_l) = F_j(1,1,\ldots,1;\mathbf{y}_l), \end{cases} \quad j = 1,\ldots,n. \tag{5.3.5}$$

Therefore, we have

$$\mathcal{H}_1^j = \{(\mathbf{y}_1,\ldots,\mathbf{y}_l):\ F_j(I_1^{(1)},\ldots,I_1^{(r_1)};\cdots;I_{l-1}^{(1)},\ldots,I_{l-1}^{(r_{l-1})};\mathbf{y}_l) = 1\} =$$

$$\left\{ \begin{pmatrix} \mathbf{y}_1 \\ \vdots \\ \mathbf{y}_l \end{pmatrix} : \begin{array}{c} (I_1^{(1)} = 0,\ldots,I_{l-1}^{(r_{l-1})} = 0; I_l^{((j-1)2^L+1)}(\mathbf{y}_l) = 1), \\ (I_1^{(1)} = 1, I_1^{(2)} = 0,\ldots,I_{l-1}^{(r_{l-1})} = 0; I_l^{((j-1)2^L+2)}(\mathbf{y}_l) = 1), \\ \cdots, \\ (I_1^{(1)} = 1,\ldots,I_{l-1}^{(r_{l-1})} = 1; I_l^{(j2^L)}(\mathbf{y}_l) = 1). \end{array} \right\}, \tag{5.3.6}$$

where $\mathcal{H}_1^j$ is the jth ($j \le n$) $\mathbf{H}_1$ *decision region* in n binary decision regions for the general fusion rule F given in Eq. (5.3.2). The performance of this fusion rule is equivalent to the performance of the fusion center, the lth sensor, using all compressed local binary messages from the first $l-1$ sensors and the uncompressed observation $\mathbf{y}_l$ to make n binary decisions. According to the method presented in Subsection 4.1.4, the corresponding two *decision polynomials* $P_{\mathbf{H}_1}^{(j)}$ and $P_{\mathbf{H}_0}^{(i)} = 1 - P_{\mathbf{H}_1}^{(j)}$ can be easily written down from the jth decision region $\mathcal{H}_1^j$ above. Thus, the n sub-fusion-rules define totally $n2^L$ binary sensor compression rules at the lth sensor: $(I_l^{(1)}(\mathbf{y}_l),\ldots,I_l^{(L)}(\mathbf{y}_l);\cdots;I_l^{((n-1)2^L+1)}(\mathbf{y}_l),\ldots,I_l^{(n2^L)}(\mathbf{y}_l))$. We can write the above n sub-fusion-rules as their following polynomial versions:

$$\begin{cases} F_1(I_1^{(1)}(\mathbf{y}_1),\ldots,I_1^{(r_1)}(\mathbf{y}_1);\cdots;I_{l-1}^{(1)}(\mathbf{y}_{l-1}),\ldots,I_{l-1}^{(r_{l-1})}(\mathbf{y}_{l-1});\mathbf{y}_l) \\ = P_{\mathbf{H}_1}^{(1)}(I_1^{(1)},\ldots,I_1^{(r_1)};\cdots;I_{l-1}^{(1)},\ldots,I_{l-1}^{(r_{l-1})};I_l^{(1)},\ldots,I_l^{(2^L)}), \\ \cdots \\ F_n(I_1^{(1)},\ldots,I_1^{(r_1)}(\mathbf{y}_1);\cdots;I_{l-1}^{(1)}(\mathbf{y}_1),\ldots,I_{l-1}^{(r_{l-1})}(\mathbf{y}_{l-1});\mathbf{y}_l) \\ = P_{\mathbf{H}_1}^{(n)}(I_1^{(1)},\ldots,I_1^{(r_1)};\cdots;I_{l-1}^{(1)},\ldots,I_{l-1}^{(r_{l-1})};I_l^{((n-1)2^L+1)},\ldots,I_l^{(n2^L)}). \end{cases} \tag{5.3.7}$$

It follows that the corresponding n $\mathbf{H}_0$ decision polynomials $P_{\mathbf{H}_0}^{(1)}, \dots, P_{\mathbf{H}_0}^{(n)}$ are also defined by

$$P_{\mathbf{H}_0}^{(j)} = 1 - P_{\mathbf{H}_1}^{(j)}, \quad \forall j \leq n. \tag{5.3.8}$$

Using the $2n$ polynomials, we define well the m-ary unified fusion rule as follows.

Firstly, define an 1-1 correspondence between integers $\{0, 1, \dots, m-1\}$ and n-tuple binary code $(d_1, d_2, \dots, d_n)$. For any k: $0 \leq k \leq m-1$, suppose that a given $(d_1, d_2, \dots, d_n)$ stands for k. Then we can obtain a polynomial

$$P_{\mathbf{H}_k}(I_1^{(1)}(\mathbf{y}_1), \dots, I_1^{(r_1)}(\mathbf{y}_1); \cdots, I_{l-1}^{(r_{l-1})}(\mathbf{y}_{l-1}); I_l^{(1)}(\mathbf{y}_l), \dots, I_l^{(n2^L)}(\mathbf{y}_l))$$

so that

$$\begin{aligned} \{(\mathbf{y}_1, \dots, \mathbf{y}_l) : F(\mathbf{y}_1, \dots, \mathbf{y}_l) = k\} = \{(\mathbf{y}_1, \dots, \mathbf{y}_l) : \\ P_{\mathbf{H}_k}(I_1^{(1)}, \dots, I_1^{(r_1)}; \cdots; I_{l-1}^{(1)}, \dots, I_{l-1}^{(r_{l-1})}; I_l^{(1)}, \dots, I_l^{(nL)}) = 1\}, \end{aligned} \tag{5.3.9}$$

where $P_{\mathbf{H}_k} = P_1 P_2 \cdots P_n$ and P_j is given by

$$P_j = \begin{cases} P_{\mathbf{H}_1}^{(j)} & \text{if} \quad d_j = 1, \\ P_{\mathbf{H}_0}^{(j)} & \text{if} \quad d_j = 0, \end{cases} \qquad \forall j \leq n. \tag{5.3.10}$$

Here, $P_{\mathbf{H}_1}^{(j)}$ and $P_{\mathbf{H}_0}^{(j)}$ are defined in (5.3.7) and (5.3.8) above.

In general, $m \neq 2^n$, i.e., $2^{n-1} < m < 2^n$. In this case, we still need compute $n2^L$ local compression rules at the lth sensor, the only modification in this case is to define some

$$P_{\mathbf{H}_k} = P_{k_1} P_{k_2} \cdots P_{k_{n-1}} = P_{k_1} P_{k_2} \cdots P_{k_{n-1}} P_{k_n} + P_{k_1} P_{k_2} \cdots P_{k_{n-1}} (1 - P_{k_n})$$

just for $2^n - m$ times so that the total number of $P_{\mathbf{H}_k}$ now still m other than 2^n. Noting the definitions of $P_{\mathbf{H}_1}^{(j)}$ and $P_{\mathbf{H}_0}^{(j)}$ in (5.3.7) and (5.3.8), it can be seen that the m decision regions determined by above m polynomials $P_{\mathbf{H}_k}$ and $(I_1^{(1)}(\mathbf{y}_1), \dots, I_1^{(r_1)}(\mathbf{y}_1); \dots; I_{l-1}^{(1)}(\mathbf{y}_{l-1}), \dots, I_{l-1}^{(r_{l-1})}(\mathbf{y}_{l-1}); I_l^{(1)}(\mathbf{y}_l), \dots, I_l^{(n2^L)}(\mathbf{y}_l))$ can be any possible m-subset partition; therefore, the derived above fusion rule is also the unified m-ary fusion rule.

In doing so, we do not increase communications. What we need to do now is to compute fairly more optimal local compression rules under the unique unified fusion rule. In practice, this is a tractable problem.

From the above argument, now we have the following theorem.

Theorem 5.3.1(a) Under the following parallel information structures

$$(\{\mathbf{y}_1^{(r_1)} // \mathbf{y}_2^{(r_2)} // \cdots // \mathbf{y}_{l-1}^{(r_{l-1})} // \mathbf{y}_l^{(n2^L)}\} \Rightarrow \mathcal{F} \rightarrow \{m\})$$

and

$$(\{\mathbf{y}_1^{(r_1)}//\mathbf{y}_2^{(r_2)}//\cdots//\mathbf{y}_{l-1}^{(r_{l-1})}\} \Rightarrow \{\mathbf{y}_l \cdot \mathcal{F}\} \rightarrow \{m\}),$$

the fusion rule that is constructed as aforementioned (5.3.2)–(5.3.10) is a unified version of all fusion rules.

For convenience of extending the result in Theorem 5.3.1(a) to the tandem and tree

networks, it can be rewritten as the following more general version.

Theorem 5.3.1(b) Suppose that a sensor, namely the lth sensor, can observe its own data $\mathbf{y}_l$ and receive K information bits $(I_1, I_2, \ldots, I_K)$ simultaneously, where I_i $(\forall 1 \leq i \leq K)$ may depend on observations $(\mathbf{y}_1, \mathbf{y}_2, \ldots, \mathbf{y}_{l-1})$ in any way, such as, no matter $(I_1, I_2, \ldots, I_K)$ comes from a parallel or tandem information structure. Then, the lth sensor fuses the both to m different digits (or compresses to n bits, i.e., $m = 2^n$ in this case). Finally, a unified (local/global) fusion rule for the lth is given by the above method.

Using the Theorem 5.3.1(b), immediately we have

Corollary 5.3.1. For multisensor m-ary decision system, suppose that a sensor, namely the lth sensor, can observe its own data $\mathbf{y}_l$ and receive K information bits $(I_1, I_2, \ldots, I_K)$ simultaneously, where I_i $(\forall 1 \leq i \leq K)$ may depend on some of observation data $(\mathbf{y}_1, \mathbf{y}_2, \ldots, \mathbf{y}_{l-1})$ in any way, then the lth sensor needs to send at most $n2^K$ bits to the fusion center, where n satisfies inequality $2^{n-1} < m \leq 2^n$.

5.4 UNIFIED FUSION RULE FOR TANDEM AND HYBRID NETWORKS

The above two theorems have shown the unified, therefore, the optimal fusion rule for the parallel network. In this section, we can view each level of a tandem or tree networks as a parallel network and obtain the unified local fusion rule at this level. Finally, the global unified fusion rule for whole tandem or tree networks can be composed of the unified local fusion rules at all levels.

As mentioned in the previous section, when $m = 2^n$, the fusion center combining a local sensor in parallel network can be viewed as an intermediate node in a sensor network. Using this point of view, we propose a unified fusion rule for the tandem network as follows.

Theorem 5.4.1. Under the following information structure

$$(\mathbf{y}_1^{(r_1)} \overset{(r_1)}{\longmapsto} \mathbf{y}_2^{(r_2)} \overset{(s_2)}{\longmapsto} \cdots \overset{(s_{l-2})}{\longmapsto} \mathbf{y}_{l-1}^{(r_{l-1})} \overset{(s_{l-1})}{\longmapsto} (\mathbf{y}_l^{(r_l)} \cdot \mathcal{F}) \rightarrow \{m\}),$$

where "$\overset{(s_{i-1})}{\longmapsto} \mathbf{y}_i^{(r_i)} \overset{(s_i)}{\longmapsto}$" means the ith sensor compresses own data $\mathbf{y}_i$ to $r_i = s_i 2^{s_{i-1}}$ (see Eq. (4.1.6)) bits for $\forall i : 2 \leq i \leq l-1$ ($s_1 \triangleq r_1$), the unified fusion rule can be constructed sequentially as follows: starting with the first sensor, at each sub-structure "$\overset{(s_{i-1})}{\longmapsto} \mathbf{y}_i^{(r_i)} \overset{(s_i)}{\longmapsto}$" (if $i = l$, "$\mathbf{y}_i^{(r_i)} \overset{(s_i)}{\longmapsto}$" is replaced by "$(\mathbf{y}_l^{(s_l)} \cdot \mathcal{F}) \rightarrow \{m\}$"), we determine a set of unified local fusion rules $(T_i^{(1)}, \ldots, T_i^{(s_i)})$ (or final fusion rule F_{tan}) by the constructive method for parallel network in Theorems 5.3.1(a) or 5.3.1(b), but noticing $m = 2^{s_i}$ when $i < l$ and replacing $(I_1^{(1)}, \ldots, I_1^{(r_1)}; \ldots ; I_{l-1}^{(1)}, \ldots, I_{l-1}^{(r_{l-1})})$ in parallel network by the bits $(T_{i-1}^{(1)}, \ldots, T_{i-1}^{(s_{i-1})})$ coming from the $i-1$th sensor now in the tandem network.

Proof. Using Theorem 5.3.1(b) , we know that the above fusion rule at each sub-structure "$\overset{(s_{i-1})}{\longmapsto} \mathbf{y}_i^{(r_i)} \overset{(s_i)}{\longmapsto}$" gives the unified local (or final when $i = l$) fusion rule. Therefore, at top level sensor the lth sensor, the final unified fusion rule $F_{tan}(T_{l-1}^{(1)}, \ldots, T_{l-1}^{(s_{l-1})}; \mathbf{y}_l)$ given in this theorem must be also the unified version of all final fusion rules. ■

Obviously, combining Theorems 5.3.1 and 5.4.1, one can construct the unified fusion rule for the tree network system.

Using Theorems 5.3.1 and 5.4.1, the fusion rules given in the previous Example 4.2.2 of the tandem system and Example 4.2.3 of the tree system are both actually unified fusion rules for the given information structures there.

5.5 PERFORMANCE COMPARISON OF PARALLEL AND TANDEM NETWORKS

In [60, 89], some performance comparison between the parallel and tandem networks are presented. Their main conclusion is that the tandem network is dominant in two sensor case, for over two sensor case, a dominant network does not exist in general, but the parallel network is asymptotically better than the tandem network as the number of sensors goes to infinity, the value of sensor number N at which the parallel network becomes superior is not known. In this section, we give a more rigorous theoretical analysis and more precise answer to the above problem.

We only consider over two sensor case because two sensor parallel and tandem networks in nature are the same while the second sensor in the parallel network is also the fusion center. To fairly compare the performances of various networks in the sequel, we assume the fusion center in the parallel network also a sensor, namely the lth sensor, and all other corresponding sensors in both networks transmit the same number of information bits to the fusion center (in the parallel network case) or the next sensor at upper level (in the tandem network case). For notational simplicity, further assume every sensor is one-bit sensor, in other words, we compare the performance of the following two decision networks:

$$(\{\mathbf{y}_1^{(1)}//\mathbf{y}_2^{(1)}//\cdots//\mathbf{y}_{l-1}^{(1)}\} \Rightarrow (\mathbf{y}_l \cdot \mathcal{F}) \rightarrow \{2\}) \tag{5.5.1}$$

and

$$(\mathbf{y}_1 \overset{(1)}{\longmapsto} \mathbf{y}_2 \overset{(1)}{\longmapsto} \cdots \overset{(1)}{\longmapsto} \mathbf{y}_{l-1} \overset{(1)}{\longmapsto} (\mathbf{y}_l \cdot \mathcal{F}) \rightarrow \{2\}). \tag{5.5.2}$$

Our analysis shows that in general, a dominance between the parallel and tandem networks does not exist no matter how large the number N is because there is no dominance between two sets of all possible decision regions generated by the parallel and tandem networks. In other words, there is no the relationship between the two sets such that one will contain another after N becomes big enough , i.e., there always exists some decision regions generated by either of the two networks, which cannot be generated by another one no matter how large the number N is.

According to the analysis from (4.1.7) to (4.1.8), we know that the above tandem network is equivalent to the the following tandem network:

$$(\mathbf{y}_1 \overset{(1)}{\longmapsto} \mathbf{y}_2^{(2)} \overset{(1)}{\longmapsto} \cdots \overset{(1)}{\longmapsto} \mathbf{y}_{l-1}^{(2)} \overset{(1)}{\longmapsto} (\mathbf{y}_l \cdot \mathcal{F}) \rightarrow \{2\}). \tag{5.5.3}$$

Hence, from the second sensor, each sensor actually can compress own observation to two bits first and then locally fuses them and the received a bit from the preceding local sensor to be a bit and transmit it out. However, in the above parallel network, every local sensor only compresses own observation into a bit; therefore, there always exists some decision regions generated by the tandem network, such as the $\mathbf{H}_1$ decision region

$$\mathcal{H}_1 = \{(I_1(\mathbf{y}_1) = 1, I_2^{(1)}(\mathbf{y}_2) = 1), (I_1(\mathbf{y}_1) = 0, I_2^{(2)}(\mathbf{y}_2) = 1)\} \tag{5.5.4}$$

and its decision polynomial

$$P_{\mathbf{H}_1} = I_1(\mathbf{y}_1)I_2^{(1)}(\mathbf{y}_2) + (1 - I_1(\mathbf{y}_1))I_2^{(2)}(\mathbf{y}_2), \tag{5.5.5}$$

which in general cannot be generated by the parallel network (5.5.1) because (5.5.4) and (5.5.5) cannot be constructed by the two bits $\{I_1(\mathbf{y}_1), I_2(\mathbf{y}_2)\}$.

On the other hand, the fusion center in the l-sensor parallel network can receive $l-1$ bits of information from other $l-1$ local sensors, and the fusion center in the tandem network can receive a bit of information totally from all other local sensors. It turns out that some of final decision regions which can be generated by the parallel network, but it is impossible for the tandem network.

For example, the $\mathbf{H}_1$ decision region

$$\mathcal{H}_1 = \begin{Bmatrix} (I_1(\mathbf{y}_1) = 1, I_2(\mathbf{y}_2) = 1, I_3(\mathbf{y}_3) = 1) \\ (I_1(\mathbf{y}_1) = 0, I_2(\mathbf{y}_2) = 1, I_3(\mathbf{y}_3) = 0) \\ (I_1(\mathbf{y}_1) = 1, I_2(\mathbf{y}_2) = 0, I_3(\mathbf{y}_3) = 0) \end{Bmatrix} \tag{5.5.6}$$

can be chosen by the parallel network, but cannot be generated by the tandem network since its third sensor can receive only a bit from the first 2 sensors and the fusion rule (5.5.6) requires two bits from the first two sensors to present the three different regions:

$$(I_1(\mathbf{y}_1) = 1, I_2(\mathbf{y}_2) = 1), (I_1(\mathbf{y}_1) = 0, I_2(\mathbf{y}_2) = 1),$$
$$(I_1(\mathbf{y}_1) = 1, I_2(\mathbf{y}_2) = 0).$$

Obviously, the above situation can not change even if the sensor number in networks goes to infinity. In practice, however, it is extremely rare that a better decision region only can be generated by the tandem network other than by the parallel network due to a much higher information compression in the tandem network than in the parallel network; therefore, in most practical situations, the parallel network outperforms the corresponding tandem network very often (see the numerical examples in Section 5.6). However, the tandem network clearly has more survivability than the parallel network, particularly, in the situation where the fusion center is very likely to be destroyed. Clearly, when a fusion center is destroyed, the damaged parallel network performs like a single sensor decision maker, but when the same situation happens in the tandem network, the network becomes a smaller tandem network which perform better than a single decision maker.

5.6 NUMERICAL EXAMPLES

In the following numerical examples, we consider three and four sensor two Gaussian signal detection systems.

5.6.1 THREE SENSOR SYSTEM

The signals and noises are

$$\mathbf{H}_0 : \mathbf{y}_1 = \nu_1, \qquad \mathbf{y}_2 = \nu_2, \qquad \mathbf{y}_3 = \nu_3,$$

$$\mathbf{H}_1 : \mathbf{y}_1 = s_1 + \nu_1, \mathbf{y}_2 = s_1 + \nu_2, \mathbf{y}_3 = s_1 + \nu_3,$$

$$\mathbf{H}_2 : \mathbf{y}_1 = s_2 + \nu_1, \mathbf{y}_2 = s_2 + \nu_2, \mathbf{y}_3 = s_2 + \nu_3,$$

where the two signals s_1, s_2 and the three sensor observation noises ν_1, ν_2 and ν_3 are all mutually independent, and

$$s_1 \sim N(2,3), \quad s_2 \sim N(-2,3), \quad \nu_1 \sim N(0,3), \quad \nu_2 \sim N(0,2), \quad \nu_3 \sim N(0,1).$$

Therefore, the three conditional Probability Density Function (PDF) given $\mathbf{H_0}$, $\mathbf{H_1}$ and $\mathbf{H_2}$ are respectively

$$p(\mathbf{y}_1, \mathbf{y}_2, \mathbf{y}_3 | \mathbf{H_0}) \sim N\left(\begin{pmatrix} 0 \\ 0 \\ 0 \end{pmatrix}, \begin{bmatrix} 3 & 0 & 0 \\ 0 & 2 & 0 \\ 0 & 0 & 1 \end{bmatrix}\right),$$

$$p(\mathbf{y}_1, \mathbf{y}_2, \mathbf{y}_3 | \mathbf{H_1}) \sim N\left(\begin{pmatrix} 2 \\ 2 \\ 2 \end{pmatrix}, \begin{bmatrix} 6 & 3 & 3 \\ 3 & 5 & 3 \\ 3 & 3 & 4 \end{bmatrix}\right),$$

$$p(\mathbf{y}_1, \mathbf{y}_2, \mathbf{y}_3 | \mathbf{H_2}) \sim N\left(\begin{pmatrix} -2 \\ -2 \\ -2 \end{pmatrix}, \begin{bmatrix} 6 & 3 & 3 \\ 3 & 5 & 3 \\ 3 & 3 & 4 \end{bmatrix}\right).$$

In all of the following examples, we take $c_{ij} = 1$ as $i \neq j$, $c_{ii} = 0$, $P_0 = 1/2$, $P_1 = P_2 = 1/4$. In this case, the Bayes cost functional is actually *decision error probability* P_e.

Example 5.6.1.

The information structure in this numerical example is

$$(\{\mathbf{y}_1^{(1)} // \mathbf{y}_2^{(1)} // \mathbf{y}_3^{(8)}\} \Rightarrow \mathcal{F} \rightarrow \{3\})$$

or

$$(\{\mathbf{y}_1^{(1)} // \mathbf{y}_2^{(1)}\} \Rightarrow (\mathbf{y}_3^{(8)} \cdot \mathcal{F}) \rightarrow \{3\}),$$

which can be yielded by adding 7 local compression rules at the third sensor in the information structure given in Example 4.2.1. Then the unified fusion rule by using the method proposed in Theorem 5.3.1 can be defined by the following 3 *decision polynomials*:

$$\begin{aligned}
P_{\mathbf{H}_0} &= (1 - I_1)(1 - I_2)(1 - I_3^{(1)}) + I_1(1 - I_2)(1 - I_3^{(2)}) \\
&\quad + (1 - I_1)I_2(1 - I_3^{(3)}) + I_1 I_2(1 - I_3^{(4)}), \\
P_{\mathbf{H}_1} &= (1 - P_{\mathbf{H}_0})[(1 - I_1)(1 - I_2)(1 - I_3^{(5)}) + I_1(1 - I_2)(1 - I_3^{(6)}) \\
&\quad + (1 - I_1)I_2(1 - I_3^{(7)}) + I_1 I_2(1 - I_3^{(8)})], \\
P_{\mathbf{H}_2} &= 1 - P_{\mathbf{H}_0} - P_{\mathbf{H}_1}.
\end{aligned}$$

It can be seen from the algorithm given in Section 4.4 and the above fusion rule that the following 6 initial values ($I_1, I_2, I_3^{(5)}, I_3^{(6)}, I_3^{(7)}, I_3^{(8)}$) are already enough to run the iterative algorithm. To evaluate the performance of the above fusion rule, we compare its final decision cost, denoted by P_{e_D}, with the centralized decision cost P_{e_C}.

$$P_{e_C} = 0.2157, \quad P_{e_D} = 0.2349 \quad \text{with initial values}$$

$$(I(\mathbf{y}_1), I(\mathbf{y}_2), I(\mathbf{y}_3), I(\mathbf{y}_3), I(\mathbf{y}_3), I(\mathbf{y}_3)),$$

$$P_{e_D} = 0.2365 \quad \text{with initial values}$$

$$(I(\mathbf{y}_1 - 1), I(\mathbf{y}_2 + 1), I(-\mathbf{y}_3), I(\mathbf{y}_3), I(-\mathbf{y}_3), I(\mathbf{y}_3)),$$

where $I(\cdot)$ here and below is a indicator function defined as before. From this example, we see that the optimal distributed decision (ODD) cost is a little bit bigger than that of the centralized decision (CD), but not too sensitive to the initial values in this Gaussian data example. Then, we exchange the positions of the first sensor and the third sensor in the information structure, i.e., in contrast with the original case, $\mathbf{y}_1$ with the largest noise power ($\sigma_1^2 = 3$) now is compressed to 8 bits and $\mathbf{y}_3$ with the smallest noise power ($\sigma_3^2 = 1$) is compressed to 1 bit. The numerical results with the same initial values as before are given as follows:

$$P_{e_D} = 0.2828 \quad \text{with initial values } (I(\mathbf{y}_1), I(\mathbf{y}_2), I(\mathbf{y}_3), I(\mathbf{y}_3), I(\mathbf{y}_3), I(\mathbf{y}_3)),$$

$$P_{e_D} = 0.2788 \quad \text{with initial values } (I(\mathbf{y}_1 - 1), I(\mathbf{y}_2 + 1), I(-\mathbf{y}_3), I(\mathbf{y}_3), I(-\mathbf{y}_3), I(\mathbf{y}_3)).$$

In comparison of the results of two cases, the performance of the first case is significantly better than the second one.

Example 5.6.2.

The information structure in this example is

$$(\mathbf{y}_1 \xmapsto{(1)} \mathbf{y}_2^{(2)} \xmapsto{(1)} (\mathbf{y}_3^{(4)} \cdot \mathcal{F}) \rightarrow \{3\}).$$

A unified fusion rule by Theorem 5.4.1 is just the fusion rule shown in Example 4.2.3. It can be seen from in Section 4.4 and the fusion rule that the following 5 initial values ($I_1, I_2^{(1)}, I_2^{(2)}, I_3^{(3)}, I_3^{(4)}$) are already enough to start running the iterative algorithm. The numerical results with the similar initial values as in Example 5.6.1 are given by

$$P_{e_D} = 0.2447 \text{ with initial values } (I(\mathbf{y}_1), I(\mathbf{y}_2), I(\mathbf{y}_2), I(\mathbf{y}_3), I(\mathbf{y}_3)),$$

$$P_{e_D} = 0.2555 \text{ with initial values } (I(\mathbf{y}_1 - 1), I(\mathbf{y}_2 + 1), I(-\mathbf{y}_2), I(\mathbf{y}_3), I(-\mathbf{y}_3)).$$

The numerical results show that the ODD costs here are still not far away from CD cost but bigger than the ODD cost in Example 5.6.1.

Then, similar to Example 5.6.1, we exchange the positions of the first sensor and the third sensor in the information structure. The numerical results with the same initial values as before are given as follows:

$$P_{e_D} = 0.3200 \text{ with initial values } (I(\mathbf{y}_1), I(\mathbf{y}_2), I(\mathbf{y}_3), I(\mathbf{y}_3), I(\mathbf{y}_3)),$$

$$P_{e_D} = 0.3700 \text{ with initial values } (I(\mathbf{y}_1 - 1), I(\mathbf{y}_2 + 1), I(-\mathbf{y}_3), I(\mathbf{y}_3), I(-\mathbf{y}_3)).$$

In comparison of the results of two cases, the performance of the first case is significantly better than the second one.

5.6.2 FOUR SENSOR SYSTEM

Now we add one more sensor again with sensor noise $\nu_4 \sim N(0, 0.5)$ to the above system. Therefore, the three pdfs given $\mathbf{H}_0$, $\mathbf{H}_0$ and $\mathbf{H}_1$ are respectively

$$p(\mathbf{y}_1, \mathbf{y}_2, \mathbf{y}_3, \mathbf{y}_4|\mathbf{H_0}) \sim N\left(\begin{pmatrix} 0 \\ 0 \\ 0 \\ 0 \end{pmatrix}, \begin{bmatrix} 3 & 0 & 0 & 0 \\ 0 & 2 & 0 & 0 \\ 0 & 0 & 1 & 0 \\ 0 & 0 & 0 & 0.5 \end{bmatrix}\right),$$

$$p(\mathbf{y}_1, \mathbf{y}_2, \mathbf{y}_3, \mathbf{y}_4|\mathbf{H_1}) \sim N\left(\begin{pmatrix} 2 \\ 2 \\ 2 \\ 2 \end{pmatrix}, \begin{bmatrix} 6 & 3 & 3 & 3 \\ 3 & 5 & 3 & 3 \\ 3 & 3 & 4 & 3 \\ 3 & 3 & 3 & 3.5 \end{bmatrix}\right),$$

$$p(\mathbf{y}_1, \mathbf{y}_2, \mathbf{y}_3, \mathbf{y}_4|\mathbf{H_2}) \sim N\left(\begin{pmatrix} -2 \\ -2 \\ -2 \\ -2 \end{pmatrix}, \begin{bmatrix} 6 & 3 & 3 & 3 \\ 3 & 5 & 3 & 3 \\ 3 & 3 & 4 & 3 \\ 3 & 3 & 3 & 3.5 \end{bmatrix}\right).$$

Example 5.6.3.

The information structure in this numerical example has been shown in Example 4.2.3, i.e.,

$$(\{\mathbf{y}_1^{(1)} // \mathbf{y}_2^{(1)}\} \Rightarrow \mathbf{y}_3^{(8)} \stackrel{(2)}{\longmapsto} (\mathbf{y}_4^{(8)} \cdot \mathcal{F}) \rightarrow \{3\}) \quad \text{structure}.$$

A set of the unified *decision polynomials* can be sequentially constructed as shown there.

It can be seen from Example 4.2.3 that the following 14 initial values for (I_1, I_2, $I_3^{(1)}$, $I_3^{(2)}$, ..., $I_3^{(8)}$, $I_4^{(5)}$, ..., $I_8^{(8)}$) are already enough to start running the iterative algorithm.

The numerical results with two different initial values are given as follows.

$$P_{e_C} = 0.1691, \quad P_{e_D} = 0.2202 \quad \text{with initial values}$$

$$(I(\mathbf{y}_1), I(\mathbf{y}_2), I(\mathbf{y}_3), I(\mathbf{y}_3), I(\mathbf{y}_3), I(\mathbf{y}_3), I(\mathbf{y}_3),$$
$$I(\mathbf{y}_3), I(\mathbf{y}_3), I(\mathbf{y}_3), I(\mathbf{y}_4), I(\mathbf{y}_4), I(\mathbf{y}_4), I(\mathbf{y}_4)),$$
$$P_{e_D} = 0.1916 \quad \text{with initial values}$$
$$(I(\mathbf{y}_1 + 1), I(\mathbf{y}_2 - 1), I(-\mathbf{y}_3), I(\mathbf{y}_3), I(\mathbf{y}_3), I(-\mathbf{y}_3), I(-\mathbf{y}_3),$$
$$I(\mathbf{y}_3), I(\mathbf{y}_3), I(\mathbf{y}_3), I(\mathbf{y}_4), I(\mathbf{y}_4), I(\mathbf{y}_4), I(\mathbf{y}_4)).$$

The numerical results show that the ODD costs here are still not far away from CD cost, and not too sensitive to the initial values in this Gaussian data example.

Example 5.6.4.

In order to compare the performance of the above information structure with that of the parallel sensor decision system, we consider the following information structure in this example:

$$(\{\mathbf{y}_1^{(1)} // \mathbf{y}_2^{(1)} // \mathbf{y}_3^{(1)}\} \Rightarrow (\mathbf{y}_4^{(16)} \cdot \mathcal{F}) \rightarrow \{3\}).$$

As shown in Example 5.6.1, the optimal fusion rule in this case can be constructed similarly; therefore, we omit it here. To run the iterative algorithm, one need 11 initial local compression rules (I_1, I_2, I_3, $I_4^{(9)}$, $I_4^{(10)}$, ..., $I_4^{(15)}$, $I_8^{(16)}$).

The numerical results with two different initial values are given as follows.

$$P_{e_D} = 0.1849 \quad \text{with initial values}$$

$$(I(\mathbf{y}_1), I(\mathbf{y}_2), I(\mathbf{y}_3), I(\mathbf{y}_4), I(\mathbf{y}_4), I(\mathbf{y}_4), I(\mathbf{y}_4), I(\mathbf{y}_4), I(\mathbf{y}_4), I(\mathbf{y}_4), I(\mathbf{y}_4)),$$
$$P_{e_D} = 0.1914 \quad \text{with initial values}$$
$$(I(\mathbf{y}_1 + 1), I(\mathbf{y}_2 - 1), I(-\mathbf{y}_3),$$
$$I(\mathbf{y}_4), I(\mathbf{y}_4), I(\mathbf{y}_4), I(\mathbf{y}_4), I(\mathbf{y}_4), I(\mathbf{y}_4), I(\mathbf{y}_4), I(\mathbf{y}_4)).$$

In this example, one needs communications of 3 bits and computation of 19 local compression rules. The performance looks better than that of the tree network decision system in Example 5.6.3, where communications of 4 bits and computation of 18 local compression rules are required.

5.7 OPTIMIZATION DESIGN OF NETWORK DECISION SYSTEMS

In the numerical examples given in Sections 3.6, 5.1, and 5.6, as well as, the analysis in Section 5.5, we mentioned some optimization problems on the design of network decision systems. Moreover, even if a network structure category has been chosen according to the requirement of practical applications, how to assign communication amount to every sensor under a constraint of total communication amount and how to allocate each sensor to an appropriate position in the network, in particular, which sensor should be built in the same station with the fusion center still play very significant role in improving system performance. Up to now, unfortunately, a rigorous theoretical analysis for above problems is very difficult therefore has not seen yet. Due to the importance of the problems, in this section, we propose some guides for optimization design of network structures from above numerical experiments and heuristic thought.

5.7.1 SELECTION OF A NETWORK STRUCTURE CATEGORY

In Section 5.5, we have shown that under the same communication amount constraint, in general, a dominance between the parallel and tandem networks does not exist no matter how large the number l of sensors is because there is no dominance between two sets of all possible decision regions generated by the parallel and tandem networks, in other words, there always exists some decision regions generated by either of the two networks, which cannot be generated by another one. On the other hand, the fusion center in a parallel network can receive much more information bits from local sensors than the corresponding tandem network with the same number of sensors can do. It turns out that many final decision regions which can be selected by the parallel network, but it is impossible to be chosen by the tandem network. Thus, in many cases, as shown by the numerical examples 5.6.3 and 5.6.4, the parallel network outperforms the corresponding tandem network very often. Therefore, we suggest that

- prefer the parallel network first in most cases;
- in the situation, where the system survivability problem is a serious problem, the tandem network may be more preferable;
- when the practical situation needs to properly balance the system performance and system survivability, the tree network may be a good choice.

5.7.2 ALLOCATION OF SENSORS' POSITIONS AND COMMUNICATION AMOUNTS

In many numerical examples in Sections 3.6, 5.1, and 5.6, we obtain some idea about this problem. It can be seen that less information bits are sent out from a sensor implies worse resolution rate of observation at this sensor. Naturally, to raise up system performance, we should choose less noisy sensor to send out more information bits. Mathematically, this makes sense. Suppose that the observation noise of the ith sensor is smaller. Then, the variances of marginal densities $p(\mathbf{y}_i|\mathbf{H}_j), j = 0, 1, \ldots, m-1$, should be smaller, moreover, the variances of joint densities $p(\mathbf{y}_1, \ldots, \mathbf{y}_l|\mathbf{H}_j), j = 0, 1, \ldots, m-1$, with respect to $\mathbf{y}_i$ should be also smaller, i.e., the variant rate of $p(\mathbf{y}_1, \ldots, \mathbf{y}_l|\mathbf{H}_j)$, $j = 0, 1, \ldots, m-1$, with respect to $\mathbf{y}_i$ should be higher. Therefore, it is more beneficial to the system performance for people to raise the resolution rate of $\mathbf{y}_i$. Since the most important to final system performance is the resolution rate of observational data for the fusion center, relevant positions of every sensors to the fusion center are more significant than the resolution rate of sensors.

Summing up, we have the following suggestions for the optimization design of the network decision systems:

- Choose the parallel network decision system in general case whenever the system survivability is not a serious concern. Otherwise, consider the tandem or tree network systems.
- Build the fusion center and the sensor with the smallest noise power at the same station. In the tandem or tree network systems, the smaller the noise power of a sensor is, the closer to the fusion center the position of this sensor is.
- Accordingly, the smaller the noise power of a sensor is, the more communications are assigned to this sensor.

Under various constraints and practical requirements, the optimization design of the communication pattern in the network decision systems is really a very interesting problem, and there have been few rigorous analytic results so far.

II

ESTIMATION FUSION

In the second part of the book, we concentrate ourselves on the multiple sensor/sub-processor estimation fusion problem. In practical data processing, it is very often that people have to deal with the multisensor multi-object estimation problem. In this situation, the first thing is to determine data of which sensors are the observations of the same object, i.e., to derive an optimal match of the observed data and the estimated objects in terms of a certain criterion. This research subject is called the *data association* that has been investigated in the existing literature, such as [3, 6, 22, 23, 41, 94] among others. This is also an N-P hard problem. This book will not cover this subject at all. After the data association is completed, our task is how to optimally estimate parameters or states of a common system based on the observations received by multiple sensors. In doing so, there are two strategies to choose. First one is usually called the *measurement fusion*. Every sensor receive and transmit data, does not do any preprocessing. Then, due to limited computer load and computation capacity, the fusion center must use available methods to real-time estimate the object. In the existing literature, most research results focus on how to linearly compress data to obtain the optimal estimate as the same as using all uncompressed sensor observations (see [3, 6, 94]). Of course, there must be some restrictive assumptions in order for the above results to hold. For example, no limit of communication bandwidth exists, and a multisensor estimation mathematical model can be formulated well. When the limitation of communication bandwidth exists, how to optimally pre-compress sensor observations before sensors' transmission to obtain a constrained optimal estimation at the fusion center or what are minimum communications under an allowed estimation error constraint seems received much less attention. These issues may have more practical values in the multisensor measurement fusion.

The second strategy is two level optimization in the estimation fusion. Every sensor is also a sub-processor. It first optimally estimate the object based on its own observation in terms of a criterion, and then transmits its estimate to the fusion center. Finally, the fusion center optimally estimate the object based on all received sensor estimates. Clearly, the second one has significant practical value, such as, it consumes less communication bandwidth, balances the local purpose with the global purpose fairly, has more survivability in a war situation. The second part of the book will pay attention only to the second category of the estimation fusion.

Chapter 6

MULTISENSOR POINT ESTIMATION FUSION

In this chapter, we are concerned with the multisensor point estimation fusion which was received most attention in the estimation fusion area in the past a few decades (see [3], [6], and [94]). A popular purpose to pursue in those research results is to find some frameworks of the distributed estimation fusion whose performance are the same as the Kalman filtering using all unprocessed sensor observations, which is usually called the *centralized* estimation fusion. Therefore, the assumptions for the centralized Kalman filtering must be a part of the assumptions of those estimation fusion results as well. This is one of the reasons why the mutual independence of all sensor observation noises becomes a common assumption in those results.

In this chapter, a general version of the linear unbiased estimation fusion rule is developed. It has the minimum variance of estimation error or least mean-square (LMS) error among all linear unbiased estimation fusion rules. It is very general – it relies only on two assumptions: **I.** the local estimators are unbiased[1] and **II.** the error covariance matrix of all local estimates is known. Not only does it include previous estimation fusion results as special cases, but it is also valid for many more general cases, including (A) coupled measurement noises across sensors; (B) sophisticated network structures or communication patterns; (C) different local dynamic models or estimator types; and (D) efficient fusion of asynchronous estimates. Therefore, this chapter gives more thorough solution to the optimal estimation fusion problem than the previous work.

[1] In Subsection 6.2.4, the unbiasedness can be relaxed to some degree.

6.1 PREVIOUS MAIN RESULTS

Although there were extensive results on this area, most of them are concerned with practical estimation methods, the rigorous theoretical results were not too many. In this section, the previous main theoretical results on the estimation fusion given in [4, 5, 22, 23, 41]) will be presented below.

Consider a linear dynamic system

$$\mathbf{x}_{k+1} = \Phi_k \mathbf{x}_k + v_k \tag{6.1.1}$$

with additive zero-mean white noise

$$E[v_k] = 0, E[v_k v_j'] = Q_k \delta_{kj}, \ \delta_{kj} = \begin{cases} 1 \text{ if } k = j, \\ 0 \text{ if } k \neq j, \end{cases}$$

and l sensor noisy measurements

$$\mathbf{y}_k^{(i)} = H_k^{(i)} \mathbf{x}_k + w_k^{(i)}, \quad \forall i \leq l, \tag{6.1.2}$$

where the measurement noise $w_k^{(i)}$ is zero-mean white noises, uncorrelated with the process noise:

$$E[w_k^{(i)}] = 0, \quad E[w_k^{(i)} w_j^{(i)\prime}] = \sigma_k^{(i)} \delta_{kj}, \quad E[v_k w_j^{(i)\prime}] = 0,$$

and $w_k^{(i)}$s are uncorrelated across sensors:

$$E[w_k^{(i)} w_k^{(j)\prime}] = 0.$$

This model and the above assumptions were used in the early work of the measurement fusion (see [94] among others) and Bar-Shalom's two sensor track-to-track fusion and Chong, Chang and Mori's multiple estimation fusion [4, 5, 22, 23, 41].

In Bar-Shalom's two sensor track-to-track fusion, two sensors' state estimates $\mathbf{x}_{k|k}^{(1)}$ and $\mathbf{x}_{k|k}^{(2)}$ need to be fused. The fusion formulae are, using the standard notation, given by

$$\mathbf{x}_{k|k} = W_k^{(1)} \mathbf{x}_{k|k}^{(1)} + W_k^{(2)} \mathbf{x}_{k|k}^{(2)}, \tag{6.1.3}$$

where the fusion weights are defined as

$$W_k^{(1)} = (P_{k|k}^{(11)} + P_{k|k}^{(22)} - P_{k|k}^{(12)} - P_{k|k}^{(21)})^{-1} (P_{k|k}^{(22)} - P_{k|k}^{(12)}) \tag{6.1.4}$$

and

$$W_k^{(2)} = (P_{k|k}^{(11)} + P_{k|k}^{(22)} - P_{k|k}^{(12)} - P_{k|k}^{(21)})^{-1} (P_{k|k}^{(11)} - P_{k|k}^{(21)}), \tag{6.1.5}$$

where $P_{k|k}^{(ij)}$ is the (ij)th sub-matrix of the estimation error covariance matrix $P_{k|k}$ for the vector $(\mathbf{x}_{k|k}^{(1)\prime} \quad \mathbf{x}_{k|k}^{(2)\prime})'$. Clearly,

$$W_k^{(1)} + W_k^{(2)} = I.$$

The analysis of the optimality of the above fusion formulae can be found in [4, 5], and an alternative proof (prove the fusion (6.1.3)–(6.1.3) to be a special case of Theorem 6.2.2) will be given by Corollary 6.2.1. In general, the performance of the above fusion worse than the centralized fusion since it uses the local estimates which are the processed information other than all of the original sensor observations.

The fusion equations presented in [22, 41] use the fusion center's one-step prediction as well as the l sensors' state estimates and their one-step predictions

$$\mathbf{x}_{k|k-1}; \quad \mathbf{x}_{k|k}^{(1)}, \quad \mathbf{x}_{k|k-1}^{(1)}; \quad \cdots; \quad \mathbf{x}_{k|k}^{(l)}, \quad \mathbf{x}_{k|k-1}^{(l)}$$

to derive a final fusion $\mathbf{x}_{k|k}$ of the l local state estimates. The fusion formulae are, using their notations, given by

$$P^{-1}(k|k) = P^{-1}(k|k-1) + \sum_{i=1}^{l} [P_i^{-1}(k|k) - P_i^{-1}(k|k-1)] \qquad (6.1.6)$$

and

$$\begin{aligned} \mathbf{x}_{k|k} = & \; P(k|k)\{P^{-1}(k|k-1)\mathbf{x}_{k|k-1} \\ & + \textstyle\sum_{i=1}^{l} [P_i^{-1}(k|k)\mathbf{x}_{k|k}^{(i)} - P_i^{-1}(k|k-1)\mathbf{x}_{k|k-1}^{(i)}]\}. \end{aligned} \qquad (6.1.7)$$

That is to say, the weighting matrices are

$$\begin{aligned} &[P(k|k)P^{-1}(k|k-1), \;\; P(k|k)P_1^{-1}(k|k), \;\; \ldots, \;\; P(k|k)P_l^{-1}(k|k), \\ &-P(k|k)P_1^{-1}(k|k-1), \;\; \ldots, \;\; -P(k|k)P_l^{-1}(k|k-1)]. \end{aligned} \qquad (6.1.8)$$

From (6.1.6) and (6.1.8), it is easy to see the weighting matrices satisfy

$$P(k|k)P^{-1}(k|k-1) + P(k|k)\sum_{i=1}^{l} [P_i^{-1}(k|k) - P_i^{-1}(k|k-1)] = I. \qquad (6.1.9)$$

In [22, 41], they proved rigorously that the above fusion formulae (6.1.6)–(6.1.8) can be converted equivalently from the corresponding centralized Kalman filtering (see (6.3.1)–(6.3.19) in Subsection 6.3.1 below). This provides an interesting message that it is possible to fuse the processed estimates without the knowledge of mean and variance of the state under estimation to achieve the performance of the linear minimum variance of estimation error with those knowledge. What is the conditions to let this happen? we will give a theoretical analysis for this question in Theorems 6.2.4 and 6.2.5. Besides, It must be emphasized that in the above two previous fusion formulae, mutual dependence of the sensor measurement noises is a necessary condition, which may not be satisfied in some practical applications. We will present the optimal estimation fusion formulae without this restriction in Subsection 6.2.

6.2 LINEAR MINIMUM VARIANCE ESTIMATION FUSION

A general version of the linear minimum variance (LMV) of estimation error[2] (other than the minimum variance of the estimator itself) fusion rule will be developed (cf. [104, 51]) in this section. It has the LMV error among all linear unbiased estimation fusion rules. It is very general — it relies only on two assumptions: (1) the local estimators are unbiased and (2) the error covariance matrix C_k of all local estimates at each time k is known. Not only does it include those fusion results in (6.1.3)–(6.1.5) and (6.1.6)–(6.1.7) as special cases, but it is also valid for many more general cases, including (a) coupled measurement noises across sensors; (b) sophisticated network structures or communication patterns; (c) different local dynamic models or estimator types; and (d) efficient fusion of asynchronous estimates.

First, we formulate the problem of the distributed estimation fusion in a general, i.e., the LMV setting. In this setting, the fused estimator is a weighted sum of local estimates with a matrix weight. We show that the set of weights is optimal if and only if it is a solution of a matrix quadratic optimization problem subject to a linear equality constraint.

Secondly, we present a general solution to the above optimization problem, which depends only on the covariance matrix C_k. We also show that the solution of the optimization problem is unique no matter C_k is nonsingular or not.

Thirdly, we present an explicit formula for the optimal weights that only depend on the block sub-matrices of C_k^{-1}. In this case where C_k is invertible, the explicit formula of the optimal weights is actually the well-known optimal weighted least squares (OWLS) estimation using all sensor estimations as the observations of a linear model.

If a priori information, the expectation and covariance, of the estimated quantity is unknown, a necessary and sufficient condition for the above LMV fusion becoming the best unbiased LMV estimation with known the above prior

[2] In some of existing papers, such as [51, 52, 102], the linear minimum variance (LMV) estimation was called the best linear unbiased estimation (BLUE).

information will be presented. We also discuss the generality and usefulness of the fusion formulae developed.

As a practical application, for a class of multisensor linear systems with coupled measurement noises we provide an off-line recursion of C_k which permits asynchronous sensor estimates.

6.2.1 FORMULATION OF THE LMV FUSION AS AN OPTIMIZATION PROBLEM

Consider an l-sensor distributed estimation system. Denote an N-dimensional state sequence of a dynamic system by $\{\mathbf{x}_k\}$, and the corresponding sequence of local unbiased estimates of the state sequence based on all received data at the kth stage by $\{\hat{\mathbf{x}}_k^{(i)},\ i = 1, \ldots, l\}$. Assume that the following estimation error covariance matrix is available

$$C_k = \begin{bmatrix} C_k^{(11)} & \cdots & C_k^{(1l)} \\ & \cdots & \\ C_k^{(l1)} & \cdots & C_k^{(ll)} \end{bmatrix}, \tag{6.2.1}$$

where l is the number of local estimates to be fused and

$$C_k^{(ij)} = E[(\hat{\mathbf{x}}_k^{(i)} - \mathbf{x}_k)(\hat{\mathbf{x}}_k^{(j)} - \mathbf{x}_k)']. \tag{6.2.2}$$

Note that $C_k^{(ij)}$ can be calculated recursively in many cases (see [4, 5] and Subsection 6.2.4 below). However, for some practical problem, especially, different local dynamic models or estimator types, $C_k^{(ij)}$ may not be calculated easily. To approach an optimal result, however, we should try all possible ways to estimate $C_k^{(ij)}$ as precisely as possible.

Given a set of the unbiased "local estimates" $\{\hat{\mathbf{x}}_k^{(1)}, \ldots, \hat{\mathbf{x}}_k^{(l)}\}$ [3], we want to find an optimal fused estimator in the LMV error sense:

$$\hat{\mathbf{x}}_k = B_k + W_k' X_k, \tag{6.2.3}$$

where B_k and W_k are both compatible constant vector and matrix, respectively, and X_k is denoted by

$$X_k = \begin{bmatrix} \hat{\mathbf{x}}_k^{(1)} \\ \vdots \\ \hat{\mathbf{x}}_k^{(l)} \end{bmatrix}, \quad W_k = \begin{bmatrix} W_k^{(1)} \\ \vdots \\ W_k^{(l)} \end{bmatrix}, \quad W_k \in \Re^{(lN)\times N}. \tag{6.2.4}$$

[3] Although the term "local estimates" is used, this set of estimates is not necessarily obtained based on all *distinct* sensor data, e.g., the local estimates could be the estimates of the same sensor at different time instants. This will be clear later.

That is, $\hat{\mathbf{x}}_k$ has the minimum mean-square error among all choices of B_k and W_k that guarantee unbiasedness. Taking expectation of (6.2.3) yields, by the unbiasedness of the fused and local estimators,

$$E(\mathbf{x}_k) = B_k + W_k' E(X_k) = B_k + \sum_{i=1}^{l} W_k^{(i)\prime} E(\mathbf{x}_k).$$

Therefore, we have

$$B_k + (\sum_{i=1}^{l} W_k^{(i)\prime} - I)E(\mathbf{x}_k) = 0.$$

In order for this equation to hold for every possible $E(\mathbf{x}_k)$, a necessary and sufficient condition is

$$B_k = 0, \quad \sum_{i=1}^{l} W_k^{(i)\prime} = I,$$

or

$$B_k = 0, \quad AW_k = I, \qquad \text{with } A = [I \cdots I]. \tag{6.2.5}$$

Thus

$$\hat{\mathbf{x}}_k = W_k' X_k, \tag{6.2.6}$$

and the LMV estimation fusion problem can reduce to one of a matrix quadratic optimization problem subject to a linear equality constraint:

$$W_k = \arg\min_{W} E[(W'X_k - \mathbf{x}_k)(W'X_k - \mathbf{x}_k)'] \tag{6.2.7}$$

subject to

$$AW_k = I \tag{6.2.8}$$

for some $W_k \in \Re^{(lN)\times N}$. Note that $W_k^{(i)}$ is the matrix-valued weight for the estimate $\hat{\mathbf{x}}_k^{(i)}$ and the above implies that the fusion weight satisfies the linear-equality constraint $AW_k = I$.

The error covariance of the fused estimate associated with a weighting matrix W_k is

$$\begin{aligned} P_k &= E[(W_k'X_k - \mathbf{x}_k)(W_k'X_k - \mathbf{x}_k)'] \\ &= E[(W_k'X_k - W_k'A'\mathbf{x}_k)(W_k'X_k - W_k'A'\mathbf{x}_k)'] \\ &= W_k' C_k W_k. \end{aligned} \tag{6.2.9}$$

Substituting this equation into (6.2.7) and (6.2.8) yields

$$\begin{aligned} W_k &= \arg\min\nolimits_{AW=I} W'C_kW \\ P_k &= \min\nolimits_{AW=I} W_k'C_kW_k. \end{aligned} \tag{6.2.10}$$

In other words, a necessary and sufficient condition of the LMV estimation fusion is that the optimal weight is a solution of the above quadratic optimization problem subject to a linear-equality constraint.

It is easy to see that previous results (6.1.3)–(6.1.5) and (6.1.6)–(6.1.9) on the distributed estimation fusion satisfy the linear equality constraint $AW_k = I$.

Note that the "local estimates" do not have to be obtained from different local filters or from entirely different set of data as shown in the fusion formulae (6.1.6)-(6.1.9) provided they are unbiased estimates of the same quantity.

6.2.2 OPTIMAL FUSION WEIGHTS

It should be recognized that the optimization problem (6.2.10) is actually a matrix linear least-squares problem subject to a linear equality constraint. A number of solution methods and algorithms are available for such problems (see e.g., [49, 8, 107, 31, 95] and the references therein). Some of them are precise and others are approximate. Some are numerically more efficient than others.

We now derive the most general version of the optimal weights given by (6.2.10) without any assumption using a method we developed for the general linear least-squares problem with linear constraints, presented in [107]. This method is based on the pseudo-inverse technique and the perfect square method.

Theorem 6.2.1. The general solution of (6.2.10) is

$$W_k = \frac{1}{l}[I - (PC_kP)^+C_k]A' + PZ, \tag{6.2.11}$$

where $P = I - \frac{1}{l}A'A$ and $Z \in \Re^{(lN)\times N}$ is an arbitrary matrix satisfying $C_k^{\frac{1}{2}\prime}PZ = 0$. The covariance of the estimation error is

$$R_{LMV} = \frac{1}{l^2}A[I - C_k(PC_kP)^+]C_kA'. \tag{6.2.12}$$

Proof. It is well-known (see [7]) that

$$W = A^+ + P\xi, \qquad \forall\xi \in \Re^{(lN)\times N} \tag{6.2.13}$$

is the general solution of the following matrix equation

$$AW = I.$$

It is straightforward to show from the basic properties of the matrix pseudo-inverse directly [7] that

$$C_k^{\frac{1}{2}\prime}P(PC_kP)^+PC_kP = C_k^{\frac{1}{2}\prime}P(C_k^{\frac{1}{2}\prime}P)^+(PC_k^{\frac{1}{2}})^+PC_k^{\frac{1}{2}}C_k^{\frac{1}{2}\prime}P = C_k^{\frac{1}{2}\prime}P. \tag{6.2.14}$$

Substituting the last three equations above into the right side of (6.2.9), we have

$$
\begin{aligned}
&[(A^{+'} + \xi' P)C_k^{\frac{1}{2}}][(A^{+'} + \xi' P)C_k^{\frac{1}{2}}]' \\
&= [\xi' + A^{+'} C_k^{\frac{1}{2}} C_k^{\frac{1}{2}'} P(PC_kP)^+] PC_kP [\xi' + A^{+'} C_k^{\frac{1}{2}} C_k^{\frac{1}{2}'} P(PC_kP)^+]' \\
&\quad + A^{+'} C_k A^+ - A^{+'} C_k^{\frac{1}{2}} C_k^{\frac{1}{2}'} P(PC_kP)^+ PC_k^{\frac{1}{2}} C_k^{\frac{1}{2}'} A^+.
\end{aligned}
$$

Clearly, minimizing the quadratic objective function amounts to making

$$[\xi' + A^{+'} C_k P(PC_kP)^+] PC_kP [\xi' + A^{+'} C_k P(PC_kP)^+]' = 0.$$

Using the pseudo-inverse technique (see e.g., [107]), it can be shown that

$$P(PC_kP)^+ = (PC_kP)^+ P = (PC_kP)^+. \tag{6.2.15}$$

Hence, we should take

$$\xi = -(PC_kP)^+ C_k A^+ + Z, \ \forall Z : C_k^{\frac{1}{2}'} PZ = 0. \tag{6.2.16}$$

The theorem thus follows from (6.2.13), (6.2.16), (6.2.15), (6.2.12), and $A^+ = \frac{1}{l} A'$. ■

Now we claim that the optimal fusion $W_k' X_k$ is actually unique although the solution of W_k given in Theorem 6.2.1 may not be unique.

Theorem 6.2.2. The optimal fusion $W_k' X_k$ is given uniquely by

$$W_k' X_k = \tfrac{1}{l} A(I - C_k(PC_kP)^+) X_k \quad \text{almost surely}. \tag{6.2.17}$$

Proof. Noticing the definition of C_k and Eqs. (6.2.16), (6.2.8), for any Z in Theorem 6.2.1, we have

$$E[Z'P(X_k - A'\mathbf{x}_k)(X_k - A'\mathbf{x}_k)'PZ] = Z'PC_kPZ = 0. \tag{6.2.18}$$

Therefore,

$$Z'P(X_k - A'\mathbf{x}_k) = 0 \quad \text{almost surely}, \tag{6.2.19}$$

moreover, using Theorem 6.2.1, identity $PA' = l(I - A^+A)A^+ = 0$, and the above equation, we have

$$
\begin{aligned}
W_k' X_k &= \tfrac{1}{l} A(I - C_k(PC_kP)^+) X_k + Z'P(X_k - A'\mathbf{x}_k + A'\mathbf{x}_k) \\
&= \tfrac{1}{l} A(I - C_k(PC_kP)^+) X_k.
\end{aligned} \tag{6.2.20}
$$

The theorem follows. ■

From the above theorem, we can view W_k as the following unique solution:

$$W_k = \frac{1}{l}(I - (PC_kP)^+C_k)A'. \tag{6.2.21}$$

By the definition of A and Theorem 6.2.2, we have a clear expression for each of its elements $W_k^{(i)}$ immediately.

Denote by $M_{ij}^{(+)}$ the (i,j)th sub-block $(N \times N)$ matrix of $\frac{1}{l}(I-(PC_kP)^+C_k)$. Then

$$W_k^{(i)} = \sum_{j=1}^{l} M_{ij}^{(+)}. \tag{6.2.22}$$

Moreover, if C_k has full rank, we have a more explicit expression of $W_k^{(i)}$ that depends only on C_k^{-1} explicitly.

Theorem 6.2.3. If C_k has full rank, we have the following explicit expression of W_k:

$$W_k = C_k^{-1}A'(AC_k^{-1}A')^{-1}. \tag{6.2.23}$$

I.e., each element $W_k^{(i)}$ of W_k is given by

$$W_k^{(i)} = \sum_{j=1}^{l} C_{ij}^{(-1)} \Big(\sum_{r,j=1}^{l} C_{rj}^{(-1)} \Big)^{-1}, \tag{6.2.24}$$

where $C_{rj}^{(-1)}$ is the (r,j)th sub-matrix of C_k^{-1}. Besides, the covariance of the final estimation error is given by

$$P_k = (AC_k^{-1}A')^{-1} = \Big(\sum_{i,j=1}^{l} C_{ij}^{(-1)} \Big)^{-1}. \tag{6.2.25}$$

Proof. First of all, we know $AC_k^{-1}A'$ is nonsingular since A has full row rank. Then we can easily verify the following identity

$$\begin{aligned} &\min_{AW=I} W'C_kW \\ \overset{(6.2.23)}{=} &\min_{AW=I} [(W_k - W)'C_k(W_k - W) + (AC_k^{-1}A')^{-1}], \end{aligned}$$

where W_k is given in (6.2.23). Clearly, $AW_k = AC_k^{-1}A'(AC_k^{-1}A')^{-1} = I$. The theorem follows. ■

In fact, we can also obtain the fusion weight given in (6.2.23) directly from (6.2.21) as follows.

Using $A'A = lA^+A = l(I - P)$ and the identities (6.2.15) and (6.2.14), we have

$$C_k(PC_kP)^+C_kP = C_kP(PC_kP)^+PC_kP = C_kP$$

and

$$\begin{aligned} & C_k \cdot \tfrac{1}{l}[I - (PC_kP)^+C_k]A' \cdot (AC_k^{-1}A') \\ =& C_k[I - (PC_kP)^+C_k](I - P)C_k^{-1}A' \\ =& C_k[I - P - (PC_kP)^+C_k + (PC_kP)^+C_kP]C_k^{-1}A' \\ =& [I - C_kPC_k^{-1} - C_k(PC_kP)^+ + C_k(PC_kP)^+C_kPC_k^{-1}]A' \\ =& [I - C_k(PC_kP)^+P]A' \\ =& A'. \end{aligned} \tag{6.2.26}$$

Thus, by (6.2.26) and (6.2.23), we know

$$\frac{1}{l}(I - (PC_kP)^+C_k)A' = C_k^{-1}A'(AC_k^{-1}A')^{-1}.$$

It is easy to verify that the optimal fusion formula in Theorem 6.2.3 is also the Optimal Weighted Least Squares (OWLS) estimation of the following linear model:

$$\begin{cases} \hat{\mathbf{x}}_k^{(1)} = \mathbf{x}_k + \hat{\mathbf{x}}_k^{(1)} - \mathbf{x}_k, \\ \quad\cdots \\ \hat{\mathbf{x}}_k^{(l)} = \mathbf{x}_k + \hat{\mathbf{x}}_k^{(l)} - \mathbf{x}_k. \end{cases} \tag{6.2.27}$$

Then, by the definitions of the two notations A and X_k in (6.2.5) and (6.2.4), the matrix version of the above model becomes

$$X_k = A'\mathbf{x}_k + \omega, \tag{6.2.28}$$

where the "noise" $\omega = X_k - A'\mathbf{x}_k$, and its covariance R_ω was denoted by C_k in (6.2.1) and (6.2.2).

We must emphasize that $\mathbf{x}_k$ here is random. When it is deterministic, there were many discussions about its estimates in the existing literature, for example, see Ben-Israel and Greville's book [7] and the references among others.

6.2.3 EFFICIENCY OF THE LMV FUSION

The above fused estimator is the LMV only based on its own available unbiased local estimates (observation X_k). In general, it is not the following best LMV estimate (see, e.g., [18], by treating X_k as the observation in the standard formulation):

$$\hat{\mathbf{x}}_{best} = E(\mathbf{x}_k) + R_{\mathbf{x}X_k} R^{+}_{X_k}(X_k - A'E(\mathbf{x}_k)), \tag{6.2.29}$$

where

$$\begin{aligned} R_{X_k} &= E[(X_k - E(X_k))(X_k - E(X_k))'] \\ &= E[(X_k - A'E(\mathbf{x}_k))(X_k - A'E(\mathbf{x}_k))'], \end{aligned} \tag{6.2.30}$$

$$R_{\mathbf{x}X_k} = E[(\mathbf{x}_k - E(\mathbf{x}_k))(X_k - A'E(\mathbf{x}_k))']. \tag{6.2.31}$$

If $E(\mathbf{x}_k)$, R_X and $R_{\mathbf{x}X}$ are all known, using (6.2.29)–(6.2.31), the best LMV fused estimator is given by

$$\hat{\mathbf{x}}_{best} = (I - \bar{W}'_k A')E(\mathbf{x}_k) + \bar{W}'_k X_k, \tag{6.2.32}$$

where the optimal weighting matrix is given by

$$\bar{W}_k = R^{+}_{X_k} R_{X_k \mathbf{x}}. \tag{6.2.33}$$

The covariance associated with the fused estimator is given by

$$P_k = Cov(\mathbf{x}_k) - \bar{W}'_k R_{X_k} \bar{W}_k = Cov(\mathbf{x}_k) - R_{\mathbf{x}X_k} R^{+}_{X_k} R_{X_k \mathbf{x}}. \tag{6.2.34}$$

Note that in many practical situations, some of $E(\mathbf{x}_k)$, $R_{X_k\mathbf{x}}$, and R_{X_k} are not known; therefore, the best LMV fusion (6.2.29) is not applicable. In general, the estimation error covariance of our LMV fusion may be larger than P_k. However, in some cases, it is possible for our LMV fusion to be the same as the P_k given in (6.2.34). For example, the fusion given by (6.1.6)–(6.1.9) is equivalent to the centralized fusion, but it is a special example of our LMV fusion that does not use $E\mathbf{x}_k$, R_{X_k} and $R_{\mathbf{x}X_k}$.

By (6.2.32)–(6.2.34) and the following (6.2.35), a sufficient condition for our LMV fusion, without using $E\mathbf{x}_k$, R_{X_k} and $R_{\mathbf{x}X_k}$, being the fusion (6.2.32)–(6.2.34) is given straightforwardly by

$$AR^{+}_{X_k} R_{X_k \mathbf{x}} = I, \tag{6.2.35}$$

since $\hat{\mathbf{x}}_{best} = \bar{W}'_k X_k$ and $A\bar{W}_k = I$. The above sufficient condition depends on the covariance matrix R_{X_k} of all local estimations that we suppose unknown. Therefore, this sufficient condition needs to be improved. In the following theorem, we present the necessary and sufficient condition depending on C_k and the cross covariance matrix $R_{\omega\mathbf{x}}$ of all local estimation errors and the true state ($R_{\omega\mathbf{x}}$ can be known in some cases, such as the fusion given by (6.1.6)–(6.1.9)).

Theorem 6.2.4. The optimal fusion given in (6.2.29)–(6.2.34) becomes our fusion (6.2.21) given in Theorem 6.2.2 if and only if

$$\frac{1}{l} A[I - C_k(PC_kP)^{+}](C_k + R_{\omega\mathbf{x}}A) = 0, \tag{6.2.36}$$

where

$$R_{\omega\mathbf{x}} = E(X_k - A'\mathbf{x}_k)(\mathbf{x}_k - E\mathbf{x}_k)'. \tag{6.2.37}$$

Proof. Since the two types of the optimal fusion are both unbiased and unique, the problem reduces to prove the covariance of their difference is zero. By (6.2.29), (6.2.17), and $Cov(a + x) = Cov(x)$, we know

$$\begin{aligned} &Cov(\hat{\mathbf{x}}_{best} - W_k' X_k) \\ &= Cov[E(\mathbf{x}_k) + R_{\mathbf{x}X_k} R_{X_k}^{+}(X_k - A'E(\mathbf{x}_k)) - \tfrac{1}{l}A(I - C_k(PC_kP)^{+})X_k] \\ &= Cov[(R_{\mathbf{x}X_k} R_{X_k}^{+} - \tfrac{1}{l}A(I - C_k(PC_kP)^{+})X_k]. \end{aligned} \tag{6.2.38}$$

Therefore, $Cov(\hat{\mathbf{x}}_k - W_k' X_k) = 0$ if and only if

$$(R_{\mathbf{x}X_k} R_{X_k}^{+} - \frac{1}{l}A(I - C_k(PC_kP)^{+})R_{X_k} = 0, \tag{6.2.39}$$

i.e.,

$$\frac{1}{l}A(I - C_k(PC_kP)^{+})R_{X_k} = R_{\mathbf{x}X_k} R_{X_k}^{+} R_{X_k} = R_{\mathbf{x}X_k}. \tag{6.2.40}$$

Thus, by the definitions of R_{X_k} and $R_{\mathbf{x}X_k}$, and the well-known identities $(PC_kP)^{+} = (PC_kP)^{+}P$ and $PA' = 0$, it is easy to verify that Eq. (6.2.40) is equivalent to Eq. (6.2.36). ■

If C_k is invertible, we can derive the following simpler necessary and sufficient conditions for the equivalence of the two types of optimal fusion.

Theorem 6.2.5. Suppose that C_k is invertible. The optimal fusion given in (6.2.29)–(6.2.34) becomes our fusion (6.2.24) given in Theorem 6.2.3 if and only if

$$AC_k^{-1}R_{\omega\mathbf{x}} = -I. \tag{6.2.41}$$

Proof. Using Theorems 6.2.2 and 6.2.3, we have

$$\frac{1}{l}A[I - C_k(PC_kP)^{+}] = (AC_k^{-1}A')^{-1}AC_k^{-1}.$$

From Theorem 6.2.4, we know that the necessary and sufficient condition here is equivalent to the following equality

$$\frac{1}{l}A[I - C_k(PC_kP)^{+}](C_k + R_{\omega\mathbf{x}}A) = 0. \tag{6.2.42}$$

Thus, combining the above two equalities yields

$$A + AC_k^{-1}R_{\omega\mathbf{x}}A = 0.$$

Noting $AA' = \frac{1}{l}I$ and the above equality, condition (6.2.42) now reduces to (6.2.41), i.e., the theorem holds. ■

Obviously, to make (6.2.41) true, $R_{\omega\mathbf{x}}$ must be of column full rank. reader can verify that the fusion given by (6.1.6)–(6.1.9) satisfies the above condition (6.2.41).

6.2.4 EXTENSION TO A MORE GENERAL MODEL

In the above the linear model (6.2.27) and (6.2.28), the sensor estimates $\{\hat{\mathbf{x}}_k^{(1)}, \ldots, \hat{\mathbf{x}}_k^{(l)}\}$ are viewed as the observations of a specific linear model. Naturally, we can extend the above fusion results to more general linear model

$$\begin{cases} \mathbf{y}^{(1)} = H_1\mathbf{x}_k + \omega_1, \\ \qquad \cdots \\ \mathbf{y}^{(l)} = H_l\mathbf{x}_k + \omega_l. \end{cases} \tag{6.2.43}$$

Then, the matrix version of the above model becomes

$$Y = H\mathbf{x}_k + \xi, \tag{6.2.44}$$

where the observation vector Y, noise vector ξ, and matrix H are the stacked $\mathbf{y}^{(i)}$, ω_i, and H_i, $i \le l$, respectively. In comparison of the model (6.2.27) and (6.2.28) with the model (6.2.43) and (6.2.44), the general observation Y, matrix H, and noise ξ replace the sensor estimates X_k, special matrix A, and estimate errors ω. Denote the covariance of noise ξ by C, $E\xi = 0$. To guarantee $W'Y$ is unbiased, the weight matrix W satisfies constraint $W'H = I$. Such W exists if and only if H has full column rank, or equivalently, $H'(H')^+ = I$ by using the properties of matrix pseudo-inverse. This assumption is reasonable from the practical point of view. To get an unbiased state estimate, the observations/local-estimates of all sensors must contain the information of all components of the state even if individual local observation/estimate might not satisfy this. Therefore, H must have full column rank. Thus, by replace A' in Theorems 6.2.1–6.2.4 by H, the four theorems can be extended to the corresponding forms for the model (6.2.44) (for details, see [51]). Therefore, the above fusion model is very general. For example, the row dimension of H_i in (6.2.43) may be less than the dimension of the state, which implies that the local estimator may only get a partial estimate of the state, which, of course, is biased. Besides, this model can also use to the measurement fusion. Clearly, $W'Y$ optimally fuses the total measurement Y of l sensors to the LMV estimate of the state.

6.2.5 PREVIOUS FUSION FORMULAE AS SPECIAL CASES

We can verify directly that the two previous optimal estimation fusion formulae in [4, 5] and [22, 41] are special cases of the LMV estimation fusion formulae in the previous section.

Corollary 6.2.1. For the two-sensor case, if $D \triangleq C_k^{11} + C_k^{22} - C_k^{12} - C_k^{21}$ is invertible, then $W_k^{(i)}$ in Theorem 6.2.2 reduces to the same form as (6.1.4) and (6.1.5) given in [4, 5], only noting that the differences of the notations in the two fusion formulae, i.e., $C_k, C_k^{(ij)}$ here are $P_{k|k}, P_{k|k}^{(ij)}$ there, respectively.

Remark 6.2.1. Since $D = [I \quad -I]C_k[I \quad -I]'$, the invertibility of C_k implies the invertibility of D. The assumption for this corollary is weaker than that given in [4, 5] for the fusion formulae (6.1.4) and (6.1.5). In other words, even if C_k is not invertible, it is still possible for D to be invertible, for example, $C_k^{11} = I$ and $C_k^{22} = C_k^{12} = C_k^{21} = 0$. Thus, when D is invertible but C_k singular, formulae (6.1.4) and (6.1.5)are not $W_k^{(i)}$ in Theorem 6.2.3 but the more general one given in Theorem 6.2.2.

Proof of Corollary 6.2.1. By simple manipulation, it is easy to verify

$$PC_kP = \frac{1}{4}\begin{bmatrix} D & -D \\ -D & D \end{bmatrix}, \; (PC_kP)^+ = \begin{bmatrix} D^{-1} & -D^{-1} \\ -D^{-1} & D^{-1} \end{bmatrix}. \tag{6.2.45}$$

Using the result in Theorem 6.2.2 and (6.2.45), we have

$$\begin{aligned} W_k &= \frac{1}{2}(I - (PC_kP)^+C_k)A' \\ &= \frac{1}{2}\left(I - \begin{bmatrix} D^{-1} & -D^{-1} \\ -D^{-1} & D^{-1} \end{bmatrix}\begin{bmatrix} C_k^{11} & C_k^{12} \\ C_k^{21} & C_k^{22} \end{bmatrix}\right)\begin{bmatrix} I \\ I \end{bmatrix} \\ &= \frac{1}{2}\begin{bmatrix} I - D^{-1}(C_k^{11} - C_k^{21}) & D^{-1}(C_k^{22} - C_k^{12}) \\ D^{-1}(C_k^{11} - C_k^{21}) & I - (C_k^{22} - C_k^{12}) \end{bmatrix}\begin{bmatrix} I \\ I \end{bmatrix} \\ &= \frac{1}{2}\begin{bmatrix} D^{-1}(D - C_k^{11} + C_k^{21} - C_k^{12} + C_k^{22}) \\ D^{-1}(D + C_k^{11} - C_k^{21} + C_k^{12} - C_k^{22}) \end{bmatrix} \\ &= \begin{bmatrix} D^{-1}(C_k^{22} - C_k^{12}) \\ D^{-1}(C_k^{11} - C_k^{21}) \end{bmatrix}. \end{aligned} \tag{6.2.46}$$

It follows that the corollary holds. ■

As a matter of fact, the fusion formulae (6.1.3)–(6.1.5) given in [4, 5] at best would be a special case of the unified LMV estimation fusion formulae in this section, in particular, given by Theorems 6.2.2 and 6.2.3. The reasons are that the latter is the best linear fusion result in the minimum variance sense with the linear constraint and is unique, the former is also a linear fusion formulae with the same constraint. The above proof just shows that the fusion formulae (6.1.3)–(6.1.5) really achieve the minimum variance of the estimation error. On the other hand, the two weighting matrices in the fusion formulae (6.1.3)–(6.1.5) are given directly by the sub-matrices of $P_{k|k}$. This is really convenient. Unfortunately, in the more than two sensor case, a similar extension has not been found even when all sensor noises are independent mutually.

The estimation fusion (6.1.6)–(6.1.9) given in [22, 41] was proved to be the same as the centralized Kalman filtering, in other words, its performance has reached the best. In general, the unified LMV fusion may not perform as best as the centralized Kalman filtering. However, the former is also a linear fusion with the constraint given in (6.2.8), as well as, the latter is the unique best linear fusion in the minimum variance sense with the same constraint. Hence, the estimation fusion given in [22, 41] must be a special case of the unified LMV fusion. In fact, we can prove the two fusion formulae are exactly identical directly. Due to the limited space of the book, we have to omit the detailed argument here.

6.2.6 DISCUSSION

In this subsection, we discuss mainly the generality and usefulness of the unified LMV fusion results obtained above.

First, it should be emphasized that the above LMV fusion results rely on two assumptions: (1) the total of local estimates can be linearly transmitted to be unbiased and (2) the error covariance matrix C_k is known. Both assumptions are fairly reasonable.

If the measurement noises are independent across sensors, the covariance C_k is not (block) diagonal only because the same *random* state is estimated by all local estimators, for example, because of the common process noise in the system dynamics on which each local estimator is based. C_k in this case and in some more general and coupled cases can be easily obtained (see Subsection 6.2.6 below). In general, C_k quantifies the coupling among local estimators. It, or its equivalent, is needed for optimal fusion. This availability assumption for C_k basically guarantees that our LMV fusion results are optimal given the coupling among local estimators. When this coupling is neither known nor obtainable, the above fusion results are not applicable directly but may facilitate the development of the corresponding optimal fusion.

Note that in the above LMV fusion, $\hat{\mathbf{x}}_k^{(i)}$ could be any unbiased estimate of $\mathbf{x}_k$. For example, one could have

$$X_k = [\mathbf{x}_{k|k-1}^{(1)\prime}, \mathbf{x}_{k|k}^{(1)\prime}, \mathbf{x}_{k|k-2}^{(2)\prime}, \mathbf{x}_{k|k}^{(2)\prime}, \ldots, \mathbf{x}_{k|k-1}^{(l)\prime}, \mathbf{x}_{k|k}^{(l)\prime}]',$$

where the superscript j denotes the estimate by the jth "local" estimator. With this understanding, most advantages of the above LMV fusion results presented below can be easily appreciated.

Comparing with existing results on the estimation fusion, the above LMV fusion formulae have at least the following advantages:

- They are valid for cases with coupled observation noises across sensors. This is useful in practice when the dynamic process is observed in a common noisy environment, such as when a target is taking an electronic countermeasure (ECM), e.g. noise jamming, or when the sensor noises are coupled because of, say, their dependence on the target state. A class of systems that fall into this category is given in Subsection 6.2.6. Another important application area is the fusion of estimates based on observations obtained over different time periods. The fusion-based optimal smoothing using measurements corrupted by autocorrelated noise is a good candidate for application of our results. Almost all previous fusion results assume that the sensor observations are conditionally independent given the target state/signal to be estimated. A formula was mentioned in [53] that is valid for fusing local estimates based on not necessarily disjoint observation data, which is a limited yet useful form of data dependence, still a special case.

- The fused estimator depends on the network structures or communication patterns only through C_k. Consequently, the fusion rule is invariant no matter if there is feedback or not, if the network has a parallel, tandem, tree or general structure, or what the communication bandwidths are. This means that in practice all we need to obtain is the coupling between every pair of local estimates. If there is no fusion center, such as for the network structure considered in [38], each sensor can use our LMV fusion formulae to obtain the best estimator based on its own observations and any information received from other sensors.

- The local estimators do not have to use the same dynamic/measurement model. The local estimates may be obtained based on different dynamic models and then be fused using our LMV fusion formulae. The use of different dynamic models for the local estimators may be necessary or more effective. It is necessary, for example, when state augmentation is needed for some local estimators with autocorrelated sensor measurement noise. It may be more effective, for example, when multiple models are used for the process to be estimated. Of course the key to a successful application of our results in such cases lies in the determination of the covariance matrix C_k. In order to have the best fusion performance, the choice of the local models is of course not arbitrary (see, e.g., [2]).

- There is no requirement on synchronism of the local estimates provided that they can be converted to the estimates of the state at the same time (based on observations up to different times). In reality, it is difficult to synchronize local estimates. However, many well-known results are valid only for synchronized local estimates and thus require artificial synchronization for the local estimates. This either increases the computational burden considerably or degrade the fusion performance. Our LMV fusion formulae provide a more convenient and efficient framework for fusing asynchronous local estimates. For example, our LMV fusion formulae are applicable (optimally) after we simply convert all local estimates $\mathbf{x}^{(i)}(t_i|t_j)$ to $\mathbf{x}^{(i)}(k|t_j)$, that is, the same point k in time at which the state is to be estimated by the fusion center, where the conversion could either be prediction if $k > t_i$ or smoothing if $k < t_i$.

- The local estimators need not be of the same type. For example, our results are valid if some local estimators are LMV estimators while others are MAP (maximum a posteriori) estimators. This flexibility is useful for some applications.

- It is quite possible in practice for C_k to be singular. Intuitively, this may be the case if there is no independent parts between any two sensor observation noises (see the dynamic system with $w_k^{(i)} \equiv 0$ given in the following subsection). However, Theorem 6.2.2 guarantees the uniqueness of our LMV fusion no matter whether C_k is singular or not.

6.2.7 RECURSIVE COMPUTATION OF ERROR COVARIANCE

It can be easily seen that the optimal weighting matrix W_k, given by Theorems 6.2.1, 6.2.2, and 6.2.3, depends only on the covariance matrix C_k and the computational burden of W_k relies mostly on the computation of C_k and its inverse. Obviously, C_k may not be calculated recursively in general. However, in many practical situations, C_k may depend only on the system coefficient matrices and known noise covariances. Hence, C_k and thus W_k can be calculated off-line. An off-line recursion for $C_k^{(i,j)}$ is presented in [4, 5] assuming that the measurement noises are independent across sensors. In this section, we extend that result to a class of linear systems having dependent measurement noises with known correlations between any two sensors.

Consider an linear dynamic process

$$\mathbf{x}_{k+1} = \Phi_k \mathbf{x}_k + v_k \tag{6.2.47}$$

with additive zero-mean white noise

$$E[v_k] = 0, E[v_k v_j'] = Q_k \delta_{kj},$$

and noisy measurement

$$\mathbf{y}_k^{(i)} = H_k^{(i)} \mathbf{x}_k + w_k^{(i)} + e_k^{(i)}, \quad \forall i \le l, \tag{6.2.48}$$

where the measurement noise is the sum of two zero-mean white noises $w_k^{(i)}$ and $e_k^{(i)}$, uncorrelated with the process noise:

$$E[w_k^{(i)}] = 0, \quad E[w_k^{(i)} w_j^{(i)\prime}] = \sigma_k^{(i)} \delta_{kj}, \quad E[e_k^{(i)}] = 0,$$
$$E[e_k^{(i)} e_j^{(i)\prime}] = S_k^{(ii)} \delta_{kj}, \quad E[v_k w_j^{(i)\prime}] = 0, \quad E[v_k e_j^{(i)\prime}] = 0.$$

However, while $w_k^{(i)}$'s are independent across sensors, $e_k^{(i)}$'s are coupled across sensors:

$$E[w_k^{(i)} w_k^{(j)\prime}] = 0, \qquad E[e_k^{(i)} e_k^{(j)\prime}] = S_k^{(ij)}.$$

Clearly, this system reduces to the one with independent measurement noise when $e_k^{(i)} \equiv 0$. As explained before, this model may be useful e.g., when a target is generating noise jamming or when the sensor noises are dependent on the target state.

Similar to the derivation in [4, 5], it can be shown using Kalman filter formulae for the above system that we have the following recursive formulae, for $k = 1$ and assuming $S_0 = 0$,

$$\begin{aligned} C_1^{(ij)} &\\ = (I - K_1^{(i)} H_1^{(i)}) Q_0 (I - K_1^{(j)} H_1^{(j)})' &+ K_1^{(i)} S_1^{(ij)} K_1^{(j)} + K_1^{(i)} \sigma_1^{(i)} \delta_{ij} K_1^{(j)\prime}, \\ i, j = 1, \ldots, l, & \end{aligned} \tag{6.2.49}$$

and for any $k > 1$,

$$\begin{aligned} C_k^{(ij)} = &(I - K_k^{(i)} H_k^{(i)}) Q_{k-1} (I - K_k^{(j)} H_k^{(j)})' \\ &+ (I - K_k^{(i)} H_k^{(i)}) \Phi_{k-1} S_{k-1}^{(ij)} F_{k-1'} (I - K_k^{(j)} H_k^{(j)})' \\ &+ K_k^{(i)} S_k^{(ij)} K_k^{(j)} + K_k^{(i)} \sigma_k^{(i)} \delta_{ij} K_k^{(j)\prime}, \quad i, j = 1, \ldots, l, \end{aligned} \tag{6.2.50}$$

where K is the Kalman filter gain.

Let

$$\lambda_k^{(i)} = (I - K_k^{(i)} H_k^{(i)})\Phi_{k-1}, \quad \Lambda_k = \text{diag}\{\lambda_k^{(1)}, \ldots, \lambda_k^{(l)}\},$$

$$K(k) = \text{diag}\{K_k^{(1)}, \ldots, K_k^{(l)}\}, \quad \Sigma_k = \text{diag}\{\sigma_k^{(1)}, \ldots, \sigma_k^{(l)}\},$$

$$S_k = \begin{bmatrix} S_k^{(11)} \cdots S_k^{(1l)} \\ \cdots \\ S_k^{(l1)} \cdots S_k^{(ll)} \end{bmatrix}, \tag{6.2.51}$$

$$M_k = [Q_{k-1}^{\frac{1}{2}\prime}(I - K_k^{(1)} H_k^{(1)})' \quad \cdots \quad Q_{k-1}^{\frac{1}{2}\prime}(I - K_k^{(l)} H_k^{(l)})']'.$$

Then, an off-line recursion of C_k is obtained by rewriting (6.2.49)–(6.2.50) in the matrix form as

$$C_k = \Lambda_k C_{k-1} \Lambda_k' + K(k)(\Sigma_k + S_k)K(k)' + M_k M_k',$$

which can be initialized by

$$C_1 = M_1 M_1' + K(1)(\Sigma_1 + S_1)K(1)'.$$

When the local estimates $\hat{\mathbf{x}}_k^{(i)}$'s are asynchronous, for example, the fusion center can know $\hat{\mathbf{x}}_j^{(i)}$ of the ith sensor at the time instant j, where $j < k$, we can use the $k - j$ step prediction $\Phi_{k-1} \cdots \Phi_j \hat{\mathbf{x}}_j^{(i)}$ as the unbiased estimation $\hat{\mathbf{x}}_k^{(i)}$. Then, using the dynamical model (6.2.47) and (6.2.48), the above recursive computation of C_k can still be implemented. Since C_k^{-1} appears in the fusion formula (6.2.24), it is more valuable that C_k^{-1} can be recursively calculated. Unfortunately, we found few cases can be done so.

6.3 THE OPTIMALITY OF KALMAN FILTERING FUSION WITH FEEDBACK

In [23], a Kalman filtering fusion with feedback was suggested, which is a heuristic modification of (6.1.6)–(6.1.8). In this section, a rigorous performance analysis is dedicated to this distributed recursive filtering fusion with feedback for dynamic systems. Under the assumption of independent observations across sensors, it will be shown that the Kalman filtering fusion with feedback is, as the fusion without feedback of (6.1.6)–(6.1.8), exactly equivalent to the centralized filtering fusion, moreover, the so called P matrices in the feedback Kalman filtering at both local filters and fusion center are still the covariance matrices of filtering errors. Although the feedback here cannot improve the performance at the fusion center, the feedback does reduce the covariance of each local filtering error. The above results can be extended to a hybrid filtering fusion with the feedback received by partial local filters.

6.3.1 PROBLEM FORMULATION

The l-sensor distributed dynamic system is given by

$$\mathbf{x}_{k+1} = \Phi_k \mathbf{x}_k + v_k, \quad k = 0, 1, ..., \tag{6.3.1}$$

$$\mathbf{y}_k^i = H_k^i \mathbf{x}_k + w_k^i, \quad i = 1, \ldots, l, \tag{6.3.2}$$

where Φ_k is a matrix of order $(r \times r)$, $\mathbf{x}_k, v_k \in \Re^r, H_k^i \in \Re^{N_i \times r}, \mathbf{y}_k^i, w_k^i \in \Re^{N_i}$. The process noise v_k and measurement noise w_k are both zero-mean random variables independent of each other temporally and are not cross-correlated.

To compare performances between the centralized and distributed filtering fusion, the stacked measurement equation is written as

$$\mathbf{y}_k = H_k \mathbf{x}_k + \omega_k, \tag{6.3.3}$$

where

$$\mathbf{y}_k = (\mathbf{y}_k^{1'}, \ldots, \mathbf{y}_k^{l'})', \quad H_k = (H_k^{1'}, \ldots, H_k^{l'})', \quad \omega_k = (w_k^{1'}, \ldots, w_k^{l'})', \tag{6.3.4}$$

and the covariance of the noise ω_k is given by

$$Cov(\omega_k) = R_k = diag(R_k^1, \ldots, \Re_k^l), \quad R_k^i = Cov(w_k^i), \quad i = 1, \ldots, l. \tag{6.3.5}$$

According to the standard results in Kalman filtering [37], [40] and [54], the local Kalman filtering at the i-th sensor is

$$\mathbf{x}_{k/k}^i = \mathbf{x}_{k/k-1}^i + K_k^i(\mathbf{y}_k^i - H_k^i \mathbf{x}_{k/k-1}^i) = (I - K_k^i H_k^i)\mathbf{x}_{k/k-1}^i + K_k^i \mathbf{y}_k^i,$$
$$K_k^i = P_{k/k}^i H_k^{i'} R_k^{i^{-1}} \tag{6.3.6}$$

with covariance of filtering error given by

$$P_{k/k}^i = (I - K_k^i H_k^i) P_{k/k-1}^i \quad (P_{k/k-1}^{i^{-1}} = P_{k/k}^{i^{-1}}(I - K_k^i H_k^i)), \tag{6.3.7}$$

or

$$P_{k/k}^{i^{-1}} = P_{k/k-1}^{i^{-1}} + H_k^{i'} R_k^{i^{-1}} H_k^i, \tag{6.3.8}$$

where

$$\mathbf{x}_{k/k-1}^i = \Phi_k \mathbf{x}_{k-1/k-1}^i,$$
$$P_{k/k}^i = E[(\mathbf{x}_{k/k}^i - \mathbf{x}_k)(\mathbf{x}_{k/k}^i - \mathbf{x}_k)'],$$
$$P_{k/k-1}^i = E[(\mathbf{x}_{k/k-1}^i - \mathbf{x}_k)(\mathbf{x}_{k/k-1}^i - \mathbf{x}_k)'].$$

Similarly, the centralized Kalman filtering with all sensor data is given by

$$\mathbf{x}_{k/k} = \mathbf{x}_{k/k-1} + K_k(\mathbf{y}_k - H_k\mathbf{x}_{k/k-1}) = (I - K_kH_k)\mathbf{x}_{k/k-1} + K_ky_k,$$
$$K_k = P_{k/k}H_k'R_k^{-1} \tag{6.3.9}$$

with covariance of filtering error given by

$$P_{k/k}^{-1} = P_{k/k-1}^{-1} + H_k'R_k^{-1}H_k = P_{k/k-1}^{-1} + \sum_{i=1}^{l} H_k^{i'}R_k^{i^{-1}}H_k^i, \tag{6.3.10}$$

or

$$P_{k/k} = (I - K_kH_k)P_{k/k-1}, \tag{6.3.11}$$

where

$$\mathbf{x}_{k/k-1} = \Phi_k\mathbf{x}_{k-1/k-1},$$
$$P_{k/k} = E[(\mathbf{x}_{k/k} - \mathbf{x}_k)(\mathbf{x}_{k/k} - \mathbf{x}_k)'],$$
$$P_{k/k-1} = E[(\mathbf{x}_{k/k-1} - \mathbf{x}_k)(\mathbf{x}_{k/k-1} - \mathbf{x}_k)'].$$

Using (6.3.8) and (6.3.10), the estimation error covariance of the centralized Kalman filtering is given by using the estimation error covariances of all local filters:

$$P_{k/k}^{-1} = P_{k/k-1}^{-1} + \sum_{i=1}^{l}(P_{k/k}^{i^{-1}} - P_{k/k-1}^{i^{-1}}). \tag{6.3.12}$$

Using (6.3.9), (6.3.4) and (6.3.5), we have

$$K_ky_k = P_{k/k}\sum_{i=1}^{l} H_k^{i'}R_k^{i^{-1}}y_k^i. \tag{6.3.13}$$

To express the centralized filtering $\mathbf{x}_{k/k}$ in terms of the local filtering, we use (6.3.6) and (6.3.13) to eliminate y_k from (6.3.9). From (6.3.6) and (6.3.7) we have

$$\begin{aligned} H_k^{i'}R_k^{i^{-1}}y_k^i &= P_{k/k}^{i^{-1}}\mathbf{x}_{k/k}^i - P_{k/k}^{i^{-1}}(I - K_k^iH_k^i)\mathbf{x}_{k/k-1}^i \\ &= P_{k/k}^{i^{-1}}\mathbf{x}_{k/k}^i - P_{k/k-1}^{i^{-1}}\mathbf{x}_{k/k-1}^i. \end{aligned} \tag{6.3.14}$$

Thus, substituting (6.3.11), (6.3.13) and (6.3.14), into (6.3.9) yields

$$P_{k/k}^{-1}\mathbf{x}_{k/k} = P_{k/k-1}^{-1}\mathbf{x}_{k/k-1} + \sum_{i=1}^{l}(P_{k/k}^{i^{-1}}\mathbf{x}_{k/k}^i - P_{k/k-1}^{i^{-1}}\mathbf{x}_{k/k-1}^i). \tag{6.3.15}$$

That is to say that the centralized filtering (6.3.15) and error matrix (6.3.12) are explicitly expressed in terms of the local filtering. Hence, the performance of the distributed Kalman filtering fusion given in (6.1.6)–(6.1.8) is the same as that of the centralized fusion.

When there is feedback, the fusion center broadcasts its latest estimate to the local sensors. Thus, for all i, the following local and global one-stage predictions were modified naturally [23] as

$$\mathbf{x}^i_{k/k-1} = \Phi_k \hat{\mathbf{x}}_{k-1/k-1} = \mathbf{x}_{k/k-1}, \tag{6.3.16}$$

$$\hat{P}^i_{k/k-1} = \hat{P}_{k/k-1}. \tag{6.3.17}$$

Naturally, the following filtering fusion with feedback was suggested

$$\begin{aligned} \hat{P}^{-1}_{k/k} &= \hat{P}^{-1}_{k/k-1} + \textstyle\sum_{i=1}^{l}(\hat{P}^{i^{-1}}_{k/k} - \hat{P}^{-1}_{k/k-1}) \\ &= \textstyle\sum_{i=1}^{l} \hat{P}^{i^{-1}}_{k/k} - (l-1)\hat{P}^{-1}_{k/k-1}, \end{aligned} \tag{6.3.18}$$

$$\begin{aligned} \hat{P}^{-1}_{k/k}\mathbf{x}_{k/k} &= \hat{P}^{-1}_{k/k-1}\hat{\mathbf{x}}_{k/k-1} + \textstyle\sum_{i=1}^{l}(\hat{P}^{i^{-1}}_{k/k}\hat{\mathbf{x}}^i_{k/k} - \hat{P}^{-1}_{k/k-1}\hat{\mathbf{x}}_{k/k-1}) \\ &= \textstyle\sum_{i=1}^{l} \hat{P}^{i^{-1}}_{k/k}\hat{\mathbf{x}}^i_{k/k} - (l-1)\hat{P}^{-1}_{k/k-1}\hat{\mathbf{x}}_{k/k-1}, \end{aligned} \tag{6.3.19}$$

where the superscript "hat" here is to show the corresponding estimates with feedback. Obviously, the performance of the filtering fusion (6.3.12) and (6.3.15) without feedback cannot be improved. Then the following questions arise: **(I)** Can the performance of the filtering fusion (6.3.16)–(6.3.19) with feedback also attain that of the centralized filtering fusion? **(II)** Are the matrices $\hat{P}^{-1}_{k/k}$ and $\hat{P}^{i^{-1}}_{k/k}$ still the covariances of the global and local filtering errors respectively? **(III)** What is the main advantage of feedback here?

Obviously, $\hat{P}^i_{k/k} \neq P^i_{k/k}$ and $\hat{\mathbf{x}}^i_{k/k} \neq \mathbf{x}^i_{k/k}$ in general. To determine the tracking gate for the data association (see, for example, [3, 6]), we have to know what is the covariance of the local estimate error. Question (II) is related to this issue. Questions (I) and (III) are the base of the filtering fusion with feedback. In this section, a rigorous analysis for the feedback filtering fusion is presented to answer the above questions. The answers for questions (I) and (II) are both positive, i.e.,

$$\hat{\mathbf{x}}_{k/k} = \mathbf{x}_{k/k}, \quad \hat{P}_{k/k} = P_{k/k}, \tag{6.3.20}$$

$$\hat{P}^i_{k/k} = E[(\hat{\mathbf{x}}^i_{k/k} - \mathbf{x}_k)(\hat{\mathbf{x}}^i_{k/k} - \mathbf{x}_k)']. \tag{6.3.21}$$

The main advantage of the feedback filtering fusion is to reduce the covariance of every local estimate error, i.e.,

$$\hat{P}^i_{k/k} \leq P^i_{k/k}, \quad i = 1, 2, \ldots, l. \tag{6.3.22}$$

6.3.2 GLOBAL OPTIMALITY OF THE FEEDBACK FILTERING FUSION

We use mathematical induction to show no global performance difference between two types of filtering fusion.

Clearly, it is reasonable to have the same initial values at both the local and the global filters no matter feedback exists or not.

$$\hat{\mathbf{x}}_{0/0} = \mathbf{x}_{0/0} = \mathbf{x}^i_{0/0} = \hat{\mathbf{x}}^i_{0/0}, \quad \hat{P}_{0/0} = P_{0/0} = P^i_{0/0} = \hat{P}^i_{0/0}. \tag{6.3.23}$$

Suppose that

$$\hat{\mathbf{x}}_{k-1/k-1} = \mathbf{x}_{k-1/k-1}, \quad \hat{\mathbf{x}}_{k/k-1} = \mathbf{x}_{k/k-1}, \tag{6.3.24}$$

$$\hat{P}_{k-1/k-1} = P_{k-1/k-1}, \quad \hat{P}_{k/k-1} = P_{k/k-1}. \tag{6.3.25}$$

Therefore, at the $k-$th stage, using the above equations and (6.3.18), we have

$$\hat{P}^{-1}_{k/k} = P^{-1}_{k/k-1} + \sum_{i=1}^{l}(\hat{P}^{i^{-1}}_{k/k} - P^{i^{-1}}_{k/k-1}). \tag{6.3.26}$$

Note the definition of $\hat{P}^i_{k/k}$ similar to that in (6.3.8), we get

$$\hat{P}^{i^{-1}}_{k/k} = P^{-1}_{k/k-1} + H^{i'}_k R^{i^{-1}}_k H^i_k. \tag{6.3.27}$$

Substituting (6.3.27) into (6.3.26) and comparing it with (6.3.10) yield

$$\hat{P}^{-1}_{k/k} = P^{-1}_{k/k}. \tag{6.3.28}$$

Then using (6.3.6) and (6.3.8) in the feedback case, and (6.3.27), we have

$$\begin{aligned}\hat{P}^{i^{-1}}_{k/k}\hat{\mathbf{x}}^i_{k/k} &= \hat{P}^{i^{-1}}_{k/k}\mathbf{x}_{k/k-1} + H^{i'}_k R^{i^{-1}}_k(\mathbf{y}^i_k - H^i_k\mathbf{x}_{k/k-1})\\ &= \hat{P}^{i^{-1}}_{k/k}\mathbf{x}_{k/k-1} + H^{i'}_k R^{i^{-1}}_k\mathbf{y}^i_k - H^{i'}_k R^{i^{-1}}_k H^i_k\mathbf{x}_{k/k-1}\\ &= \hat{P}^{i^{-1}}_{k/k}\mathbf{x}_{k/k-1} + H^{i'}_k R^{i^{-1}}_k\mathbf{y}^i_k - \hat{P}^{i^{-1}}_{k/k}\mathbf{x}_{k/k-1} + P^{-1}_{k/k-1}\mathbf{x}_{k/k-1}\\ &= H^{i'}_k R^{i^{-1}}_k\mathbf{y}^i_k + P^{-1}_{k/k-1}\mathbf{x}_{k/k-1}.\end{aligned} \tag{6.3.29}$$

Similarly, from (6.3.9) and (6.3.10), we have

$$\begin{aligned}P^{-1}_{k/k}\mathbf{x}_{k/k} &= H'_k R^{-1}_k\mathbf{y}_k + P^{-1}_{k/k-1}\mathbf{x}_{k/k-1}\\ &= \textstyle\sum_{i=1}^{l} H^{i'}_k R^{i^{-1}}_k\mathbf{y}^i_k + P^{-1}_{k/k-1}\mathbf{x}_{k/k-1}.\end{aligned} \tag{6.3.30}$$

Hence, using (6.3.19), (6.3.25) and (6.3.28)–(6.3.30), we have

$$\hat{x}_{k/k} = P_{k/k}[P_{k/k-1}^{-1}\mathbf{x}_{k/k-1} + \sum_{i=1}^{l} H_k^{i'} R_k^{i^{-1}} \mathbf{y}_k^i] = x_{k/k}. \tag{6.3.31}$$

The result is natural because (a) (6.3.18) and (6.3.19) are optimal for the feedback case since they follow from Kalman filtering; (b) each local estimate $\hat{x}_{k/k}^i$ can never be worse than $\mathbf{x}_{k/k}^i$ since more information is used.

6.3.3 LOCAL ESTIMATE ERRORS

Although $\hat{P}_{k/k}^{i-1}$ has been defined in Eq. (6.3.27), we need to verify if it is still the covariance of local estimate error.

From the last equality of (6.3.29) and ((6.3.27)), we know

$$\begin{aligned}
\hat{\mathbf{x}}_{k/k}^i &= \hat{P}_{k/k}^i H_k^{i'} R_k^{i^{-1}} \mathbf{y}_k^i + \hat{P}_{k/k}^i P_{k/k-1}^{-1}\mathbf{x}_{k/k-1} \\
&= \hat{P}_{k/k}^i H_k^{i'} R_k^{i^{-1}} w_k^i + \hat{P}_{k/k}^i H_k^{i'} R_k^{i^{-1}} H_k^i \mathbf{x}_k + \hat{P}_{k/k}^i P_{k/k-1}^{-1}\mathbf{x}_{k/k-1} \\
&= \hat{P}_{k/k}^i H_k^{i'} R_k^{i^{-1}} w_k^i + \hat{P}_{k/k}^i (\hat{P}_{k/k}^{i-1} - P_{k/k-1}^{-1})\mathbf{x}_k + \hat{P}_{k/k}^i P_{k/k-1}^{-1}\mathbf{x}_{k/k-1} \\
&= \hat{P}_{k/k}^i H_k^{i'} R_k^{i^{-1}} w_k^i + \hat{P}_{k/k}^i P_{k/k-1}^{-1}(\mathbf{x}_{k/k-1} - \mathbf{x}_k) + \mathbf{x}_k.
\end{aligned} \tag{6.3.32}$$

It follows that

$$\begin{aligned}
&E[(\hat{\mathbf{x}}_{k/k}^i - \mathbf{x}_k)(\hat{\mathbf{x}}_{k/k}^i - \mathbf{x}_k)'] \\
&= E\{[\hat{P}_{k/k}^i H_k^{i'} R_k^{i^{-1}} w_k^i + \hat{P}_{k/k}^i P_{k/k-1}^{-1}(\mathbf{x}_{k/k-1} - \mathbf{x}_k)] \\
&\quad \cdot [\hat{P}_{k/k}^i H_k^{i'} R_k^{i^{-1}} w_k^i + \hat{P}_{k/k}^i P_{k/k-1}^{-1}(\mathbf{x}_{k/k-1} - \mathbf{x}_k)]'\} \\
&= \hat{P}_{k/k}^i H_k^{i'} R_k^{i^{-1}} H_k^i \hat{P}_{k/k}^i + \hat{P}_{k/k}^i P_{k/k-1}^{-1} \hat{P}_{k/k}^i \\
&= \hat{P}_{k/k}^i \hat{P}_{k/k}^{i-1} \hat{P}_{k/k}^i = \hat{P}_{k/k}^i.
\end{aligned} \tag{6.3.33}$$

6.3.4 THE ADVANTAGE OF THE FEEDBACK

Although the feedback cannot improve global performance as shown in Subsection 6.3.2, we would like to know what is the advantage of the feedback. From (6.3.8) and (6.3.27),

$$\begin{aligned}
\hat{P}_{k/k}^{i-1} - P_{k/k}^{i-1} = P_{k/k-1}^{-1} - P_{k/k-1}^{i-1} &= (\Phi_{k-1}P_{k-1/k-1}\Phi_{k-1}' + Cov(v_{k-1}))^{-1} \\
&\quad -(\Phi_{k-1}P_{k-1/k-1}^i\Phi_{k-1}' + Cov(v_{k-1}))^{-1} \geq 0,
\end{aligned} \tag{6.3.34}$$

where the last inequality comes from

$$P_{k-1/k-1} \leq P_{k-1/k-1}^i, \quad k = 2, 3, \ldots, \tag{6.3.35}$$

which can be proved recursively by (6.3.8), (6.3.10) and (6.3.23) as follows:

$$P_{1/0}^{-1} - P_{1/0}^{i-1} = (\Phi_0 P_{0/0} \Phi_0' + Cov(v_0))^{-1} - (\Phi_0 P_{0/0}^i \Phi_0' + Cov(v_0))^{-1} = 0,$$

$$P_{1/1}^{-1} - P_{1/1}^{i-1} = P_{1/0}^{-1} - P_{1/0}^{i-1} + \sum_{j \neq i} H_1^{j'} R_1^{j^{-1}} H_1^j \geq 0,$$

$$P_{2/1}^{-1} - P_{2/1}^{i-1} = (\Phi_1 P_{1/1} \Phi_1' + Cov(v_1))^{-1} - (\Phi_1 P_{1/1}^i \Phi_1' + Cov(v_1))^{-1} \geq 0,$$

$$\cdots$$

$$P_{k-1/k-1}^{-1} - P_{k-1/k-1}^{i-1} = P_{k-1/k-2}^{-1} - P_{k-1/k-2}^{i-1} + \sum_{j \neq i} H_{k-1}^{j'} R_{k-1}^{j^{-1}} H_{k-1}^j \geq 0. \tag{6.3.36}$$

Hence,

$$\hat{P}_{k/k}^i \leq P_{k/k}^i. \tag{6.3.37}$$

Furthermore, assuming that Φ_k is invertible for all k, by (6.3.36), the necessary and sufficient condition for the above inequality (6.3.37) being strict is that there exists some $n \leq k$ such that $P_{n/n} < P_{n/n}^i$, in other words, once the above inequality become true,

$$\hat{P}_{k/k} < P_{k/k}^i \quad \text{for all } k \geq n.$$

A sufficient condition for the inequality above is

$$\sum_{j \neq i} H_r^{j'} R_r^{j^{-1}} H_r^j > 0 \quad \text{some } r \leq k-1. \tag{6.3.38}$$

Therefore, by the results in Subsections 6.3.3 and 6.3.4, the feedback here can improve all local filtering performances.

6.3.5 EXTENSION TO A HYBRID FILTERING FUSION

In some practical distributed filtering systems, it is impossible sometimes for a local filter to wait feedback from the fusion center, in addition, the communications between the fusion center and local filters may fail sometimes, such as in a war situation. In those cases, only partial local filters could receive feedback from the fusion center.

A hybrid filtering fusion for the above situation can be modelled as follows:

$$\hat{\mathbf{x}}_{k/k-1}^i = \begin{cases} \Phi_k \hat{\mathbf{x}}_{k-1/k-1} = \hat{\mathbf{x}}_{k/k-1}, & \text{if feedback received at the } i\text{th tracker,} \\ \mathbf{x}_{k/k-1}^i, & \text{if no feedback received at the } i\text{th tracker,} \end{cases}$$

$$\hat{P}_{k/k-1}^i = \begin{cases} \hat{P}_{k/k-1}, & \text{if feedback received at the } i\text{th tracker,} \\ P_{k/k-1}^i, & \text{if no feedback received at the } i\text{th tracker,} \end{cases}$$

as well as, the following filtering fusion with feedback received by partial local filters is given by

$$\begin{aligned}\hat{P}_{k/k}^{-1} &= \hat{P}_{k/k-1}^{-1} + \sum_{\mathcal{U}_f(k)} (\hat{P}_{k/k}^{i-1} - \hat{P}_{k/k-1}^{-1}) + \sum_{\mathcal{U}_{no}(k)} (P_{k/k}^{i-1} - P_{k/k-1}^{i-1}) \\ &= \sum_{\mathcal{U}_{no}(k)} (P_{k/k}^{i-1} - P_{k/k-1}^{i-1}) + \sum_{\mathcal{U}_f(k)} \hat{P}_{k/k}^{i-1} - (|\mathcal{U}_f(k)| - 1)\hat{P}_{k/k-1}^{-1},\end{aligned}$$

$$\begin{aligned}\hat{P}_{k/k}^{-1}\hat{\mathbf{x}}_{k/k} =& \hat{P}_{k/k-1}^{-1}\hat{\mathbf{x}}_{k/k-1} + \sum_{\mathcal{U}_f(k)} (\hat{P}_{k/k}^{i-1}\hat{\mathbf{x}}_{k/k}^i - \hat{P}_{k/k-1}^{-1}\hat{\mathbf{x}}_{k/k-1}) \\ &+ \sum_{\mathcal{U}_{no}(k)} (P_{k/k}^{i-1}\mathbf{x}_{k/k}^i - P_{k/k-1}^{i-1}\mathbf{x}_{k/k-1}^i) \\ =& \sum_{\mathcal{U}_{no}(k)} (P_{k/k}^{i-1}\mathbf{x}_{k/k}^i - P_{k/k-1}^{i-1}\mathbf{x}_{k/k-1}^i) \\ &+ \sum_{\mathcal{U}_f(k)} \hat{P}_{k/k}^{i-1}\hat{\mathbf{x}}_{k/k}^i - (|\mathcal{U}_f(k)| - 1)\hat{P}_{k/k-1}^{-1}\hat{\mathbf{x}}_{k/k-1},\end{aligned}$$

where $\mathcal{U}_f(k)$ is the set of all local filters receiving feedback at the kth stage, $\mathcal{U}_{no}(k)$ is the set of all other local filters, $|\mathcal{U}_f(k)|$ is the number of elements of set $\mathcal{U}_f(k)$.

Noticing that all arguments in Subsections 6.3.2–6.3.4 proceed at one step iteration, it is easy to extend them to the above hybrid filtering fusion, that is to say, the results in Subsections 6.3.2–6.3.4 still hold.

Those results in Section 6.3 in fact make a theoretical base for the distributed filtering fusion with feedback. If the sensor noises are dependent across sensors, the situation becomes more complicated. Using the unified estimation fusion formula in Subsection 6.2.2 and the results given in in Subsection 6.2.5, one can extend the performance analysis results here to distributed estimation system with dependent sensor noises given in Subsection 6.2.5.

6.4 FUSION OF THE FORGETTING FACTOR RLS ALGORITHM

We present two types of fusion methods for the distributed recursive state estimators of dynamic systems without knowledge of noise covariances. This estimator at every local sensor is to embed the dynamic matrix and the forgetting factor into the Recursive Least Squares (RLS) method to remedy the lack of knowledge of noises, which was developed before (see, [97]) and called the Extended Forgetting Factor Recursive Least Squares (EFRLS) estimator. We will prove that the aforementioned fusion methods are exactly equivalent to the centralized EFRLS that uses all measurements from local sensors directly. Therefore, the two fusion methods have the same advantages as the centralized EFRLS does. For example, they can perform slightly worse than precisely specified Kalman filter and still well even if there exists unknown cross-correlation between sensors and/or between the process and measurement noise sequences in time and space (cf. numerical simulations in [97] and this section).

6.4.1 FORGETTING FACTOR RLS ALGORITHM

An l-sensor distributed dynamic system is described as in (6.3.1)–(6.3.5), but the transit matrix Φ_k is a full-rank square matrix and no knowledge of noises is given.

An important practical problem in the above system is to find an optimal estimator of $\mathbf{x}_k$ given the observations $\mathbf{y}_1, \ldots, \mathbf{y}_j, j \geq 1$.

A well-known method to estimate the unknown state $\mathbf{x}_k$ in (6.3.1) is to derive a linear, unbiased, and minimum error variance (LMV) estimator. To derive such an estimator it is necessary to know the second-order statistical quantities, covariances R_{v_j} and R_{w_j} of the dynamic process noise $\{v_j\}$ and the measurement noise $\{w_j\}$, $j \leq k$. Kalman filtering is the best known recursive LMV algorithm to optimally estimate the unknown state of a dynamic system. When the processing center can receive all measurements from the local sensors in time, the Kalman filtering equations are given by (6.3.9)–(6.3.11).

In spite of its widespread applicability, Kalman filtering still has some practical disadvantages. First, it requires knowledge of the covariances R_{v_k} and R_{w_k} at all times. When these are not known, it is necessary to estimate them. In some practical applications (e.g., when the dimension of the state is very large), however, it is hard to do this, and incorrect variance estimates significantly degrade the performance of the resulting state estimates. In addition, if there is any unknown time correlation in the noise terms, or unknown cross-correlated between v_k and w_k, Kalman filter may perform very poorly (see the simulation examples in [97]). In order to overcome these difficulties of Kalman filtering, a recursive estimation procedure that does not require knowledge of the noise variances was developed in [97] by incorporating the dynamic matrix and the forgetting factor into the LS method, resulting in the Extended Forgetting Factor RLS–EFRLS method.

In this section, we develop two types of fusion methods for the EFRLS estimators, which are both exactly equivalent to their corresponding centralized LS estimate that uses all measurements from local sensors. These two methods differ in that one ignores the choices of the forgetting factors at the local sensors and the fusion center can select its own favorite forgetting factor, while the other believes the choices of the forgetting factors at the local sensors and the fusion center can recursively fuse the sensor estimates, which is equivalent to using all sensor measurements with their own forgetting factors. Obviously, these two methods may be suitable for different practical situations. When the local selections of the forgetting factors at the sensors are not good on average, one may use the first method; if the local selections are all good enough, we should choose the second one. The simulation results show that the fused EFRLS can perform much better than single sensor EFRLS and only slightly worse than optimal but often unrealistic Kalman filtering with exact noise statistics. It was illustrated in [97] via simulations that EFRLS could perform still well even if sensor or process noise sequences are autocorrelated or there is cross-correlation between the sensor and process noises. In order to illustrate the use of the multisensor EFRLS fusion methods and understand practical performance issues, we have done a number of simulations comparing Kalman filtering and the EFRLS at each local sensor as well as their fusion methods. These simulations were done for a dynamic system modelled as an object moving in a circular trajectory at constant speed with disturbances and measurement noises. The simulation results show that the fused EFRLS algorithms can track the moving object satisfactorily. The correctly specified Kalman filter outperforms the EFRLS algorithms as expected. However, it has been shown in [97] that when there is time correlation in the noise terms the EFRLS actually performs better than an incorrectly specified Kalman filter. In addition, the fused EFRLS algorithm consistently outperforms the single sensor EFRLS algorithm.

6.4.2 TWO TYPES OF DISTRIBUTED EFRLS FUSION METHODS

The EFRLS algorithm estimates the state of the dynamic system at the ith sensor as follows (cf. [97]) :

$$\mathbf{x}^i_{k/k} = \Phi_{k-1}[\mathbf{x}^i_{k-1/k-1} + L^i_k(\mathbf{y}^i_k - H^i_k\Phi_{k-1}\mathbf{x}^i_{k-1/k-1})], \tag{6.4.1}$$

$$L^i_k = P^i_{k-1/k-1}\Phi'_{k-1}(H^i_k)'(\lambda_i I + H^i_k\Phi_{k-1}P^i_{k-1/k-1}\Phi'_{k-1}(H^i_k)')^{-1}, \tag{6.4.2}$$

$$P^i_{k/k} = \lambda_i^{-1}\Phi_{k-1}(I - L^i_k H^i_k\Phi_{k-1})P^i_{k-1/k-1}\Phi'_{k-1}, \tag{6.4.3}$$

where $0 < \lambda_i \leq 1$ is the forgetting factor. Similarly, when the fusion center can receive all observational data $\mathbf{y}_k = (\mathbf{y}^{1'}_k, \ldots, \mathbf{y}^{l'}_k)'$, the centralized EFRLS algorithm estimates the state of the dynamic system at the fusion center as follows:

$$\mathbf{x}_{k/k} = \Phi_{k-1}[\mathbf{x}_{k-1/k-1} + L_k(\mathbf{y}_k - H_k\Phi_{k-1}\mathbf{x}_{k-1/k-1})], \tag{6.4.4}$$

$$L_k = P_{k-1/k-1}\Phi'_{k-1}H'_k(\lambda_i I + H_k\Phi_{k-1}P_{k-1/k-1}\Phi'_{k-1}H'_k)^{-1}, \tag{6.4.5}$$

$$P_{k/k} = \lambda_{-1}\Phi_{k-1}(I - L_k H_k\Phi_{k-1})P_{k-1/k-1}\Phi'_{k-1}, \tag{6.4.6}$$

where $H_k = (H^{1'}_k, \ldots, H^{l'}_k)'$.

Now we develop two different methods to fuse all the local data $\mathbf{x}^i_{k/k}$ and $\mathbf{x}^i_{k-1/k-1}$ $(i = 1, 2, \ldots, l)$ so as for the two types of fusion to be the same as the centralized ERLS with two different forgetting factors.

Fusion I

Let

$$\mathbf{x}^i_{k/k-1} = \Phi_{k-1}\mathbf{x}^i_{k-1/k-1}, \quad \mathbf{x}_{k/k-1} = \Phi_{k-1}\mathbf{x}_{k-1/k-1}, \tag{6.4.7}$$

$$K^i_k = \Phi_{k-1}L^i_k, \quad K_k = \Phi_{k-1}L_k, \tag{6.4.8}$$

$$\begin{aligned} P^i_{k/k-1} &= \lambda_i^{-1}\Phi_{k-1}P^i_{k-1/k-1}\Phi'_{k-1}, \\ P_{k/k-1} &= \lambda_f^{-1}\Phi_{k-1}P_{k-1/k-1}\Phi'_{k-1}. \end{aligned} \tag{6.4.9}$$

Using (6.4.7)-(6.4.9), and the following definition of $P_{k/k}$ (see [97])

$$P_{k/k} = (X'_{k/k}\Lambda_k X_{k/k})^{-1}, \tag{6.4.10}$$

where

$$X_{k/k} = \begin{pmatrix} H_1\Phi_1^{-1}\cdots\Phi_{k-1}^{-1} \\ \vdots \\ H_{k-1}\Phi_{k-1}^{-1} \\ H_k \end{pmatrix}, \tag{6.4.11}$$

$$\Lambda_k = \begin{pmatrix} \lambda_f^{k-1}I & & & 0 \\ & \ddots & & \\ & & \lambda_f I & \\ 0 & & & I \end{pmatrix}, \tag{6.4.12}$$

we have (see [97])

$$\begin{aligned} K_k &= \lambda_f P_{k/k-1}H'_k(\lambda_f I + H_k\Phi_{k-1}P_{k-1/k-1}\Phi'_{k-1}H'_k)^{-1} \\ &= P_{k/k-1}H'_k(I + H_kP_{k/k-1}H'_k)^{-1} \\ &= P_{k/k-1}H'_k[I - (I + H_kP_{k/k-1}H'_k)^{-1}H_kP_{k/k-1}H'_k] \\ &= [P_{k/k-1} - P_{k/k-1}H'_k(I + H_kP_{k/k-1}H'_k)^{-1}H_kP_{k/k-1}]H'_k \\ &= P_{k/k}H'_k \end{aligned} \tag{6.4.13}$$

and

$$P_{k/k}^{-1} = P_{k/k-1}^{-1} + H'_kH_k. \tag{6.4.14}$$

Similarly, from (6.4.1)-(6.4.3), the definition of $P^i_{k/k}$ similar to $P_{k/k}$, we have

$$\begin{aligned} K^i_k &= \lambda_i P^i_{k/k-1}(H^i_k)'(\lambda_i I + H^i_k \Phi_{k-1} P^i_{k-1/k-1}\Phi'_{k-1}(H^i_k)')^{-1} \\ &= P^i_{k/k-1}(H^i_k)'(I + H^i_k P^i_{k/k-1}(H^i_k)')^{-1} \\ &= P^i_{k/k}(H^i_k)' \end{aligned} \tag{6.4.15}$$

and

$$(P^i_{k/k})^{-1} = (P^i_{k/k-1})^{-1} + (H^i_k)'H^i_k. \tag{6.4.16}$$

Thus, by the definition of H_k and Eqs. (6.4.14) and (6.4.16),

$$P^{-1}_{k/k} = P^{-1}_{k/k-1} + \sum_{i=1}^{l}[(P^i_{k/k})^{-1} - (P^i_{k/k-1})^{-1}]. \tag{6.4.17}$$

Then by (6.4.7), (6.4.4), (6.4.14) and (6.4.16), we have

$$\begin{aligned} (P^i_{k/k})^{-1}\mathbf{x}^i_{k/k} &= [(P^i_{k/k-1})^{-1} + (H^i_k)'H^i_k]\mathbf{x}^i_{k/k-1} + (H^i_k)'(\mathbf{y}^i_k - H^i_k\mathbf{x}^i_{k/k-1}) \\ &= (P^i_{k/k-1})^{-1}\mathbf{x}^i_{k/k-1} + (H^i_k)'\mathbf{y}^i_k, \end{aligned} \tag{6.4.18}$$

$$\begin{aligned} P^{-1}_{k/k}\mathbf{x}_{k/k} &= (P^{-1}_{k/k-1} + H'_k H_k)\mathbf{x}_{k/k-1} + H'_k(\mathbf{y}_k - H_k\mathbf{x}_{k/k-1}) \\ &= P^{-1}_{k/k-1}\mathbf{x}_{k/k-1} + H'_k\mathbf{y}_k \\ &= P^{-1}_{k/k-1}\mathbf{x}_{k/k-1} + \textstyle\sum_{i=1}^{l}(H^i_k)'\mathbf{y}^i_k \\ &= P^{-1}_{k/k-1}\mathbf{x}_{k/k-1} + \textstyle\sum_{i=1}^{l}[(P^i_{k/k})^{-1}\mathbf{x}^i_{k/k} - (P^i_{k/k-1})^{-1}\mathbf{x}^i_{k/k-1}]. \end{aligned} \tag{6.4.19}$$

After the derivation of the fusion formulae (6.4.17) and (6.4.19), we present the following two remarks.

Remark 6.4.1. Fusion I is of the same form as the Kalman filtering fusion proposed in (6.1.6)–(6.1.8) except using two different notations K_k and L_k as the gain factors in the recursive formulae (6.3.9), (6.4.1) and (6.4.4), although the matrices $P_{k/k}$ here represents the inverses of parameter matrices $X'_{k/k}\Lambda_k X_{k/k}$.

Remark 6.4.2. It is interesting that the forgetting factors λ_f at the fusion center and λ_i at each sensor may not be the same, that is, they may use their own favorite forgetting factors since they may know more about their own measurement errors. The fusion center also uses its own favorite forgetting factor λ_f. In spite of the variety of the local forgetting factors, the fused filtering from all local filtering is still equivalent to the centralized EFRLS with the fusion center's forgetting factor λ_f. In other words, the choices of the local forgetting factors do not influence the performance of the final fusion, only the forgetting factor selected by the fusion center matters. This property can guarantee Fusion I to give attention to both the local goal and global goal simultaneously therefore is very useful in practice. The reason for the property being true is that although in the fusion formulae (6.4.17) and (6.4.19), the local matrices $P^i_{k/k}$ and $P^i_{k/k-1}$, the local estimates $\mathbf{x}^i_{k/k}$ and $\mathbf{x}^i_{k/k-1}$ appear and are all dependent on λ^i, the really useful information for fusion is

$$(P^i_{k/k})^{-1} - (P^i_{k/k-1})^{-1} = (H^i_k)' H^i_k$$

and

$$(P^i_{k/k})^{-1}\mathbf{x}^i_{k/k} - (P^i_{k/k-1})^{-1}\mathbf{x}^i_{k/k-1} = (H^i_k)'\mathbf{y}^i_k.$$

However, the measurement matrices H^i_k and the observation $\mathbf{y}^i_k$ are both independent of λ_i.

Fusion II

Obviously, when local sensors choose good forgetting factors, the above Fusion I cannot take advantage of this. Hence, we now want to develop another recursive fusion method which is equivalent to the RLS estimate that uses all local measurements with their own forgetting factors. This Fusion II is the recursive algorithm of the following weighted LS problem

$$\mathbf{x}_{k/k} = \operatorname*{argmin}_x (Y_k' - x' X_{k/k}') \Lambda_k (Y_k - X_{k/k} x), \tag{6.4.20}$$

where $X_{k/k}$ is defined as (6.4.11).

$$Y_k = \begin{pmatrix} \mathbf{y}_1 \\ \vdots \\ \mathbf{y}_k \end{pmatrix}, \tag{6.4.21}$$

Λ_k is a $(k \times k)$ block diagonal matrix as follows.

$$\Lambda_k = \begin{pmatrix} \lambda^{k-1} & & & 0 \\ & \ddots & & \\ & & \lambda & \\ 0 & & & I \end{pmatrix}, \tag{6.4.22}$$

where I is an $(M \times M)$ unit matrix, $M = \sum_{i=1}^{l} N_i$ and λ is another diagonal matrix of the local forgetting factors $(\lambda_1, \ldots, \lambda_l)$, which is different from the scalar forgetting factor λ_f in Fusion I.

In the aforementioned case where all local forgetting factors $(\lambda_1, \ldots, \lambda_l)$ are good enough, it can be expected that Fusion II outperforms Fusion I. The simulations below support the above expectation.

We present Fusion II as follows:

$$\mathbf{x}_{k/k} = P_{k/k} \sum_{i=1}^{l} Q_k^i, \tag{6.4.23}$$

where

$$P_{k/k}^{-1} = \sum_{i=1}^{l} (P_{k/k}^i)^{-1}, \tag{6.4.24}$$

$$Q_k^i = \lambda_i (\Phi_{k-1}^{-1})' Q_{k-1}^i + [(P_{k/k}^i)^{-1} \mathbf{x}_{k/k}^i - (P_{k/k-1}^i)^{-1} \mathbf{x}_{k/k-1}^i], \tag{6.4.25}$$

$$Q_1^i = H_1 y_1^i. \tag{6.4.26}$$

The proof of the equivalence of the above recursive fusion algorithm and the solution of the previous weighted LS problem (6.4.20) can be completed via a sequence of lammas. It is easy to verify that if all forgetting factors in Fusion I are identical, i.e., $\lambda_f = \lambda_1 = \cdots = \lambda_l$, $P_{k/k}$ in Fusion II is actually $P_{k/k}$ in Fusion I, and the two fusion methods are the same. Besides, the communications between the local sensors and the fusion center for Fusion I and II are the same.

To prove the equivalence of the centralized LS solution (6.4.20)–(6.4.22) and Fusion II (6.4.23)–(6.4.26), rewrite

$$X_{k/k} = \begin{pmatrix} H_1 \Phi_1^{-1} \cdots \Phi_{k-1}^{-1} \\ \vdots \\ H_{k-1} \Phi_{k-1}^{-1} \\ 0 \end{pmatrix} + \begin{pmatrix} 0 \\ \vdots \\ 0 \\ H_k \end{pmatrix} \triangleq h_1 + h_2. \tag{6.4.27}$$

Using the the standard result for the weighted LS problem, we have the solution to Eq. (6.4.20) is

$$\begin{aligned} \mathbf{x}_{k/k} &= (X_{k/k}' \Lambda_k X_{k/k})^{-1} (X_{k/k}' \Lambda_k Y_k) \\ &= P_{k/k} \textstyle\sum_{i=1}^{l} Q_k^i, \end{aligned} \tag{6.4.28}$$

where

$$Q_k^i = X_{k/k}^i \Lambda_k^i Y_k^i. \tag{6.4.29}$$

Let

$$P_k = (h_1' \Lambda_k h_1)^{-1}. \tag{6.4.30}$$

Thus,

$$P_{k/k}^{-1} = P_k^{-1} + H_k' H_k. \tag{6.4.31}$$

Then, we have the following lammas.

Lemma 6.4.1. $P_k = [\sum_{i=1}^{l} (P_{k/k-1}^i)^{-1}]^{-1}$.

Proof. Similar to (6.4.27) and (6.4.22), denote

$$X_{k/k}^i = \begin{pmatrix} H_1^i \Phi_1^{-1} \cdots \Phi_{k-1}^{-1} \\ \vdots \\ H_{k-1}^i \Phi_{k-1}^{-1} \\ H_k^i \end{pmatrix}, \quad k = 1, 2, ..., \tag{6.4.32}$$

$$\Lambda_k^i = \begin{pmatrix} \lambda_i^{k-1} I & & & 0 \\ & \ddots & & \\ & & \lambda_i I & \\ 0 & & & I \end{pmatrix}. \tag{6.4.33}$$

Thus,

$$P_{k/k}^i = (X_{k/k}^i \Lambda_k^i X_{k/k}^i)^{-1}. \tag{6.4.34}$$

Using the above (6.4.30) and Eqs. (6.4.22) and (6.4.32)–(6.4.34) yields

$$\begin{aligned} P_k &= [(\Phi_{k-1}^{-1})' X_{k-1/k-1}' diag(\lambda, \ \ldots, \ \lambda) \Lambda_{k-1} X_{k-1/k-1} \Phi_{k-1}^{-1}]^{-1} \\ &= [(\Phi_{k-1}^{-1})' \textstyle\sum_{i=1}^{l} (X_{k-1/k-1}^i)' \lambda_i \Lambda_{k-1}^i X_{k-1/k-1}^i \Phi_{k-1}^{-1}]^{-1} \\ &= [\textstyle\sum_{i=1}^{l} (P_{k/k-1}^i)^{-1}]^{-1}. \end{aligned} \tag{6.4.35}$$

■

Lemma 6.4.2. $Q_k^i = \lambda_i (\Phi_{k-1}^{-1})' Q_{k-1}^i + [(P_{k/k}^i)^{-1} \mathbf{x}_{k/k}^i - (P_{k/k-1}^i)^{-1} \mathbf{x}_{k/k-1}^i]$.

Proof.

$$\begin{aligned} Q_k^i &= X_{k/k}^i \Lambda_k^i Y_k^i \\ &= (\Phi_{k-1}^{-1})' (X_{k-1/k-1}^i)' \lambda_i \Lambda_{k-1}^i Y_{k-1}^i + H_k^i \mathbf{y}_k^i \\ &= \lambda_i (\Phi_{k-1}^{-1})' Q_{k-1}^i + H_k^i \mathbf{y}_k^i \\ &= \lambda_i (\Phi_{k-1}^{-1})' Q_{k-1}^i + [(P_{k/k}^i)^{-1} \mathbf{x}_{k/k}^i - (P_{k/k-1}^i)^{-1} \mathbf{x}_{k/k-1}^i]. \end{aligned} \tag{6.4.36}$$

■

Lemma 6.4.3. $P_{k/k}^{-1} = \sum_{i=1}^{l}(P_{k/k}^{i})^{-1}$.

Proof. Noting Lemma 6.4.1, Eq. (6.4.16) and $P_{k/k}^{-1} = P_k^{-1} + H_k' H_k$, we have

$$\begin{aligned} P_{k/k}^{-1} &= \sum_{i=1}^{l}(P_{k/k-1}^{i})^{-1} + \sum_{i=1}^{l}(H_k^i)' H_k^i \\ &= \sum_{i=1}^{l}(P_{k/k}^{i})^{-1}. \end{aligned} \qquad (6.4.37)$$

■

From (6.4.28) and Lemmas 6.4.2–6.4.3, the LS estimate $\mathbf{x}_{k/k}$ given in (6.4.20)–(6.4.22) can be computed recursively from sensor estimates $\mathbf{x}_{k/k}^i$, $\mathbf{x}_{k/k-1}^i$ and matrices $P_{k/k}^i$, $P_{k/k-1}^i$ via (6.4.23)–(6.4.26).

6.4.3 SIMULATIONS

The simulations were done for a dynamic system modelled as an object moving in a circular trajectory at constant speed with process noise and measurement noise. The simulation results support the analysis in the last Subsection and show that the fused EFRLS algorithms can track the object satisfactorily. We also compare the fused EFRLS algorithms with a fused Kalman filter in tracking the moving object. The simulations show that the EFRLS performs slightly worse than the Kalman filter when all the assumptions for the Kalman filter are fulfilled. However, it has been shown in [97] that when the noise terms are time-correlated, the performance of the EFRLS can be better than Kalman filter with knowledge of the precise covariances of both process noise and measurement noise but without knowledge of the noise autocorrelation.

The object dynamics and measurement equations are modelled as follows.

$$\mathbf{x}_{k+1} = \Phi \mathbf{x}_k + v_k, \qquad (6.4.38)$$

$$y_k^i = H^i \mathbf{x}_k + w_k^i, \quad i = 1, 2, 3. \qquad (6.4.39)$$

The position in the plane, is chosen as the state. We assume, without loss of generality, that the circle is about the origin $(0 \quad 0)'$ of the plane.

$$\Phi = \begin{pmatrix} \cos(2\pi/300) & \sin(2\pi/300) \\ -\sin(2\pi/300) & \cos(2\pi/300) \end{pmatrix} \qquad (6.4.40)$$

is a constant rotational matrix and the initial state $\mathbf{x}_0 = (50 \quad 0)'$. Given all these, in the absence of noise the object would move in a circle of radius 50 about the origin of the coordinate plane. The measurement coefficient matrices H^i are also constant and given by

$$H^i = \begin{pmatrix} 1 & i \\ 1 & -i \end{pmatrix}, \quad i = 1, 2, \quad H^3 = \begin{pmatrix} 1 & 3 \\ -1 & 3 \end{pmatrix}. \qquad (6.4.41)$$

The initial value $P^i_{0/0}$ for the EFRLS procedures is $((H^i)'H + 0.1I)^{-1}$. The covariances R_{v_k} and R_{w_k} of the noises are diagonal, given by

$$R_v = \begin{pmatrix} 2 & 0 \\ 0 & 2 \end{pmatrix}, \quad R_{w^i} = \begin{pmatrix} 1 & 0 \\ 0 & 1 \end{pmatrix}, \quad i = 1, 2, 3. \tag{6.4.42}$$

We ran the Kalman filtering fusion and our two methods for 50 times, where we chose the forgetting factors at each sensor as well as possible, that is, the second fusion method may be used in this case. We show the average square of track errors of each sensor and their fusion for Kalman filtering and the two EFRLS fusion methods in the following Figs. 6.4.1, 6.4.2 and 6.4.3.

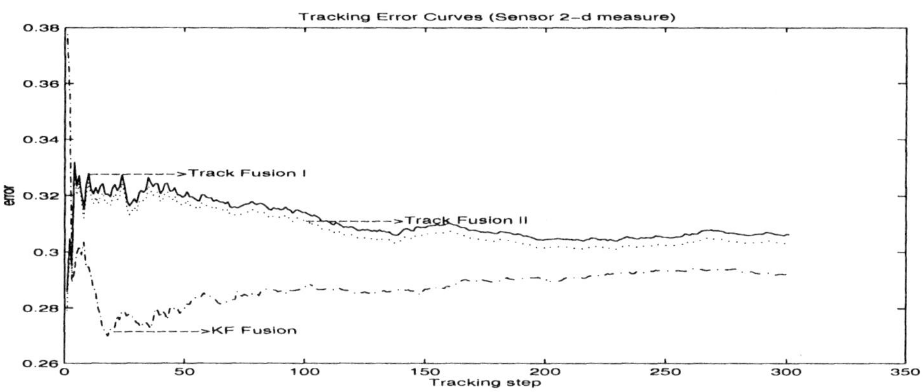

Figure 6.4.1. Three track fusion results with 2D measurements.

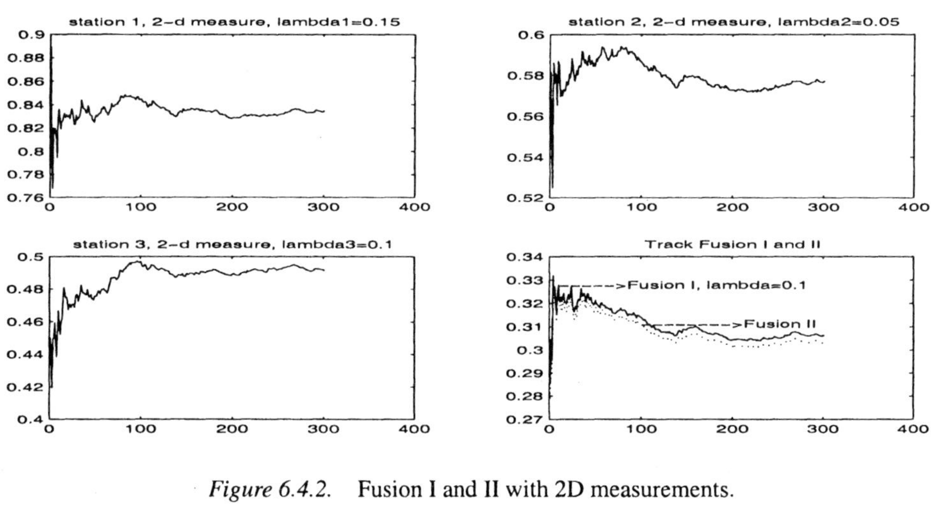

Figure 6.4.2. Fusion I and II with 2D measurements.

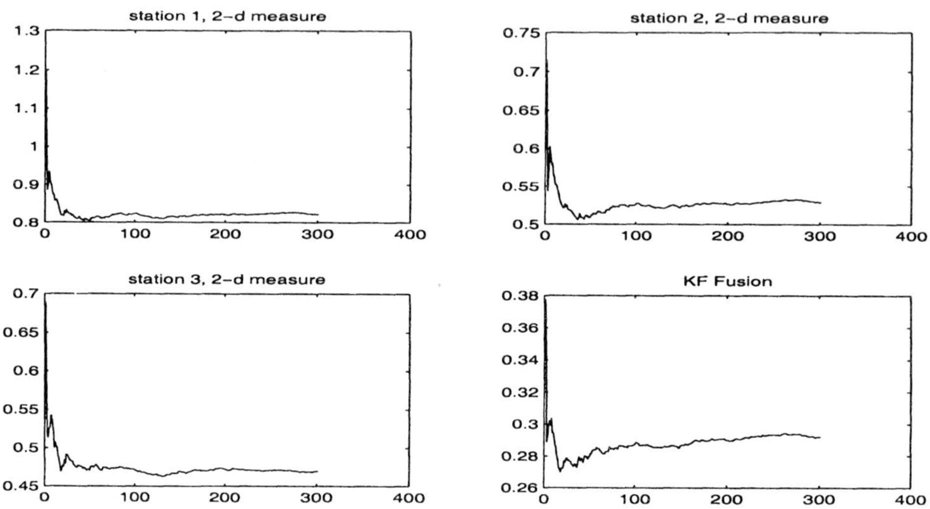

Figure 6.4.3. Kalman filter fusion with 2D measurements.

Then, we consider one dimensional measurement equations as follows:

$$H^1 = [1 \quad 2]; \; H^2 = [2 \quad 3]; \; H^3 = [3 \quad 4]; \; R_{w^i} = [1 \quad 0]. \qquad (6.4.43)$$

The performance curves are shown in Figs. 6.4.4, 6.4.5 and 6.4.6.

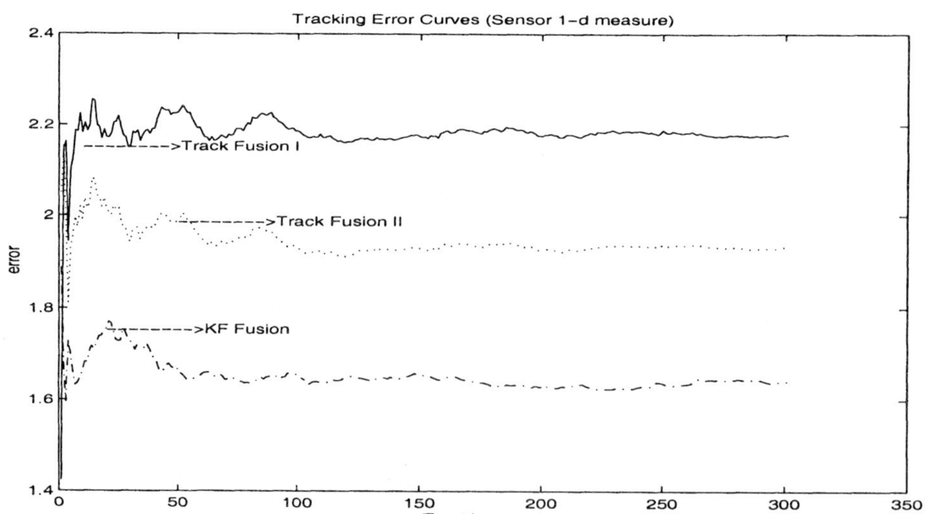

Figure 6.4.4. Three track fusion results with 1D measurements.

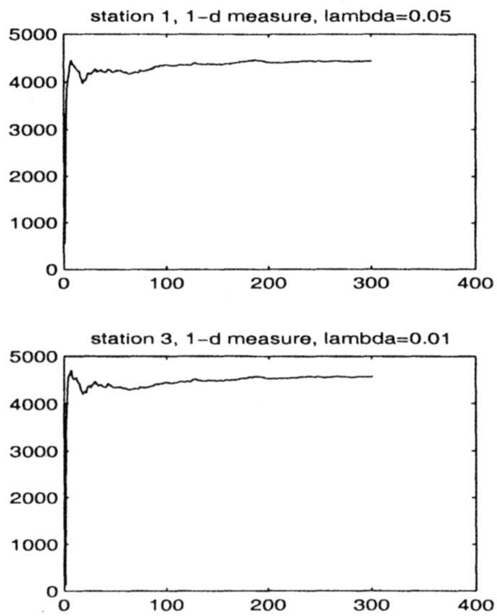

Figure 6.4.5. Fusion I and II with 1D measurements.

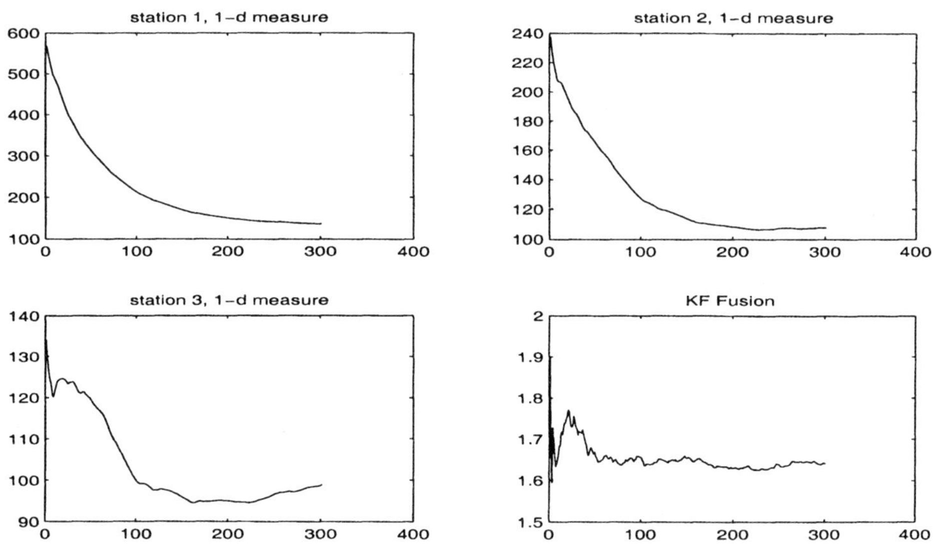

Figure 6.4.6. Kalman filter fusion with 1D measurements.

It can be seen that the sensor tracking errors are too big when the sensor measurement is 1-dimensional, even for the Kalman filtering. However, the fused tracks can provided much improved performance.

Chapter 7

MULTISENSOR INTERVAL ESTIMATION FUSION

In this chapter, we consider the multisensor interval estimation fusion which is different from the point estimation fusion. In the point estimation problems, a popular criterion is to minimize the distance between the estimate and the true value under a proper metric, such as the minimum variance estimation. In spite of the true value unknown, the error distance is still able to be minimized via given the second moments of model noises, such as the parameter estimation for a linear model, or Kalman filtering. In the interval estimation problems, although there still exist some metrics to measure error distance between intervals and true value, such as Hausdorff metric (for example, see [36]), minimizing the error distance has no much practical value. Therefore, from practical point of view, the interval estimation problem of true value is much more in favor of the interval coverage probability (confidence degree) and the length of the covering interval. On the other hand, there may be some connections between the two types of estimations. When one derives a minimum error variance point estimation, the neighborhood of an estimate should usually have greater coverage probability of the true value. This leads us to simplify interval estimation fusion in some cases (see Subsection 7.1.4).

In this chapter, we still consider the two level optimization strategy, i.e., each sensor first derives its own interval estimate in terms of local purpose, and then the fusion center uses all received local messages to get a final interval estimation. In doing so, there are two types of the available sensor messages: (1) sensor statistics and their joint distribution; (2) sensor interval outputs and their confidence degrees. Since sensor statistical interval estimations are in fact functions of sensor statistics, the former has more direct knowledge of sensor observations and more sensor information than the latter. Therefore, we can deal with the optimal fusion problem for the former in Section 7.1 based classical statistical interval estimation methods. The latter have no the knowledge of sensor observations, only the sensor outputs: intervals and their confidence degrees are available. Hence, to optimally fuse these intervals and confidence degrees, we will develop a combination method of sensor outputs in Section 7.2. In Section 7.3, we briefly introduce some of the present results on the fault-tolerant interval estimation fusion, and use the combination method proposed in Section 7.2 to deal with the fault-tolerant interval estimation fusion.

7.1 STATISTICAL INTERVAL ESTIMATION FUSION USING SENSOR STATISTICS

In this section, we deal with the fusion problem of multiple statistical interval estimators for the a unknown parameter θ by using sensor statistics. We are concerned mainly with the interval estimation fusion via pivotal quantity method. A multisensor convex linear statistic fusion model for the optimal interval estimation fusion is established. A Gauss-Seidel iteration algorithm for searching for the optimal fusion weights is suggested. In particular, the convex linear (error) minimum variance (LMV) fusion that is similar to the point LMV estimation fusion given in Chapter 6 and reduces computation of the fusion weights and yields nearly optimal estimate performance generally. Numerical examples are provided and give additional support to the analytic results. Besides, in Subsection 7.1.2, we also briefly introduce the interval estimation fusion by inverting a hypothesis testing.

7.1.1 PROBLEM FORMULATION

Many practical problems involve multiple sample data to estimate a true parameter θ. While many results do represent major progress on the point estimate fusion, there exist few results on the interval estimation[1] fusion. After reviewing and developing some results on interval mathematics and interval analysis, the point Kalman filtering was extended to the interval Kalman filtering in [16]. In many practical applications, however, one may be more interested in finding an interval covering θ with a required probability than a guess of a single value as the value of θ. For example, it is not necessary to guarantee a missile to hit target exactly, it is enough for the effective explosion region of the missile to cover the target with a required probability.

There exists an l-sensor distributed system, where every sensor observes data $\mathbf{x}_i = \{x_1^i, \ldots, x_n^i\}$ $(i \leq l)$. After each local sensor/estimator, namely the ith estimator, observes data $\mathbf{x}_i$, chooses a proper statistic $T_i(\mathbf{x}_i)$, and then gives its interval estimate for local purpose, a significant problem is how to utilize the

[1] If θ is a vector, the interval estimate is extended to a set estimate. Without loss of generality, only the interval estimate will be considered in this book.

available information from all local estimators for the purpose of obtaining a best interval of covering θ.

Suppose that the local estimators can transmit the local statistic $T_i(\mathbf{x}_i)$ (as well known, two ends of a interval estimate are actually functions of the statistic). The remaining problem now is how to construct a good final statistic from $\{T_1(\mathbf{x}_1), \ldots, T_l(\mathbf{x}_l)\}$ and an interval estimation method to obtain the best estimation performance in terms of a certain criterion.

Obviously, the optimal interval estimate fusion in general depends on not only the criterion, but also the conditional distributions $F(\mathbf{x}_1, \ldots, \mathbf{x}_l|\theta)$ and the local statistics $\{T_1(\mathbf{x}_1), \ldots, T_l(\mathbf{x}_l)\}$. Without knowledge of them, it is impossible generally to yield the optimal statistical fusion. In this section, we concentrate on the following type of interval estimation.

Suppose that all local statistics $\{T_1(\mathbf{x}_1), \ldots, T_l(\mathbf{x}_l)\}$ are unbiased estimates of θ, or more generally, a known function $F_i(T_i(\mathbf{x}_i))$ $(i \leq l)$ is an unbiased estimate of θ (i.e., $EF_i(T_i(\mathbf{x}_i)) = E\theta$, denote $F_i(T_i)$ still by T_i). This assumption is valid in many practical applications, e.g., sample mean, sample variance, and particularly Kalman filtering. Besides, we have the following example.

Example 7.1.1. Suppose that $\{x_1, \ldots, x_n\}$ is an i.i.d. sample from a uniform $(0, \theta)$ population, the statistic is $T(\mathbf{x}) = \max\{x_1, \ldots, x_n\}$. Then we know $\frac{n+1}{n}T(\mathbf{x})$ is an unbiased estimator of θ (cf. Example 7.3.5, page 320 in [11]).

To make an interval estimate of θ at the ith sensor, usually, we use the pivotal quantity method to find the confidence interval of θ as follows.

First, let $S_i(T_i(\mathbf{x}_i), \theta)$ be a function of $T_i(\mathbf{x}_i)$ and θ, and $S_i(T_i(\mathbf{x}_i), \theta)$ satisfies the following conditions :

1) The distribution of $S_i(T_i(\mathbf{x}_i), \theta)$ does not depend on θ.

2) For $\forall a < b$, inequality $a \leq S_i(T_i(\mathbf{x}_i), \theta) \leq b$ can be converted to $A \leq \theta \leq B$, A and B are determined only by $S_i(T_i(\mathbf{x}_i, \theta)$, a, and b.

Second, when the observation $\mathbf{x}_i)$ is received, using the above properties of the pivotal quantity $S_i(T_i(\mathbf{x}_i), \theta)$, we can calculate an interval estimate for θ as $\theta_L^i(T_i(\mathbf{x}_i)) \leq \theta \leq \theta_U^i(T_i(\mathbf{x}_i))$ from $a \leq S_i(T_i(\mathbf{x}_i), \theta) \leq b$. The ends $\theta_L^i(T_i(\mathbf{x}_i))$ and $\theta_U^i(T_i(\mathbf{x}_i))$ of the interval are functions of $T_i(\mathbf{x}_i)$. Hence, for interval estimation of a single sensor, a important thing is to find a good pivotal quantity. But this is difficult sometimes, and there is no general method to do this in the traditional interval estimation.

In the distributed system, the fusion center collects $T_i(\mathbf{x}_i)$ from all local sensors, denoted by

$$T(\mathbf{x}) = [T_1(\mathbf{x}_1)\ T_2(\mathbf{x}_2)\ \cdots\ T_l(\mathbf{x}_l)]' \tag{7.1.1}$$

and then use them to obtain a fused optimal interval estimation in some sense. Thus, to reach this goal, there are more difficulties than the single sensor case. First, when we are combining the data $\{T_1(\mathbf{x}_1), \cdots, T_l(\mathbf{x}_l)\}$, we must simultaneously consider whether or not we can find a pivotal quantity of the fused data. Second, even if we find the pivotal quantity that satisfies the above conditions 1) and 2), the distribution of the pivotal quantity must be available, at least, in numerical form. Third, to derive the optimal fusion, i.e., the optimal combination coefficients, the computation should be intractable (cf. Subsection 7.1.3). Sometimes, a nearly optimal solution is required to reduce the computation (for example, see Subsection 7.1.4).

To formulate a criterion for a general interval estimator $[L(T(\mathbf{x})),\ U(T(\mathbf{x}))]$ of θ, denote the *coverage probability* of $[L(T(\mathbf{x})),\ U(T(\mathbf{x}))]$ by

$$P_\theta(\theta \in [L(T(\mathbf{x})),\ U(T(\mathbf{x}))]),$$

and the *confidence degree* of $[L(T(\mathbf{x})),\ U(T(\mathbf{x}))]$ is the infimum of the coverage probabilities,

$$\inf_\theta P_\theta(\theta \in [L(T(\mathbf{x})),\ U(T(\mathbf{x}))]).$$

In practice, Neyman criterion for the optimal interval is used quite popularly, i.e., one minimizes the the length of the covering interval given the confidence degree β.

$$\min\{c : \inf_\theta P_\theta(\theta \in [L(T(\mathbf{x})),\ U(T(\mathbf{x}))]) = \beta, U - L = c\}. \tag{7.1.2}$$

Remark 7.1.1. As many other multisensor data fusion problems involving precise computation of some probabilities in evaluating fusion performances, such as decision fusion, a joint probability distribution of all sensor data is necessary here; that is, this assumption is not restrictive theoretically. On the other hand, this is not always available in practice, particularly, when $T_1, \ldots, T_l$ are not independent of each other and the correlations among them are unknown. This is a restriction of the statistical interval estimation fusion. However, if one relaxes the requirement of precise coverage probability, only maximizes the lower bound of the coverage probability in terms of the Chebyshev inequality, then only the covariance matrix of all T_i $(i \le l)$ is needed (see Subsection 7.1.4 below). It is more realistic sometimes in practice that the fusion center can receive only sensor outputs: intervals and their confidence levels. This type of the interval estimation fusion will be discussed in Section 7.2.

7.1.2 OPTIMAL CONVEX LINEAR FUSION

The optimal fusion depends on a chosen class of functions of $T(\mathbf{x}) = (T_1'(\mathbf{x}_1), \ldots, T_l'(\mathbf{x}_l))'$. If there was no restriction on this, it would be impossible in general to solve the optimal fusion problem. In this subsection, we restrict ourselves to optimize the fused statistic T_f only over convex linear functions of $(T_1(\mathbf{x}_1), \ldots, T_l(\mathbf{x}_l))$. More precisely speaking, to find an available pivotal quantity for the combination of $(T_1(\mathbf{x}_1), \ldots, T_l(\mathbf{x}_l))$, we may choose a linear combination of the functions F_i of $T_i(\mathbf{x}_i))$, $i \leq l$, namely, a linear combination of $F_i(T_i(\mathbf{x}_i))$, $i \leq l$, but they still must be unbiased estimates and their distributions must be known. In this case, we can view $F_i(T_i(\mathbf{x}_i))$ as $T_i(\mathbf{x}_i)$.

As mentioned in Remark 7.1.1, suppose that the fusion center knows the joint distribution of $(T_1(\mathbf{x}_1), \ldots, T_l(\mathbf{x}_l))$. There are two ways to construct pivotal quantity of the fused $(T_1(\mathbf{x}_1), ..., T_l(\mathbf{x}_l))$: **I)** directly based on the joint distribution of $(T_1(\mathbf{x}_1), \ldots, T_l(\mathbf{x}_l))$, **II)** based on a known joint distribution of $(S_1(T_1(\mathbf{x}_1), \theta), \ldots, S_l(T_l(\mathbf{x}_l), \theta))$, where $S_i(T_i(\mathbf{x}_i), \theta)$ is the pivotal quantity of the ith sensor. When $S_i(T_i(\mathbf{x}_i), \theta)$ are all unbiased, we view S_i as T_i; therefore, the two ways become one way.

Assume that all local statistics $\{T_1(\mathbf{x}_1), \ldots, T_l(\mathbf{x}_l)\}$ are unbiased estimates of θ. To maintain the unbiasedness, naturally, the fused final statistic should be the following convex combination

$$\begin{array}{c} T_f(T_1(\mathbf{x}_1), \ldots, T_l(\mathbf{x}_l)) = \sum_{i=1}^{l} w_i T_i(\mathbf{x}_i)) = WT \\ \text{subject to } \sum_{i=1}^{l} w_i = 1, \end{array} \tag{7.1.3}$$

where $W = (w_1 \quad w_2 \quad \ldots \quad w_l)$. Here we do not need the weights to be nonnegative. This is a generalized convex combination such that the fused statistic T_f is still an unbiased estimate of θ. In fact, it is possible for the optimal weights to be negative as shown in the following example.

Example 7.1.2. Suppose that the two sensor statistics T_1 and T_2 have the following joint Gaussian Distribution:

$$N\left(\begin{pmatrix}\theta\\ \theta\end{pmatrix}, \begin{pmatrix}1 & 2\\ 2 & 6\end{pmatrix}\right).$$

According to the result and formula (7.1.22) in Subsection 7.1.4 below, it is easy to get the two optimal weights $w_1 = 4/3 \; w_2 = -1/3$.

After getting the fused statistic T_f, we need to construct a final pivotal quantity $S(T_f, \theta)$.

Similarly, if we know all pivotal quantities $\{S_1(T_1(\mathbf{x}_1)), \ldots, S_l(T_l(\mathbf{x}_l))\}$, then the fused pivotal quantity should be the following convex combination,

$$\begin{array}{c} S(S_1(T_1(\mathbf{x}_1)), \ldots, S_l(T_l(\mathbf{x}_l))) = \sum_{i=1}^{l} w_i S_i(T_i(\mathbf{x}_i))) \\ \text{subject to } \sum_{i=1}^{l} w_i = 1 \text{ and } w_i \geq 0, \; i \leq l. \end{array} \tag{7.1.4}$$

To keep the fused S monotonic with respect to θ, we may not fuse T_i or S_i directly but fuse their properly transformations (see Example 7.1.5 below).

To make final statistical interval estimate, it is necessary to compute distribution of the fused statistic or their functions. For the statistical interval estimation, only continuous distributions are considered usually.

Suppose that $(T_1(\mathbf{x}_1), \ldots, T_l(\mathbf{x}_l))$ (or $(S_1(T_1(\mathbf{x}_1)), \ldots, S_l(T_l(\mathbf{x}_l))))$ have joint density $f(y_1, \ldots, y_l|\theta)$, then we need to find the density $f_{T_f}(t|\theta)$ of the linear fused statistic $T_f(T_1(\mathbf{x}_1), \ldots, T_l(\mathbf{x}_l))$.

For the most general joint density, using a direct consequence of the multi-dimensional change of variable formula for integral (Proposition 35.1 in [64]), and the following changes of variables

$$t_1 = \sum_{i=1}^{l} w_i y_i, \quad t_2 = y_2, \; \ldots, \; t_l = y_l, \tag{7.1.5}$$

and noting that the above mapping from $\Re^l \to \Re^l$ is one to one and the inverse Jacobian determinant

$$\det(\frac{dy}{dt}) = \begin{vmatrix} \frac{1}{w_1} & -\frac{w_2}{w_1} & \cdots & -\frac{w_l}{w_1} \\ 0 & 1 \;\; 0 \cdots & & 0 \\ & \cdots & \cdots & \\ 0 & \cdots\cdots & 0 & 1 \end{vmatrix} = 1/w_1, \tag{7.1.6}$$

we thus have the required fused density

$$f_{T_f}(t|\theta) = \frac{1}{w_1} \int f(\frac{1}{w_1}t - \frac{w_2}{w_1}t_2 - \cdots - \frac{w_l}{w_1}t_l, t_2, \ldots, t_l|\theta) dt_2 \cdots dt_l. \tag{7.1.7}$$

In particular, if $T_1(\mathbf{x}_1), \ldots, T_l(\mathbf{x}_l)$ are conditionally independent mutually given θ, the fused density is simply

$$f_{T_f}(t|\theta) = \frac{1}{w_1} \int f_{T_1}(\frac{1}{w_1}t - \frac{w_2}{w_1}t_2 - \cdots - \frac{w_l}{w_1}t_l|\theta) f_{T_2}(t_2|\theta) \cdots f_{T_l}(t_l|\theta) dt_2 \cdots dt_l. \tag{7.1.8}$$

In practice, it is not always necessary to calculate the above density. For example, when all local statistics T_i, $i = 1, \ldots, l$, have Gaussian distributions, the fused statistic T_f has Gaussian distribution with the same mean θ. In this case, it suffices to calculate the covariance of T. Since the joint density $f(y_1, \ldots, y_l|\theta)$ is given, it is easy to get $Cov(T)$.

Now we have the distribution function of the fused statistic T. We then need to find a suitable pivotal quantity in order to calculate interval of θ. After the pivotal quantity of the fused statistic T_f is determined, the optimal fusion in terms of the criterion (7.1.2) reduces to optimizing the convex combination weights $\{w_i\}_{i=1}^l$.

7.1.3 COMPUTATION OF THE OPTIMAL WEIGHTS

In general, it is impossible to get a closed optimal solution of the weights in terms of (7.1.2). Therefore, we can now define an iterative algorithm in terms of criteria (7.1.2) by considering the following Gauss-Seidel iterative process.

$$\begin{aligned}\min_{W^{(k+1)}}\{c : \inf_\theta P_\theta(\theta \in [L(\textstyle\sum_{j=1}^i w_j^{(k+1)} T_j(\mathbf{x}_j) + \sum_{j=i+1}^l w_j^{(k)} T_j(\mathbf{x}_j)),\\ U(\textstyle\sum_{j=1}^i w_j^{(k+1)} T_j(\mathbf{x}_j) + \sum_{j=i+1}^l w_j^{(k)} T_j(\mathbf{x}_j))])\\ = \beta,\ c = U - L\}\end{aligned} \tag{7.1.9}$$

with the initial value $W^{(0)} = (w_1^{(0)},..., w_l^{(0)})$ being given, where β is a required confidence degree.

Obviously, the above objective function is monotonic as k goes to infinity. On the other hand, it is upper bounded and lower bounded, respectively. Hence, the objective converges. Two simple termination criteria of the iteration process are to stop when

$$\begin{aligned}\inf_\theta P_\theta(\theta \in [L(\textstyle\sum_{i=1}^l w_i^{(k+1)} T_i(\mathbf{x}_i)),\quad U(\sum_{i=1}^l w_i^{(k+1)} T_i(\mathbf{x}_i))])\\ - \inf_\theta P_\theta(\theta \in [L(\textstyle\sum_{i=1}^l w_i^{(k)} T_i(\mathbf{x}_i)), U(\sum_{i=1}^l w_i^{(k)} T_i(\mathbf{x}_i))]) \leq \epsilon,\end{aligned} \tag{7.1.10}$$

or

$$c^{(k+1)} - c^{(k)} \leq \epsilon',$$

($c^{(j)}$ is the jth step solution for the following problem

$$\begin{aligned}\arg\min_{W^{(j)}}\{ c^{(j)} : \inf_\theta P_\theta(\theta \in [L(\textstyle\sum_{i=1}^l w_i^{(j)} T_i(\mathbf{x}_i)), U(\sum_{i=1}^l w_i^{(j)} T_i(\mathbf{x}_i))])\\ = \beta,\quad c^{(j)} = U - L\}),\end{aligned} \tag{7.1.11}$$

where $\epsilon, \epsilon' > 0$ are both pre-specified tolerance parameters.

To facilitate computer implementation of the above process, we need to discretize the variables. Let the discretization of $w_1, w_2, \ldots, w_l$ be given by

$$\{w_{11}, w_{12}, \ldots, w_{1N_1}\}, \{w_{21}, w_{22}, \ldots, w_{2N_2}\}, \ \ldots, \ \{w_{l1}, w_{l2}, \ldots, w_{lN_l}\},$$

respectively. In addition, since the above optimization involves integral calculation, and it is impossible in general to obtain a close solution, we need to discretize the integral variables to get numerical solution as well. Since the objective becomes now discrete, bounded and monotonic, the above algorithm can achieve an equilibrium (may be local minimum) after a finite number of iterations no matter if the weight w converges.

Although the above algorithm is tractable computationally, it has two main drawbacks. First, Gauss-Seidel iteration does not guarantee the algorithm to converge to the global optimal solution, which strongly depends on the initial values $(w_1^{(0)}, \ldots, w_l^{(0)})$. Second, it costs much computation due to multi-dimensional integral calculation; therefore, may not be applicable for some real-time processing, such as for state tracking for dynamical systems.

Besides, although the above algorithm is established generally, in practice, the derivation of confidence interval depends strongly on pivotal quantity. It is not easy sometimes to get a pivotal quantity of the statistic T_f and parameter θ.

Because of the above problems, in the following subsection, a promising approach to determining initial values is presented, which can reduce the computation greatly.

7.1.4 NEARLY OPTIMAL LINEAR FUSION

It is well-known that the *Chebyshev inequality* is an approximate coverage probability estimate:

$$P(|\sum_i^l w_i T_i(\mathbf{x}_i) - \theta| \le c) \ge 1 - \frac{E(\sum_i^l w_i T_i(\mathbf{x}_i) - \theta)^2}{c^2} \qquad (7.1.12)$$

and

$$\begin{aligned} &\{\theta : |\textstyle\sum_i^l w_i T_i(\mathbf{x}_i) - \theta| \le c\} \\ &= \{\theta : \theta \in [\textstyle\sum_i^l w_i T_i(\mathbf{x}_i) - c, \quad \sum_i^l w_i T_i(\mathbf{x}_i) + c]\}. \end{aligned} \qquad (7.1.13)$$

When the confidence degree β is given, we can find c by solving the following equation

$$\beta = 1 - \frac{E(\sum_i^l w_i T_i(\mathbf{x}_i) - \theta)^2}{c^2} \qquad (7.1.14)$$

provided that

$$E(\sum_{i=1}^l w_i T_i(x_i) - \theta)^2$$

is obtained. Then, we get the coverage interval, where the length of interval is $2c$. It follows that the criterion (7.1.2) can be reduced approximately to the following minimization problem:

$$\min_{\sum_i w_i = 1} E(\sum_i^l w_i T_i(\mathbf{x}_i) - \theta)^2, \qquad (7.1.15)$$

which is in fact the LMS estimate of the weights under convex linear constraint. Assume that the following covariance matrix is given (for example, see Example

7.1.4 below).

$$\Sigma = \begin{bmatrix} \sigma_{11} \cdots \sigma_{1l} \\ \cdots \\ \sigma_{l1} \cdots \sigma_{ll} \end{bmatrix}, \tag{7.1.16}$$

where l is the number of local estimates to be fused and

$$\sigma_{ij} = E[(T_i(\mathbf{x}_i) - \theta)(T_j(\mathbf{x}_j) - \theta)'].$$

We rewrite Σ as its matrix form as follows:

$$\Sigma = E(T - A'\theta)(T - A'\theta)', \tag{7.1.17}$$

where A is a l-dimensional row vector $(1 \cdots 1)$.

Thus, the LMS fusion problem becomes one of a matrix quadratic optimization problem subject to a linear equality constraint:

$$W = \arg\min_W E(WT - \theta)^2 = \arg\min_W WE(T - A'\theta)(T - A'\theta)'W' \tag{7.1.18}$$

subject to

$$AW' = 1. \tag{7.1.19}$$

Substituting this equation and (7.1.17) into (7.1.18) and (7.1.19) yields

$$W = \arg\min_{AW'=1} W\Sigma W', \tag{7.1.20}$$

which is also an LMV estimate of the weights under convex linear constraint as discussed in Section 6.2. Applying Theorem 6.2.2 there, we have the following propositions:

Proposition 7.1.1. The optimal fusion WT is given uniquely with probability one by

$$WT = \tfrac{1}{l}A(I - \Sigma(P\Sigma P)^{+})T \quad \text{almost everywhere.} \tag{7.1.21}$$

Moreover, if Σ is of full rank, we have a more explicit expression of w_i that are expressed only by sub-matrices of Σ^{-1} (see Theorem 6.2.3 in Subsection 6.2.2).

Proposition 7.1.2. If Σ is of full rank, we have the following explicit expression of each element w_i:

$$w_i = \sum_{j=1}^{l} \sigma_{ij}^{(-1)} / \sum_{r,j=1}^{l} \sigma_{rj}^{(-1)} \tag{7.1.22}$$

and

$$E(WT - \theta)^2 = \sum_{r,j=1}^{l} \sigma_{rj}^{(-1)}, \tag{7.1.23}$$

where $\sigma_{rj}^{(-1)}$ is the (r, j)th element of Σ^{-1}.

Using Proposition 7.1.2, the optimal weights in Example 7.1.2 can be derived easily.

If there exists a solution of the above minimization problem that is independent of θ, such solution can be used as a set of near optimal weights; if not, we can take the solution of

$$\sup_{\theta} \min_{\sum_i w_i=1} E(\sum_i^l w_i T_i(\mathbf{x}_i) - \theta)^2$$

as a set of the nearly optimal weights.

It is worth noting that the solution of (7.1.15) is not exactly the solution of (7.1.2) generally, but the above inequality (7.1.12) guarantee the LMV solution to be a good approximate solution, particularly when the required distribution information is unknown and we only have the knowledge of the second moments for the statistics (at least, it is much easier to estimate second moment than distribution from samples). In addition, for some probability distributions, such as Gaussian distribution, the LMV solution is the exact solution of (7.1.2). Even if the two solutions are not the same, the above LMV solution is certainly a good initial value of Gauss-Seidel iterations and can greatly reduce the searching range of the weights. Hence, solving the minimization problem Eq. (7.1.15) is significant, and we have solve it in Subsection 6.2.2.

7.1.5 NUMERICAL EXAMPLES

In this subsection, we provide four examples for interval estimation fusion: three two-sensor examples and a three-sensor example. In the three two-sensor examples, there are two where the two sensor have the same types of Gaussian and uniform distributions respectively, and one where the two sensor have Gaussian and unform distributions respectively. Those examples use the sensor statistic fusion and the sensor pivotal quantity fusion methods. In the three sensor example, two sensors have Gaussian distributions and one has uniform distribution. Due to much computation for this case, this example uses nearly optimal linear fusion method.

Example 7.1.3. Suppose that two sensors observe data to estimate a parameter θ. The first sensor observes i.i.d. $\{x_1, \ldots, x_{10}\}$ coming from a Gaussian distribution $N(\theta, 1)$, and its statistic is $T_1(x) = \sum_i x_i/10$. The distribution of T_1 is $N(\theta, 1/10)$. At the second sensor, its i.i.d. sample $\{y_1, \ldots, y_{10}\}$ comes from a $N(\theta, \gamma^2)$, and its statistic $T_2(y) = \sum_i y_i/10$, the distribution of T_2 is $N(\theta, \gamma^2/10)$. Assume x and y are independent of each other.

Clearly, the distribution of fused statistic T_f is Gaussian distribution. We still consider $T_f = w_1T_1 + w_2T_2$. Using the pivotal quantity method and Gauss-Seidel iterative algorithm, we can obtain the confidence interval of θ with confidence degree 0.9.

Suppose that $\theta = 2$, $\gamma = 2$.

Sample data of first sensor:

[2.6232, 2.7990, 2.9409, 1.0079, 2.2120, 2.2379, 0.9922, 1.2580, 3.0823, 1.8685].

Sample data of second sensor:

[2.7798, 2.1760, 0.7291, 0.8809, 2.8873, 0.1002, 3.5624, 3.1379, 0.3566, 1.4688].

The numerical result for this example is given in Table 7.1.1.

Table 7.1.1. Interval fusion for two Gauss observation sensors.

	Confidence interval of θ	optimal weights
Convex Linear fusion	$\theta \in [1.5741,\ 2.5302]$	$w_1 = 0.7, w_2 = 0.3$
First sensor	$\theta \in [1.5678,\ 2.6366]$	$w_1 = 1, w_2 = 0$
Second sensor	$\theta \in [0.7390,\ 2.8767]$	$w_1 = 0, w_2 = 1$

Example 7.1.4. Suppose that two sensors observe data to estimate a parameter θ, where the first sensor observes $\{x_1, \ldots, x_{10}\}$ coming from a uniformly distributed population $(0, \theta)$, and its statistic is $T_1(x) = \max(x_1, \ldots, x_{10})$. We know $\frac{11}{10}T_1(x)$ is an unbiased estimate of θ(cf. example 7.3.5, page 320 [11]), and the density function of $\frac{11}{10}T_1(x)$ is $f_{T_1}(t_1) = \left(\frac{11}{10}\right)^{10} 10t_1^9/\theta^{10}, 0 \leq t_1 \leq \frac{11}{10}\theta$. Set $S_1(T_1(x), \theta) = T_1(x)/\theta$. Clearly, the $S_1(T_1(x), \theta)$ is a pivotal quantity of $T_1(x)$ for the interval estimate of θ. The second sensor observes $\{y_1, \ldots, y_{10}\}$ coming from a uniformly distributed population $(0, \gamma\theta), \gamma > 0$, and its statistic is $T_2^*(y) = \max(y_1, \ldots, y_{10})$, $\frac{11}{10}T_2^*(y)\big/\gamma$ is an unbiased estimate of θ, denote $T_2(y) = \frac{11}{10}T_2^*(y)\big/\gamma$, the density function of $T_2(y)$ is $f_{T_2}(t_2) = \left(\frac{11}{10}\right)^{10} 10t_2^9/\theta^{10}$, $0 \leq t_2 \leq \frac{11}{10}\theta$. Set $S_2(T_2(y), \theta) = T_2(y)/\theta$. Clearly, $S_2(T_2(y), \theta)$ is a pivotal quantity of $T_2(y)$ as well. Assume x and y are independent of each other.

We use the method of the convex linear fusion based on local pivotal quantities. The fused pivotal quantity is

$$W'T = w_1S_1(T_1(x), \theta) + w_2S_2(T_2(y), \theta) = w_1T_1(x)/\theta + w_2T_2(y)/\theta.$$

Assume $T_1(x)/\theta$ and $T_2(y)/\theta$ are independent of each other and their density functions are known, we can calculate their joint density function. Then we get the density function of T by using (7.1.7). Using the pivotal quantity method and the Gauss-Seidel iterative algorithm (7.1.9), we can obtain the confidence interval of θ with confidence degree 0.9.

Suppose that $\theta = 2$, $\gamma = 2$.

Sample data of the first sensor:

[0.8307, 0.6100, 1.7487, 0.0300, 1.5359, 1.9417, 1.9802, 1.5777, 0.8773, 0.9966].

Sample data of the second sensor:

[1.7117, 5.1479, 2.5603, 7.6808, 5.8131, 3.2956, 5.9565, 2.1436, 3.5194, 7.4670].

The numerical result for this example is given in Table 7.1.2.

Table 7.1.2. Interval fusion for two uniform observation sensors.

	Confidence interval of θ	optimal weights
Convex Linear fusion	$\theta \in [1.9502,\ 2.3383]$	$w_1 = w_2 = 0.5$
First local sensor	$\theta \in [1.9802,\ 2.4783]$	$w_1 = 1, w_2 = 0$
Second local sensor	$\theta \in [1.9202,\ 2.4033]$	$w_1 = 0, w_2 = 1$

Due to the independence of $S_1(T_1(x), \theta)$ and $S_2(T_2(y), \theta)$ are now independently and identically distributed, the fused weights is $w_1 = w_2 = 0.5$.

Example 7.1.5. Suppose that two sensors observe data to estimate a parameter θ, where the first sensor observes $\{x_1, \ldots, x_{10}\}$ coming from a $N(\theta, 1)$, and its statistic is $T_1(x) = \sum_i x_i/10$, the distribution of T_1 is $N(\theta, 1/10)$. Set $S(T_1(x), \theta) = \sqrt{10}(T_1(x) - \theta)$. The distribution of $S(T_1(x), \theta)$ is $N(0, 1)$. Clearly, $S_1(T_1(x), \theta)$ is a pivotal quantity of T_1 to estimate interval of θ. At the second sensor, its i.i.d. sample $\{y_1, \ldots, y_{10}\}$ comes from a uniformly distributed population $(0, \theta)$, and its statistic $T_2(y) = \max(y_1, \ldots, y_{10})$. We know the density function of $T_2(y)$ is $f_{T_2}(t_2) = 10t_2^9/\theta^{10}$, $0 \le t_2 \le \theta$. Set $S_2(T_2(y), \theta) = T_2(y)/\theta$, clearly, $S(T_2(y), \theta)$ is a pivotal quantity of $T_2(y)$ to estimate interval of θ. But, if we use fusion based on the local pivotal quantities directly,

$$S = w_1 S_1(T_1(x), \theta) + w_2 S_2(T_2(y), \theta) = w_1\sqrt{10}(T_1(x) - \theta) + w_2 T_2(y)/\theta$$

is a nonlinear function of θ; therefore, the solution of the inequality of θ is complicated. To simplify this, we transform the local pivotal quantity $S_2(T_2(y), \theta)$, and its mean equals that of $S_1(T_1(x), \theta)$. We have

$$h(S_2(T_2(y), \theta)) = \frac{\theta}{T_2(y)} - \frac{10}{9},$$

which is a linear function of θ and therefore still a monotonic pivotal quantity of θ, and we can calculate its distribution from that of $T_2(y)/\theta$. The fusion of the two local pivotal quantities is then given by:

$$S = w_1 S(T_1(x), \theta) + w_2 h(S(T_2(y), \theta)) = w_1\sqrt{10}(T_1(x) - \theta) + w_2(\frac{\theta}{T_2(y)} - \frac{10}{9}).$$

We can obtain the confidence interval of θ with confidence degree 0.9.

Suppose that $\theta = 10$. Sample data of first sensor:
[9.6001, 10.6900, 10.8156, 10.7119, 11.2902, 10.6686, 11.1908, 8.7975, 9.9802, 9.8433].

Sample data of second sensor:
[0.1527, 7.4679, 4.4510, 9.3181, 4.6599, 4.1865, 8.4622, 5.2515, 2.0265, 6.7214].

The numerical result for this example is given in Table 7.1.3.

Table 7.1.3. Interval fusion for a uniform observation sensor and a Gaussian sensor.

	Confidence interval of θ	optimal weights
Convex Linear fusion	$\theta \in [9.8896,\ 10.8313]$	$w_1 = 0.991, w_2 = 0.009$
First sensor	$\theta \in [9.8244,\ 10.8933]$	$w_1 = 1, w_2 = 0$
Second sensor	$\theta \in [9.3181,\ 11.1818]$	$w_1 = 0, w_2 = 1$

From the above results, it can be seen that the interval estimate of fusion center is the best of all above interval estimates, and the estimate of the first sensor is better than that of the second sensor. In fact, the variance of $T_1(x)$ is $1/10$, and the variance of $T_2(y)$ is $\theta^2/120$. When $\theta \geq \sqrt{12}$, the variance of statistic of the first sensor is smaller than the variance of statistic of the second sensor. Therefore, $w_1 > w_2$ makes sense.

Example 7.1.6. Suppose that three sensors observe data to estimate a parameter $\theta \leq 11$. When the first sensor observes $\{x_1, \ldots, x_{10}\}$ coming from uniformly distributed population $(0, \theta)$, and its statistic is $T_1^*(x) = \max(x_1, \ldots, x_{10})$, we know $\frac{11}{10}T_1^*(x)$ is an unbiased estimate of θ. Denote $T_1(x) = \frac{11}{10}T_1^*(x)$. The density function of $T_1(x)$ is $f_{T_1}(t_1) = \left(\frac{10}{11}\right)^{10} 10t_1^9/\theta^{10}$, $0 \leq t_1 \leq \frac{11}{10}\theta$. The second sensor observes $\{\mathbf{y}_1, \ldots, \mathbf{y}_{10}\}$ coming from $N(\theta, 1)$, and its statistic is $T_2(y) = \sum_i \mathbf{y}_i/10$. The distribution of T_2 is $N(\theta, 1/10)$. The third sensor observes $\{z_1, \ldots, z_{10}\}$ coming from $N(\theta, \gamma^2)$, and its statistic is $T_3(z) = \sum_i z_i/10$. The distribution of T_3 is $N(\theta, \gamma^2/10)$. Assume x, y and z are mutually independent.

Because of fusing the three local sensors, we require a great deal of computation by using pivotal quantity method. We use nearly optimal linear fusion.

After a simple calculation, we have $E(T_1-\theta)^2=\frac{\theta^2}{120}$. Because T_1, T_2 and T_3 are independent of each other, using Proposition 7.1.2, we have

$$w_1=\frac{120}{\theta^2}\Big/(\frac{120}{\theta^2}+10+10\big/\gamma^2),\quad w_2=10\Big/(\frac{120}{\theta^2}+10+10\big/\gamma^2),$$

$$w_3=\frac{10}{\gamma^2}\Big/(\frac{120}{\theta^2}+10+10\big/\gamma^2)$$

and

$$E(w_1T_1+w_2T_2+w_3T_3-\theta)^2=(\frac{120}{\theta^2}+10+10\big/\gamma^2)^{-1}$$

Because of $\theta\leq 11$ and suppose $\gamma=0.2$, the solution of $\sup\min_\theta E(w_1T_1+w_2T_2+w_3T_3-\theta)^2$ is $w_1=0.0038$, $w_2=0.0383$ and $w_3=0.9579$, $E(w_1T_1+w_2T_2+w_3T_3-\theta)^2=0.0038$.

Sample data of the first sensor:
[2.9741, 0.4916, 6.9318, 6.5011, 9.8299, 5.5267, 4.0007, 1.9879, 6.2520, 7.3336].

Sample data of the second sensor:
[9.1095, 10.1391, 9.7639, 9.9245, 9.6414, 7.9224, 9.8565, 11.3933, 10.6518, 9.6229].

Sample data of the third sensor:
[9.8677, 10.0498, 9.9233, 9.8943, 10.0111, 10.2508, 9.4960, 10.1170, 9.7984, 10.1889].

The numerical result is given in the following table:

Table 7.1.4. Interval fusion for a uniform observation sensor and two Gaussian sensors.

	Confidence interval of θ	Optimal weights
Nearly opt. fusion	$\theta\in[9.7575,\ 10.1489]$	$w_1=.0038, w_2=.0383, \omega_3=.9579$
First sensor	$\theta\in[9.8299,\ 11.7959]$	$w_1=1, w_2=0, \omega_3=0$
Second sensor	$\theta\in[9.2681, 10.3369]$	$w_1=0, w_2=1, \omega_3=0$
Third sensor	$\theta\in[9.7207,\ 10.1987]$	$w_1=0, w_2=0, \omega_3=1$

In this example, the confidence degree is 0.9, but the confidence degree of nearly optimal linear fusion may be in fact larger than 0.9. Therefore, to obtain shorter estimate interval, this method is suitable to more local sensor case. From above example, the length of confidence interval in the fusion center is less than the length of confidence interval of any local sensors, where the interval estimate of third sensor is only slightly worse than the fused interval estimation because its statistic variance is very small ($<\gamma^2/10\leq 0.0004$).

7.1.6 INVERTING A HYPOTHESIS TESTING

In this subsection, we briefly introduce the interval estimation fusion by inverting a hypothesis testing. There are very strong correspondence between interval estimation and hypothesis testing (see Chapter 9 in [12]). In fact, we can say generally that every confidence set corresponds to a hypothesis test and vice versa. Therefore, using our well-known results on the hypothesis test fusion, we can give another way to implement the interval estimation fusion. The following theorem gives a formal version of this correspondence.

Proposition 7.1.3. (The first part of Theorem 9.2.2 in [12]). For each $\theta_0 \in \Theta$, let $A(\theta_0)$ be acceptance region of a level α test of $\mathbf{H}_0 : \theta = \theta_0$ (i.e., $P_f \leq \alpha$). For each $\mathbf{x}$, define a set $C(\mathbf{x})$ in the parameter space by

$$C(\mathbf{x}) = \{\theta_0 : \mathbf{x} \in A(\theta_0)\}. \tag{7.1.24}$$

Then the random set C(**X**) (X is a sample data) is a $1 - \alpha$ confidence set.

Proof. Since $A(\theta_0)$ is the acceptance region of a level α test,

$$P_f = P(X \notin A(\theta_0)|\mathbf{H}_0) \leq \alpha.$$

Hence

$$P(X \in A(\theta_0)|\mathbf{H}_0) = 1 - P_f \geq 1 - \alpha.$$

Since θ_0 is arbitrary, write θ instead of θ_0. The above inequality, together with (7.1.24), shows that the coverage probability of the set $C(X)$ is given by

$$P(\theta \in C(X)|\mathbf{H}_0) = P(X \in A(\theta)|\mathbf{H}_0) \geq 1 - \alpha.$$

Hence, $C(X)$ is a $1 - \alpha$ confidence set. ∎

Clearly, it is a relatively easy task to construct a level α acceptance region. The difficult task is constructing a confidence set. So the method of obtaining a confidence set by inverting an acceptance region is quite useful. Note, however, that we carefully used the word *set* rather than *interval*. This is because there is no guarantee that the confidence set obtained by test inversion will be an interval. This is a restriction of this method.

All of the techniques we have developed in Subsections 7.1.1–7.1.3 for the optimal linear fusion of sensor statistics can still be used. Only the derivation of the final covering interval at the fusion center comes from Proposition 7.1.3. We now describe this interval estimation fusion briefly.

Step I. Every sensor gives its own acceptance region $A_i(\theta_0)$ of a level α test of $\mathbf{H}_0 : \theta = \theta_0$. Write θ instead of θ_0. Invert $A(\theta)$ to a confidence set $C(X_i)$

of a level $1 - \alpha \triangleq \beta$. Then communicate their statistics $(X_1, \ldots, X_l)$ to the fusion center.

Step II. Compute the distribution of $(W_1X_1 + \cdots + W_lX_l)$. Using Proposition 7.1.3, derive a confidence interval $C(W_1X_1 + \cdots + W_lX_l)$ of a level β.

Step III. In terms of the criterion of (7.1.2), compute the optimal weights by using Gauss-Seidel algorithm (7.1.9), and obtain the corresponding shortest confidence interval of a level β.

7.2 INTERVAL ESTIMATION FUSION USING SENSOR ESTIMATES

In the previous section, the interval estimation fusion was based on the classical interval estimation methods. Thus, the various restrictions in the classical interval estimation methods are unavoidable. Besides, when local sensors do not want to share all intimate details of their own systems, or types of multiple sensor messages, e.g., image message, voice message, language message, and digital message, are too distinct to obtain the joint distribution of all sensor observations, the available information for the fusion center is only the local interval estimates and their confidence degrees given by all sensors. In these two cases, without statistics of sensors and their joint distribution, the fusion methods given in Section 7.1 cannot be implementable. In this section, under the assumption of independent estimates across sensors, we will develop a combination method for the fusion center to optimally give a fused interval estimate and its confidence degree only based on the available sensor outputs: local interval estimates with their confidence degrees. When the fusion center has extra information, a conditional combination method can be derived straightforwardly. As an application, the fault-tolerant interval estimation fusion will be also presented.

7.2.1 OUTPUTS OF SENSORS

Consider the following case: there are l sensors to estimate an unknown parameter θ. Each sensor yields an estimated closed interval[2] of θ and their confidence degrees of covering θ, denoted as

$$I_i = [a_i,\ b_i],\ a_i \le b_i, \quad i \le l, \tag{7.2.1}$$

and

$$\beta_i = P_i(\theta \in I_i), \quad i \le l. \tag{7.2.2}$$

[2]More generally, the interval I_i could be $(a_i,\ b_i]$, $[a_i,\ b_i)$, and $(a_i,\ b_i)$.

In fact, the confidence degrees here is similar to the confidence degrees defined in Section 7.1.1, a minor difference is that the latter is an infimum of coverage probability over a set of θ. From the above message, we know that the true θ is covered by either I_i or its complementary set $I_i^c \triangleq (-\infty,\ a_i) \cup (b_i,\ \infty)$, which we still call *interval* and confidence degree of I_i^c for covering θ is

$$P_i(\theta \in I_i^c) = 1 - \beta_i, \quad i \leq l. \tag{7.2.3}$$

In practical applications, the two ends $-\infty$ and ∞ of I_i^c could be finite real numbers L_i and U_i, i.e., $I_i^c \triangleq [L_i,\ a_i) \cup (b_i,\ U_i]$. Even if so, we are still not able to use the two intervals $[L_i,\ a_i)$ and $(b_i,\ U_i]$ individually because we have no information on their own confidence degrees. Thus, each sensor actually outputs two intervals

$$I_i^1 \triangleq I_i, \quad I_i^0 \triangleq I_i^c \tag{7.2.4}$$

and their confidence degrees

$$\beta_i^1 \triangleq \beta_i, \quad \beta_i^0 \triangleq 1 - \beta_i. \tag{7.2.5}$$

Then, a local message received by the fusion center is an *interval list* of the outputs of all sensors

$$\{I_1^{r_1},\ \ldots,\ I_l^{r_l}\}, \quad r_i = 0, 1$$

and its corresponding *confidence degree list* of the outputs of all sensors

$$\{\beta_1^{r_1},\ \ldots,\ \beta_l^{r_l}\}, \quad r_i = 0, 1.$$

Then, both lists are called *sensor outputs.*

7.2.2 COMBINATION RULE OF SENSOR OUTPUTS

Using sensor outputs, under the assumption of independent estimations across sensors, we propose the following combination rule for the combined intervals and their confidence degrees at the fusion center.

The Combined Intervals

The combined intervals at the fusion center are all possible intersections of the sensor interval lists and their all possible connected unions.

Firstly, consider the following nonempty intersections of all different sensor interval lists

$$I_{\{1^{r_1},\ldots,l^{r_l}\}} = \bigcap_{i=1}^{l} I_i^{r_i}, \quad r_i = 0, 1. \tag{7.2.6}$$

Then, various connected unions of these nonempty intersections yield more intervals.

$$I_{\cup_{r=1}^{N}\{1^{r_1},\ldots,l^{r_l}\}} = \bigcup_{r=1}^{N} \bigcap_{i=1}^{l} I_i^{r_i}, \quad r_i = 0, 1. \tag{7.2.7}$$

The union of all of these nonempty intersections is the entire straight line $\Re$.

Since the two ends of l sensor interval outputs partition the straight line at most into $2l$ intervals (take the unconnected $I_{\{1^0,\ldots,l^0\}}$ as an interval), the upper bound of possible numbers of all nonempty intersections of sensor intervals is $2l$. That is to say, the number of all possible nonempty intersections of sensor intervals grows polynomially as the number of sensors increases.

The Combined Confidence Degrees

The combined confidence degrees of the nonempty intervals at the fusion center are defined by

$$P(\theta \in I_{\{1^{r_1},\ldots,l^{r_l}\}}|\mathcal{C}) = \frac{1}{c}\prod_{i=1}^{l} \beta_i^{r_i}, \quad r_i = 0, 1, \tag{7.2.8}$$

where the symbol $\mathcal{C}$ is defined as

$$\mathcal{C} = \{I_{\{1^{r_1},\ldots,l^{r_l}\}} : \; I_{\{1^{r_1},\ldots,l^{r_l}\}} \neq \emptyset\}. \tag{7.2.9}$$

Obviously, $\mathcal{C}$ is never empty. Therefore, the parameter c is given by

$$c = \sum_{I_{\{1^{r_1},\ldots,l^{r_l}\}} \neq \emptyset} \prod_{i=1}^{l} \beta_i^{r_i}, \tag{7.2.10}$$

and

$$P(\theta \in I_{\{1^{r_1},\ldots,l^{r_l}\}}|\mathcal{C}) = 0 \quad \text{if } I_{\{1^{r_1},\ldots,l^{r_l}\}} = \emptyset. \tag{7.2.11}$$

Summabilty of the Combined Confidence Degrees

Since I_i^0 and I_i^1 for any i are disjoint, all of these nonempty intersections are also disjoint, and the confidence degrees of nonempty intersections are summable, i.e.,

$$P(\theta \in \bigcup_{r=1}^{N} I_{\{1^{r_1},\ldots,l^{r_l}\}}|\mathcal{C}) = \frac{1}{c}\sum_{r=1}^{N}\prod_{i=1}^{l} \beta_i^{r_i}, \tag{7.2.12}$$

the union of all of the nonempty intersections is the entire straight line $\Re$, and its confidence degree is

$$P(\theta \in \bigcup_{I_{\{1^{r_1},\ldots,l^{r_l}\}} \neq \emptyset} I_{\{1^{r_1},\ldots,l^{r_l}\}}|\mathcal{C}) = 1, \tag{7.2.13}$$

that is to say, the combined sets and their confidence degrees are summable.

Now, we give an example to show how to apply the above combination rule.

Example 7.2.1. We have the following sensor interval estimates and their confidence degrees

$$\{[8,\ 11],\ [9,\ 12],\ [10,\ 13]\},\ \{0.8,\ 0.83,\ 0.85\}.$$

Therefore, all possible sensor outputs are

$$I_1^1 = [8,\ 11],\ I_1^0 = (-\infty,\ 8) \cup (11,\ \infty),\ I_2^1 = [9,\ 12],$$

$$I_2^0 = (-\infty,\ 9) \cup (12,\ \infty),\ I_3^1 = [10,\ 13],\ I_3^0 = (-\infty,\ 10) \cup (13,\ \infty)$$

and

$$\beta_1^1 = 0.8,\ \beta_1^0 = 0.2;\ \beta_2^1 = 0.83,\ \beta_2^0 = 0.17;\ \beta_3^1 = 0.85,\ \beta_3^0 = 0.15.$$

Using the above combination rule (7.2.6)–(7.2.10), the six fused nonempty intervals and their confidence degrees are given in Table 7.2.1.

Table 7.2.1. Fused interval outputs of Example 7.2.1.

Intervals	$(-\infty, 8) \cup (13, \infty)$	$[8, 9)$	$[9, 10)$	$[10, 11]$	$(11, 12]$	$(12, 13]$
Confi. degree	0.0059	0.0237	0.1159	0.6567	0.1642	0.0336

Since the intervals and their confidence degrees are summable, from Table 7.2.1, we can further obtain more intervals and their confidence degrees. For example, we have the original intervals estimated by three sensor and their confidence degrees:

$$[8,\ 11] : 0.7963;\ [9,\ 12] : 0.9368;\ [10,\ 13] : 0.8545.$$

Comparing them with single sensor outputs, the modifications of confidence degrees are given by the combination rule and are reasonable intuitively. The confidence degree of [9, 12] is improved significantly because [9, 12] is overlapped by [8, 11] and [10, 13], and the length sum of the two overlapped parts [9, 11] and [10, 12] are totally 4, but the corresponding length sums for other two intervals are both totaly 3. In other words, the estimate of the second sensor receives the largest support from other two sensors. Since the confidence degree of [8, 11] is the least one in the three intervals, and the significant improvement of the confidence degree of [9, 12] gives negative impact to other intervals, the fused confidence degree of [8, 11] is a little bit smaller than that of single sensor case. Similarly, we can explain why the confidence degree of [10, 13] is improved a little.

It is possible that the intersection $I_{\{1^1,\dots,l^1\}}$ of all sensor interval estimates are empty. In this case, some (generalized intervals) of $I_{\{1^{r_1},\dots,l^{r_l}\}}$ may consist of several unconnected intervals. We consider the following example.

Example 7.2.2. The following are sensor interval estimates and their confidence degrees

$$\{[8,\ 11] : 0.8;\ [12,\ 14] : 0.83;\ [13,\ 19] : 0.85\}.$$

Therefore, all possible sensor outputs are

$$I_1^1 = [8,\ 11],\ I_1^0 = (-\infty,\ 8) \cup (11,\ \infty),\ I_2^1 = [12,\ 14],$$

$$I_2^0 = (-\infty,\ 12) \cup (14,\ \infty),\ I_3^1 = [13,\ 19],\ I_3^0 = (-\infty,\ 13) \cup (19,\ \infty)$$

and

$$\beta_1^1 = 0.8,\ \beta_1^0 = 0.2;\ \beta_2^1 = 0.83,\ \beta_2^0 = 0.17;\ \beta_3^1 = 0.85,\ \beta_3^0 = 0.15.$$

Using the above combination rule (7.2.6)–(7.2.10), the five fused nonempty intervals and their confidence degrees are given in Table 7.2.2.

Table 7.2.2. Fused interval outputs of Example 7.2.2.

Intervals	$(-\infty, 8) \cup (11, 12) \cup (13, \infty)$	$[8, 11)$	$[12, 13)$	$[13, 14]$	$(14, 19]$
Confi. degree	0.0231	0.0926	0.1130	0.6402	0.1311

Then, the confidence degrees of the original three intervals estimated by the sensors now become

$$[8,\ 11] : 0.0926;\ [12,\ 14] : 0.7532;\ [13,\ 19] : 0.7713.$$

In addition, we can use any two ends of sensor intervals to generate more connected intervals to estimate θ. However, we might not have precise confidence degrees for some of them, but at least we know their lower bounds. For example,

$$[8,\ 14] : \geq 0.8458;\ [8,\ 19] : \geq 0.9769.$$

These intervals with their unprecise confidence degrees can still be useful for the optimal interval estimation fusion (see Example 7.2.4 below). Thus, the combined outputs at the fusion center have three advantages:

I. Deriving more intervals with their confidence degrees, i.e., higher resolution rate of intervals;

II. Deriving more reasonable coverage probability distribution over the entire real number space since we properly take advantage of more information coming from multiple sensors;

III. Easy extension to higher dimensional set estimation fusion problems.

Remark 7.2.1. A special case of our combination rule is $\beta_i^1 = 1$ for all $i \leq l$, this interval estimation fusion was well known before.

Remark 7.2.2. The assumption of independent sensor estimates is necessary for the derivation of the combined confidence degrees given in (7.2.8). In practice, if θ is deterministic, the sensor estimates could be independent of each other when sensor errors are independent of each other; otherwise, if θ is random, the sensor estimates are usually dependent on each other. When sensor estimates are significantly correlated, to get the confidence degrees of the combined intervals, the joint probabilities $P(\theta \in I_{\{1^{r_1},\ldots,l^{r_l}\}})$ $(r_i = 0, 1)$ of all possible sensor interval outputs are required, where all intervals $I_i^{r_i}$ are fixed. If they can be known from specific experiment or historical data, we still have a similar combination rule.

Remark 7.2.3. The basic idea of this combination rule is similar to the combination rule given by the Dempster-Shafer theory the basic definitions and combination rule of the Dempster-Shafer evidence theory is provided below (for details, see (7.2.14), (7.2.15), and [28, 69]). However, a key difference between the two combination rules is that the confidence degrees (*basic probability assignment*) defined in the Dempster-Shafer theory are not summable, and unsuitable for the interval estimation problem. For example, according to the Dempster-Shafer theory, the coverage probability of union of two disjoint intervals is not necessarily equal to the sum of confidence degrees of the two intervals, even might be zero, e.g., when equalities (7.2.2) and (7.2.3) hold, by the definition of the Dempster-Shafer theory (see Definition 7.2.1 below), one must have $P_i(\theta \in \Re) = 0$. Intuitively, this is unacceptable.

The basic definitions and combination rule of the Dempster-Shafer evidence theory is provided below. It has the following three basic notions:

Definition 7.2.1. Basic Probability Assignment: For a finite set Ξ, denote the set of its subsets by ϑ. Let $m : \vartheta \to [0, 1]$ be a function such that 1) $m(\emptyset) = 0$, 2) $m(E) \geq 0, \forall\ E \in \vartheta$, 3) $\sum m(E) = 1$. Then, $m(\cdot)$ is called as a basic probability assignment on ϑ.

Definition 7.2.2. Belief Function: Given a probability assignment $m(\cdot)$, the belief function of a set A is $Bel(A) = \sum_{B \subseteq A} m(B), \forall\ A \in \vartheta$.

Definition 7.2.3. Plausibility Function: Given a probability assignment $m(\cdot)$, the plausibility function is $Pla(A) = 1 - Bel(\bar{A}), \forall\ A \in \vartheta$, where $\bar{A}$ is the complement of A.

$m(\cdot)$ actually defines probability on ϑ. It quantifies the degree of uncertainty in one's belief. It can be easily shown that $Bel(A) \leq Pla(A)$ — in fact, belief function $Bel(A)$ and plausibility function $Pla(A)$ are the minimum and maximum amount of evidence that supports A, respectively. In this way, $[Bel(A),\ Pla(A)]$ is naturally referred to as the belief interval of A.

The following combination rule that combines multiple m_i into a single $m(\cdot)$ is fundamental for most applications of the Dempster-Shafer theory.

Combination Rule. From l different $m_1, \ldots, m_l$, we have a combined basic probability assignment $m_1 \oplus \cdots \oplus m_l(\cdot)$ as follows:

$$m(A) = m_1 \oplus \cdots \oplus m_l(A) = c^{-1} \sum_{\cap_{i=1}^{l} A_{r_i} = A} \prod_{i=1}^{l} m_i(A_{r_i}), \quad \forall A \neq \emptyset, \tag{7.2.14}$$

where c is a normalization factor:

$$c = \sum_{\cap_{i=1}^{l} A_{r_i} \neq \emptyset} \prod_{i=1}^{l} m_i(A_{r_i}). \tag{7.2.15}$$

Note that it is possible that $c = 0$ in (7.2.15). In this case, $m_1 \oplus \cdots \oplus m_l(\cdot)$ is not defined. Clearly, this never happens for the combination rule proposed in this subsection.

7.2.3 OPTIMIZATION CRITERIA

After the combination of sensor outputs, according to various practical requirements, we can choose a proper interval estimate fusion in terms of an optimization criterion.

The most popular criteria in practice are the following two:

(A) minimizing interval length under confidence degree constraint;

(B) maximizing confidence degree under interval length constraint.

Example 7.2.3. The sensor outputs are the same as Example 7.2.1. In terms of Criteria (A) and (B), the optimal interval estimation fusion is given in Table 7.2.3.

Table 7.2.3. Optimal fusion under Criteria (A) and (B) for Example 7.2.3.

Confi. degree constraint	≥ 0.6	≥ 0.8	≥ 0.9
Interval Length constraint	≤ 1	≤ 2	≤ 3
Optimal interval	[10, 11]	[10, 12]	[9, 12]
Confidence degree	0.6567	0.8209	0.9368
Interval length	1	2	3

Example 7.2.4. The sensor outputs are the same as Example 7.2.2. In terms of Criteria (A) and (B), the optimal interval estimation fusion is given in Table 7.2.4.

Table 7.2.4. Optimal fusion under Criteria (A) and (B) for Example 7.2.4.

Confi. degree constraint	≥ 0.6	≥ 0.8	≥ 0.95
Interval Length constraint	≤ 1	≤ 7	≤ 11
Optimal interval	[13, 14]	[8, 14]	[8, 19]
Confidence degree	0.6567	0.8458	0.9769
Interval length	1	6	11

7.3 FAULT-TOLERANT INTERVAL ESTIMATION FUSION

Fault tolerance is an important issue in network design because sensor networks must work in a dynamic and uncertain situation. In this section, a brief introduction is dedicated to the previous fault-tolerant interval estimation fusion methods proposed by Marzullo [58] and Prasad et al. [65] among others. Then, using the combination rule proposed in Section 7.2, we propose new fault-tolerant interval estimation fusion, which gives the fused interval outputs and their confidence degrees. In terms of the two optimization criteria given in Subsection 7.2.3, the fused interval estimates have smaller lengths than those given by the previous methods.

7.3.1 WITHOUT KNOWLEDGE OF CONFIDENCE DEGREES

In this section, still consider l sensor interval estimation model as in the previous section, but all sensors only deliver the fusion center their interval estimates without knowledge of confidence degrees. From previous statistical data, or other information resource, the fusion center knows at most how many sensor estimates in the l sensor interval estimates do not contain θ, but does not know which sensor estimates are faulty. Hence, the fusion center must tolerate possible faults and give a result that is as feasible as possible.

Suppose that the fusion center receives a list $\{I_1, I_2, \ldots, I_l\}$ of sensor interval estimates, where $I_i = [a_i, b_i]$, $a_i \le b_i$, and at most f $(0 \le f < l)$ out of the l interval estimates are assumed to be faulty; the fusion center takes the list $\{I_1, I_2, \ldots, I_l\}$ as an input and outputs a closed interval, $I = [a, b], a \le b$, representing the final estimate of θ.

Let

$$I_{\{i_1, i_2, \ldots, i_{(l-f)}\}} = \bigcap_{j=1}^{l-f} I_{i_j} \neq \emptyset \tag{7.3.1}$$

and the list

$$\{I_{\{i_1, i_2, \ldots, i_{(l-f)}\}} : \{i_1, i_2, \ldots, i_{(l-f)}\} \subseteq \{1, 2, \ldots, l\}\} \tag{7.3.2}$$

represents all the nonempty intersections of $(l - f)$ closed intervals in $\{I_1, I_2, \ldots, I_l\}$. The final interval $I = [a, b]$ given by Marzullo's method is the convex hull of the union of $I_{\{i_1, i_2, \ldots, i_{(l-f)}\}}$. We give Marzullo's method a functional representation, called Marzullo function, namely

$$I = M(\{I_1, I_2, \ldots, I_l\}) = \overline{\bigcup I_{\{i_1, i_2, \ldots, i_{(l-f)}\}}}. \tag{7.3.3}$$

Obviously, Marzullo function gives the shortest interval that is guaranteed to contain θ. However, this method has an obvious drawback. A gap, if any, between all connected intervals $I_{\{i_1, i_2, \ldots, i_{(l-f)}\}}$ obviously cannot contain θ but is maintained as a subinterval in the output interval I. If such gaps are too many or too large, the output interval I will be very large even if every $I_{\{i_1, i_2, \ldots, i_{(l-f)}\}}$ is short. Were the output interval I too long, it would have little value for application. For example, such an interval (region) estimate have little value for weapon control if it exceeds the kill zone of a weapon very much, regardless of its correctness.

Prasad et al. [65] presented a fault-tolerant interval integration function on the basis of Marzullo function. It can reduce the length of the output interval of Marzullo function under the assumption that all false intervals are located close to the unknown true value θ. Their output does not guarantee to cover θ and a reliable belief level of the output has not been shown.

Besides, it is desirable that a fault-tolerant interval fusion algorithm should be stable in the sense that a slight change in the input results has an only slight change in the output, in view of the unavoidable error in information acquisition and processing. Unfortunately, Marzullo's method is not stable in the able sense — it exhibits irregular behavior in that a slight difference in the input may produce a vastly different output. This was formalized by Lamport [48] as violation of the Lipschitz condition with respect to a certain metric on intervals. Indeed, if an intersection $I_{\{i_1, i_2, \ldots, i_{(l-f)}\}}$ is a single point as well as the right end of Marzullo's ouput interval I, and a gap between it and any other intersection exists, then no matter how small disturbance to sensor interval estimates, it is always possible to make this $I_{\{i_1, i_2, \ldots, i_{(l-f)}\}}$ empty, then, the right end of the present Marzullo's output interval I would jump over the length of the gap to be a point on the left side of the original right end. Intuitively, this case violates the Lipschitz condition.

To address this unstable behavior in sensor fusion process, Schmid and Schossmaier [70] introduced a fault-tolerant interval integration function called Schmid-Schossmaier function, represented as

$$I = S(\{I_1, I_2, \ldots, I_l\}) = [a, \ b], \tag{7.3.4}$$

where

$$a = \max_{(f+1)}\{a_i : i = 1, \ldots, l\}, \ b = \min_{(f+1)}\{b_i : i = 1, \ldots, l\}. \tag{7.3.5}$$

The main distinctive feature of Schmid-Schossmaier function is that it satisfies Lipschitz condition, which makes its algorithm globally stable. However, $M(\{I_1, I_2, \ldots, I_l\}) \subseteq S(\{I_1, I_2, \ldots, I_l\})$ and under some conditions $M(\{I_1, I_2, \ldots, I_l\})$ is a proper subset of $S(\{I_1, I_2, \ldots, I_l\})$. As a result, although the Schmid-Schossmaier algorithm has good stability, it suffers from the other drawback of the Marzullo function mentioned above — it includes gaps between connected intervals $I_{\{i_1, i_2, \ldots, i_{(l-f)}\}}$ — and its output interval is even larger than Marzullo's output.

7.3.2 WITH KNOWLEDGE OF CONFIDENCE DEGREES

When every sensor delivers its own interval estimate with confidence degree to the fusion center, and the fusion center also knows that at most f sensor interval estimates are faulty, we can extend the results on the combined interval outputs and their confidence degrees given in the previous section to obtain an optimal fault-tolerant interval estimation fusion in terms of Criterion (A) or (B).

Assume that all sensor estimates are independent of each other with the knowledge that at most f sensor interval estimate are faulty, but we do not know which of them are faulty. However, we are sure that at least $l - f$ sensor estimates are correct with probability one, of course, still not sure which sensor estimates they are. Therefore, by the definition of r_i given in (7.2.4), we only consider the following set of intersections of interval outputs:

$$\mathcal{C}_2 = \{I_{\{1^{r_1},\ldots,l^{r_l}\}} : I_{\{1^{r_1},\ldots,l^{r_l}\}} \neq \emptyset, \sum_{i=1}^{l} r_i \geq l - f\}. \quad (7.3.6)$$

Clearly, when $f = l$, the condition "$\sum_{i=1}^{l} r_i \geq l - f = 0$" vanishes, the condition set $\mathcal{C}_2$ becomes $\mathcal{C}$, which was defined in Subsection 7.2.2.

Thus, the conditional confidence degrees $I_{\{1^{r_1},\ldots,l^{r_l}\}}$ for all intervals in $\mathcal{C}_2$ are defined as

$$P(\theta \in I_{\{1^{r_1},\ldots,l^{r_l}\}}|\mathcal{C}_2) = \frac{1}{c_0}\prod_{i=1}^{l} \beta_i^{r_i}, \quad r_i = 0, 1, \quad (7.3.7)$$

where the parameter c_0 is given by

$$c_0 = \sum_{\substack{I_{\{1^{r_1},\ldots,l^{r_l}\}} \neq \emptyset \\ \sum_{i=1}^{l} r_i \geq l-f}} \prod_{i=1}^{l} \beta_i^{r_i}, \quad (7.3.8)$$

and

$$P(\theta \in I_{\{1^{r_1},\ldots,l^{r_l}\}}|\mathcal{C}_2) = 0 \quad \text{if } I_{\{1^{r_1},\ldots,l^{r_l}\}} = \emptyset, \text{ or} \sum_{i=1}^{l} r_i < l - f. \quad (7.3.9)$$

Example 7.3.1. The sensor outputs are the same as those given in Example 7.2.1, and the fusion center knows that at most f sensors are faulty. The fused interval outputs are given in Table 7.3.1.

Table 7.3.1. Fused fault-tolerant interval outputs with at most f faulty sensors in Example 7.3.1.

f	$(-\infty, 8) \cup (13, \infty)$	$[8, 9)$	$[9, 10)$	$[10, 11]$	$(11, 12]$	$(12, 13]$
3	0.0059	0.0237	0.1159	0.6567	0.1642	0.0336
2	0	0.0239	0.1166	0.6606	0.1651	0.0338
1	0	0	0.1237	0.7010	0.1753	0

It can be seen from Table 7.3.1 and the following tables that the smaller f is, the more the confidence degrees are concentrated on the less intervals and the more reliable these intervals are. This makes sense because the less f is, the more reliable sensor estimates are.

In terms of Criteria (A) and (B), the optimal interval fusion is given in Tables 7.3.2 and 7.3.3.

Table 7.3.2. Optimal fusion with $f = 1$ under Criteria (A) and (B) in Example 7.3.1.

Confi. degree constraint	≥ 0.7	≥ 0.8	≥ 0.9
Interval Length constraint	≤ 1	≤ 2	≤ 3
Optimal interval	[10, 11]	[10, 12]	[9, 12]
Confidence degree	0.7010	0.8763	1
Interval length	1	2	3

Table 7.3.3. Optimal fusion with $f = 2$ under Criteria (A) and (B) in Example 7.3.1.

Confi. degree constraint	≥ 0.6	≥ 0.8	≥ 0.9
Interval Length constraint	≤ 1	≤ 2	≤ 3
Optimal interval	[10, 11]	[10, 12]	[9, 12]
Confidence degree	0.6606	0.8257	0.9423
Interval length	1	2	3

Remark 7.3.1. When the confidence degrees $\beta_i^1 \equiv 1$ for any $i \leq l$, and the criterion for the optimal interval estimation fusion is Criterion (A) with the constraint of confidence degree being one, our optimal fault-tolerant interval fusion developed here becomes Marzullo's method. In other words, Marzullo's method is a special case of our method.

Remark 7.3.2. In fact, when one knows the fused intervals and their confidence degrees based on sensor interval estimates and our combination rule, further development can be done whenever extra information is known. So-called fault-tolerant interval fusion is a special example of this type of problems, which can be viewed as a conditional confidence degree problem.

For another example, if the fusion center knows by its experience that at least f sensor interval estimates are faulty, similar to (7.3.6)–(7.3.9), new conditional confidence degrees can be derived, the only difference is that $\sum_{i=1}^{l} r_i \geq l - f$ in (7.3.6), (7.2.10), and (7.3.9) now should be replaced by $\sum_{i=1}^{l} r_i \leq l - f$. Since this extra information implies that sensor estimates are less reliable than that without this extra information, and the larger f is, the less reliable sensor estimates are, it is expected (see Table 7.3.4) that confidence degrees of the fused intervals will be assigned more to those intersections $I_{\{1^{r_1},\ldots,l^{r_l}\}}$ where more r_is of $\{r_1, \ldots, r_l\}$ are equal to 0 (see Table 7.3.4) as f becomes large.

Example 7.3.2. The sensor outputs are the same as those given in Example 7.2.1, and the fusion center knows that at least f sensors are faulty. The fused interval outputs are given in Table 7.3.4.

Table 7.3.4. Fused fault-tolerant interval outputs with at least f faulty sensors in Example 7.3.2.

f	$(-\infty,8)\cup(13,\infty)$	[8,9)	[9,10)	[10,11]	(11,12]	(12,13]
0	0.0059	0.0237	0.1159	0.6567	0.1642	0.0336
1	0.0173	0.0691	0.3374	0	0.4781	0.0980
2	0.0938	0.3750	0	0	0	0.5312

7.3.3 EXTENSION TO SENSORS OUTPUTTING MULTIPLE INTERVALS

When sensors can output multiple interval estimates with their confidence degrees, namely the ith sensor outputs intervals $I_i^1, \ldots, I_i^{r_i}$ with their confidence degrees $\beta_i^1, \ldots, \beta_i^{r_i}$, we can extend the results given in previous sections to this more general case. Still assume the estimates $I_i^1, \ldots, I_i^{r_i}$ are independent of each other. Without loss of generality, intervals $I_i^1, \ldots, I_i^{r_i}$ are disjoint. If originally intervals I_i^j and I_i^k are joint, then define a new interval $I_i^{r_i+1} = I_i^j \cap I_i^k$ with the confidence degree $\beta_i^{r_i+1} = \beta_i^j\beta_i^k$. By the summability of the confidence degrees, redefine

$$I_i^j := I_i^j - I_i^{r_i+1} \quad \text{with the confidence degree} \quad \beta_i^j := \beta_i^j - \beta_i^{r_i+1}$$

and

$$I_i^k := I_i^k - I_i^{r_i+1} \quad \text{with the confidence degree} \quad \beta_i^k := \beta_i^k - \beta_i^{r_i+1}.$$

Thus, I_i^j, I_i^k and $I_i^{r_i+1}j$ become disjoint. Similarly, define a generalized interval

$$I_i^0 = (\bigcup_{j=1}^{r_i} I_i^j)^c \quad \text{with the confidence degree} \quad \beta_i^0 = 1 - \sum_{j=1}^{r_i} \beta_i^j.$$

Then, we can present the corresponding combination rule, its properties, and conditional combination rules.

7.3.4 CONCLUSION

In the last two section, the interval estimation fusion method based on sensor interval estimates and their confidence degrees has been developed. When sensor estimates are independent of each other, to merge sensor estimates and their confidence degrees, a combination rule has been proposed. Moreover, two popular optimization criteria: minimizing interval length with an allowable minimum confidence degree, or maximizing confidence degree with an allowable maximum interval length for the interval estimation are suggested. In terms of the two criteria, an optimal interval estimation fusion can be obtained based on the combined intervals and their confidence degrees. When the fusion center not only receives interval estimate with confidence degree to the fusion center, but also knows that at most f sensor interval estimates are faulty, we can extend the results on the combined interval outputs and their confidence degrees to obtain a conditional combination rule and the corresponding optimal fault-tolerant interval estimation fusion in terms of the two criteria. It is easy to see that Marzullo's fault-tolerant interval estimation fusion [58] is a special case where the first criterion is employed and the allowable minimum confidence degree is one, or the second criterion is applied and no constraint for the allowable maximum interval length is required. More generally, for any extra information, the corresp0onding conditional combination rule for the interval estimation fusion can be derived. We also point out that in some sense, our combination rule is similar to the combination rule in Dempster-Shafer evidence theory, but the confidence degree there is not summable; therefore, not applicable to the interval estimation fusion. When sensor estimates are dependent on each other, a similar interval estimation fusion method cannot be derived unless the joint probabilities of those sensor interval output lists can be known.

References

[1] Al-Khafaji, A.W. and Tooley, J.R. *Numerical Methods in Engineering Practice*. Holt, Reinhart and Winston, Inc., 1986.

[2] Alouani, A.T., Rice, T.R., and Helmick, R.E. On Sensor Track Fusion. *Proc. 1994 American Control Conference*, pp. 1042–1048, Baltimore, MD, June 1994.

[3] Bar-Shalom, Y. (Ed.) *Multitarget-Multisenaor Tracking: Advances and Applications*. vol. 1 & 2, Norwood, MA: Artech, 1990 and 1992.

[4] Bar-Shalom, Y. On the Track-to-Track Correlation Problem. *IEEE Trans. Automatic Control*, AC-26: 571-572, 1981.

[5] Bar-Shalom, Y. The Effect of the Common Process Noise on the Two-sensor Fused-Track Covariance. *IEEE Trans. Aerospace and Electronic Systems*, AES-22: 803-805, 1986.

[6] Bar-Shalom, Y and Li, X.R. *Multitarget-Multisenaor Tracking: Principles and Techniques*. Norwood, MA: Artech, 1995.

[7] Ben-Israel, A. and Greville, T.N.E., *Generalized Inverses: Theory and Applications* (2-nd edition). New York: John Wiley & Sons, 2002.

[8] Björck, A. *Numerical Methods for Least Squares Problems*. Philadelphia, PA: SIAM, 1996.

[9] Blum, R. and Kassam, S.A. Optimum Distributed Detection of Weak Signals in Dependent Sensors. *IEEE Transaction on Information Theory*, IT-36: 1066–1079, 1992.

[10] Blum, R., Kassam S.A., and Poor, H.V. Distributed Detection with Multiple Sensors: Part II-Advanced Topics. *Proceedings of the IEEE*, vol. 85: 64–79, 1997.

[11] Casella, G. and Berger, R.L. *Statistical Inference*. Wadsworth & Brooks/Cole, California, 1990.

[12] Casella, G. and Berger, R.L. *Statistical Inference* (2nd Ed.). Wadsworth & Brooks/Cole, California, 2002.

[13] Chair, Z. and Varshney, P.K. Optimal Data Fusion in Multiple Sensor Detection Sytems. *IEEE Transaction on Aerospace and Electronic Systems*, AES-22: 98–101, 1986.

[14] Chau, Y.A. and Geraniotis Delic, E. Distributed Multisensor Parameter Estimation in Dependent Noise. *IEEE Transaction on Communications,* 40: 373–384, 1992.

[15] Chaudhuri, S., Crandall A. and Reidy, D. Multisensor Data Fusion for Mine Detection. *Proc. SPIE–Sensor Fusion III: 3–D Perception and Recognition*, International Society of Optical Engineering, Bellingham, Washington, 1306: 187–204, 1990.

[16] Chen, G., Wang, J., and Shieh, L.S. Interval Kalman Filtering. *IEEE Trans. Aerospace and Electronic Systems*, AES-33: 250–259, 1997.

[17] Chen, P.N. and Papamarcou, A. Likelihood Ratio Partition for Distributed Signal Detection in Correlated Gaussian Noise. *Proc. IEEE Intl. Symp. on Inf. Th.* p.118, 1995.

[18] Chen, H.F. *Recursive Estimation and Control for Stochastic Systems*. New York, Wiley, 1985.

[19] Chen, P.N. Large Deviations Approaches to Performance Analysis of Distributed Detection Systems. Ph.D. dissertation, University of Maryland, 1994.

[20] Chen, X.R. *Introduction to Mathematical Statistics*. Beijing: Science Publishers, 1985.

[21] Cherikh, M. and Kantor, P.B. Counter Examples in Distributed Detection. *IEEE Transactions on Information Theory*, IT-38: 162-165, 1992.

[22] Chong, C.Y., Chang, K.C., and Mori, S. Distributed Tracking in Distributed Sensor Networks. *Proc. 1986 American Control Conf.*, Seattle, WA, June 1986.

[23] Chong, C.Y., Mori, S. and Chang, K.C. Distributed Multitarget Multisensor Tracking. *Multitarget-Multisensor Tracking: Advanced Applications* (Bar-Shalom, ed.), vol. 1, Atech House, Norwood, MA, 1990.

[24] Crow, R.W. and Schwartz, S.C. Quickest Detection for Sequential Decentralized Decision Systems. *IEEE Trans. Aerospace and Electronic Systems*, AES-32: 267–283, 1996.

[25] Dasarathy, B.V. *Decision Fusion*. Los Alamitos, CA: IEEE Computer Society Press, 1994.

[26] Delic, H., Papantoni-Kazakos and Kazakos, D. Fundamental Structures and Asymptotic Performance Criteria in Decentralized Binary Hypothesis Testing. *IEEE Transaction on Communications,* 43: 32–43, 1995.

[27] Delic, H. and Papantoni-Kazakos. Robust Decentralized Detection by Asymptotically Many Sensors. *Signal Processing*, 33: 223–233, 1993.

[28] Dempster, A.P. Upper and Lower Probabilities Induced by a Multi-valued Mapping. *Ann. Mathematical Statistics*, 38: 325–339, 1967.

[29] Drakopoulos, E. and Lee, C.C. Optimum Multisensor Fusion of Correlated Local Decisions. *IEEE Transaction on Aerospace and Electronic Systems*, AES-27: 593–606, 1991.

[30] Durrant-Whyte, H.F. Sensor Models and Multisensor Integration. *Int.J.Robot.Res.*, 7: 97–113, 1988.

[31] Farebrother, R. *Linear Least Squares Computations*. New York: Marcel Dekker, 1988.

[32] Ferguson, T. *Mathematical Statistics: a Decision Theoretic Approach*. New York, Academic Press, 1968.

[33] Freedman, D.D. and Smyton, P.A. Overview of Data Fusion Activities. *Proc. Am. Control Conf.*, San Francisco, June, 1993.

[34] Geraniotis, E. and Chau, Y.A. Robust Data Fusion for Multisensor Detection Systems. *IEEE,Trans. Inform. Theory*, IT-36, 1265–1279, 1990.

[35] Ghosh, B.K. and Sen, P.K. *Handbook of Sequential Analysis*. Marcel Dekker, Inc. 1991.

[36] Goodman, I.R., Mahler R.P.S., and Nguyen, H.T. *Mathematics of Data Fusion*. Kluwer Academic Publishers, 1997.

[37] Goodwin G.C. and Payne, R. *Dynamic System Identification: Experimental Design and Data Analysis*. New York: Academic Press, 1977.

[38] Grime S. and Durrant-Whyte, H.F. Data Fusion in Decentralized Sensor Networks. *Control Engineering Practice*, 2: 849–863, 1994.

[39] Han, J., Varshney P.K., and Vannicola, V.C. Distributed Detection of Moving Optical Objects. *Proc. SPIE–Sensor Fusion III: 3–D Perception and Recognition*, International Society of Optical Engineering, Bellingham, Washington, 1306: 115–124, 1990.

[40] Haykin, S. *Adaptive Filter Theory*. Englewood Cliffs, NJ: Prentice-Hall, 1996.

[41] Hashmipour, H.R., Roy, S., and Laub, A.J. Decentralized Structure for Parallel Kalman Filtering. *IEEE Transactions on Automatic Control*, AC-33: 88–93, 1988.

[42] Hoballah, I.Y. and Varshney, P.K. Neyman-Pearson Detection with Distributed Sensors. *Proc. 25th IEEE Conf. Decision and Control*, pp. 237–241, Athens, Greece, 10-12 December 1986.

[43] Hoballah, I.Y. and Varshney, P.K. Distributed Bayesian Signal Detection. *IEEE Transactions on Information Theory*, IT-35: 995–1000, 1989.

[44] Irving, W.W. and Tsitsiklis, J.N. Some Properties of Optimal Thresholds In Decentralized Decision. *IEEE Transaction on Automatic Control*, AC–39: 835–838, 1994.

[45] Kam, M., Zhu, Q., Gray, W.S., and Steven, W. Optimal Data Fusion of Correlated Local Decisions in Multiple Sensor Detection Systems. *IEEE Transactions on Aerospace and Electronic Systems*, AES-28: 916–920, 1992.

[46] Klein, L.A. *Sensor and Data Fusion Concepts and Applications*. SPIE Press, 1993.

[47] Kushner, H.J. and Pacut, A. A Simulation Study of a Decentralized Decision Problem. *IEEE Transaction on Automatic Control*, AC–27: 1116–1119, 1982.

[48] Lamport, L. Synchronizing Time Servers. *Technical Report* 18, Digital System Research Center, 1987.

[49] Lawson C.L. and Hanson, R.J. *Solving Least Squares Problems*. Philadelphia, PA: SIAM, 1995.

[50] Lehmann, E.L. *Testing Statistical Hypotheses*. New York: Wiley, 1959.

[51] Li, X.R., Zhu, Y.M. and Han, C.Z. Unified Optimal Linear Estimation Fusion–Part II: Unified Model and Fusion Rule. *Proceedings of the International Conference on Multisource-Multisensor Information Fusion (FISION'2000)*, pp. MoC2-10–MoC2-17, Paris, July 2000.

[52] Li, X.R., and Zhang, K.S. Optimal Linear Estimation Fusion–Part IV: Optimality and Efficiency of Distributed Fusion. *Proceedings of the International Conference on Multisource-Multisensor Information Fusion (FISION'2001)*, pp. Web1-19–Web1-26, Montreal, Aug. 2001.

[53] Liggins, M., Chong, C.Y., Kadar, I., Alford, M.G., Vannicola, V., and Thomopoulos, S. Distributed Fusion Architectures and Algorithms for Target Tracking. *Proceedings of the IEEE*, 85: 95–107, 1997.

[54] Ljung, L. *System Identification: Theory for the User*. Englewood Cliffs, NJ: Prentice Hall, 1987.

[55] Luo, R.C. and Kay, M.G. Multisensor Integration and Fusion in Intelligent Systems. *IEEE Transaction on Systems, Man, and Cybernetics*, SMC-19: 901–931, 1989.

[56] Luo, Z.Q., Wong, K.M., Zhu Y.M. and Sun, Y. Optimum Local Binary Decisions for a Fixed Fusion Rule in a Distributed Multi-Sensor System. *Technical Report*, Communications Research Laboratory, McMaster University, Canada, 1996.

[57] Mangasarian, O.L. *Nonlinear Programming*. New York, McGraw-Hill Book Company, 1969.

[58] Marzullo, K. Tolerating Failures of Continuous-Valued Sensors. *ACM Transaction on Computer System,* 8: 284-304, 1990.

[59] Oden, J.T. *Applied Functional Analysis*. Prentice-Hall, 1979.

[60] Papastavrou, J.D. and Athans, M. Distributed Detection by a Large Team of Sensors in Tandem. *IEEE Trans. Aerospace and Electronic Systems,* AES-28: 639–652, 1992.

[61] Papastavrou, J.D. and Athans, M. On Optimal Distributed Decision Architecture in a Hypothesis Testing Environment. *IEEE Trans. Automatic Control,* AC-37: 1154–1169, 1992.

[62] Polychronopoulos, G. and Tsitsiklis, J.N. Explicit Solutions for Some Simple Decentralized Detection Problems. *IEEE Trans. Aerospace and Electronic Systems,* AES-26: 789–796, 1990.

[63] Poor, H. V. *An Introduction to Signal Detection and Estimation - Second Edition.* New York : Springer-Verlag, 1994.

[64] Port, S.C. *Theoretical Probability for Applications*. John Wiley & Son Inc., 1994.

[65] Prasad, L., Iyengar, S.S., Kashyap, R.L., and Madan, R.N. Functional Characterization of Fault Tolerant Integration in Distributed Sensor Networks. *IEEE Transactions Systems, Man, and Cybernetics*, SMC-21, no.5, 1991.

[66] Reibman, A.R. and Nolte, L.W. Optimal Detection and Performance of Distributed Sensor Systems. *IEEE Trans. Aerospace and Electronic Systems,* AES-23: 24–30, 1987.

[67] Reibman, A.R. and Nolte, L.W. Design and Performance Comparison of Distributed Detection Networks. *IEEE Trans. Aerospace and Electronic Systems,* AES-23: 282–291, 1990.

[68] Samarasooriya, V.N.S. and Varshney, P.K. Sequential Approach to Asynchronous Decision Fusion. *Optical Engineering*, 35: 625–633, 1996.

[69] Shafer, G.A., *Mathematical Theory of Evidence.* Princeton, N.J., Princeton University Press, 1976.

[70] Schimd U. and Schossmaier, K. How to Reconcile Fault-Tolerant Interval Intersection with The Lipschiz Condition. Technische Universitat Wien, Department of Automation, Vienna, September 6,1999.

[71] Siegmund, D. Error Probabilities and Average Sample Number of the Sequential Probability Ratio Test. *J. Roy. Statist. Soc. Ser. B*, pp. 394–401, 1975.

[72] Siegmund, D. *Sequential Analysis: Test and Confidence Intervals.* Springer, 1985.

[73] Tang, Z.B., Pattipati K.R. and Kleinman, D.K. A Distributed M-ary Hypothesis Testing Problem with Correlated Observations. *IEEE Transactions on Automatic Control*, AC-37: 1042–1046, 1992.

[74] Tang, Z.B., Pattipati K.R. and Kleinman, D.K. An Algorithm for Determining the Decision Threshold in a Distributed Detection Problem, *IEEE Transaction on Systems, Man, and Cybernetics,* SMC-21: 231–237, 1991.

[75] Tang, Z.B., Pattipati K.R. and Kleinman, D.K. Optimization of Distributed Detection Networks: Part I–Tandem Structure. *IEEE Transaction on Systems, Man, and Cybernetics,* SMC-21: 1044–1059, 1991.

[76] Tang, Z.B., Pattipati K.R. and Kleinman, D.K. Optimization of Distributed Detection Networks: Part II–Tree Structure. *IEEE Transaction on Systems, Man, and Cybernetics,* SMC-23: 211–221, 1993.

[77] Tenney, R.R. and Sandell, N.R.Jr. Detection with Distributed Sensors. *IEEE Transaction on Aerospace and Electronic Systems*, AES-17: 501–510, 1981.

[78] Thomopoulos, S.C.A. Visvanathan R., and Bougoulias, D.C. Optimal Distributed Decision Fusion. *IEEE Transaction on Aerospace and Electronic Systems*, AES-25: 761–765, 1989.

[79] Tsitsiklis, J.N. Decentralized Detection by a Large Number of Sensors. *Mathematics of Control, Signals, and Systems,* 1: 167–182, 1988.

[80] Tsitsiklis, J.N. Decentralized Detection. *Advances in Statistical Signal Processing, vol. 2: Signal Detection* (H.V. Poor and J.B. Thomas, Ed.), Greenwich, CT, JAI Press, 1990.

[81] Tsitsiklis, J.N. Extremal Properties of Likelihood-Ratio Quantizers. *IEEE Transaction on Communications,* 41: 550–558, 1993.

[82] Tsitsiklis, J.N and Athans, M. On the Complexity of Decentralized Decision Making and Detection Problems. *IEEE Transaction on Automatic Control*, AC-30: 440–446, 1985.

[83] Van Trees, H.L. *Detection, Estimation, and Modulation Theory*. vol. 1, New York, Wiley, 1968.

[84] Varshney, P.K. *Distributed Detection and Data Fusion*. Springer, 1997.

[85] Veeravalli, V., Basar T., and Poor, H.V. Minimax Robust Decentralized Detection. *IEEE Transaction on Information Theory*, IT-38: 35–40, 1994.

[86] Veeravalli V., Basar T., and Poor, H.V. Decentralized Sequential Detection with a Fusion Center Performing the Sequential Test. *IEEE Transaction on Information Theory*, IT-37: 433–442, 1993.

[87] Vender, R. Team Decision Problems. *Ann. Math. Stat.*, 33: 857–881, 1962.

[88] Venugopal, V., Veeravalli, V., Basar T., and Poor, H.V. Decentralized Sequential Detection with a Fusion Center Performing the Sequential Test. *IEEE Transaction on Information Theory*, IT-37: 433–442, 1993.

[89] Viswanathan, R. and Varshney, P.K. Distributed Detection with Multiple Sensors: Part I-Fundamentals. *Proceedings of the IEEE*, 85: 54–63, 1997.

[90] Wald, A. *Sequential Analysis*. New York, John Wiley & Sons, 1947.

[91] Wald, A. and Wolfowitz, J. Optimum Character of Sequential Probability Ratio Test. *Ann. Math. Statist.*, 19: 326–339, 1948.

[92] Willett, P.K. and Warren, D. The Suboptiamlity of Randomized Test in Distributed and Quantized Detection Systems. *IEEE Transaction on Information Theory,* IT-38: 355–361, 1992.

[93] Wijsman, R. Stopping Times, Termination, Moments, Distribution. *Handbook of Sequential Analysis* (Ghosh, B.K. and Sen, P.K., eds.), Marcel Dekker, Inc., 1991.

[94] Willner, D., Chang, C.B., and Dunn, K.P. Kalman Filter Configurations for Multiple Radar Systems. *Technical Note 1976-21*, Lincoln Laboratory, MIT, 14 April, 1976.

[95] Wismer D.A. and Chattergy, R. *Introduction to Nonlinear Optimization*. New York: Elsevier, North-Holland, 1978.

[96] Willett, P.K. and Alford, M. The Case of Like–Sensor Predetection Fusion. *IEEE Trans. Aerospace and Electronic Systems,* AES-30: 986–999, 1994.

[97] Zhu, Y.M. Efficient Recursive State Estimator for Dynamic Systems without Knowledge of Noise Covariances. *IEEE Trans. Aerospace and Electronic Systems*, AES-35: 102–114, 1999.

[98] Zhu, Y.M., Blum, R.S., Luo, Z.Q., and Wong, K.M. Unexpected Properties and Optimum Distributed Sensor Detectors for Dependent Observation Cases. *IEEE Transactions on Automatic Control*, AC-45: 62–72, 2000.

[99] Zhu, Y.M. Optimum Fusion in Distributed Multisensor Network Decision Systems. *The Proceedings of 14th IFAC World Congress*, Beijing, July 1999.

[100] Zhu, Y.M., Liu C.Y., and Gan, Y. Multisensor Distributed Neyman-Pearson Decision with Correlated Sensor Observations. *Proceedings Of the International Conference on*

Multisource-Multisensor Information Fusion (FUSION'98), 1: 35–39, Las Vegas, July, 1998.

[101] Zhu, Y.M., Gan,Y., Zhang K.S., and Liu, C.Y. Multisensor Neyman-Pearson Type Sequential Decision with Correlated Sensor Observations. *Proceedings Of the International Conference on Multisource-Multisensor Information Fusion (FUSION'98)*, 2: 787–793, Las Vegas, July, 1998.

[102] Zhu, Y.M. and Li, X.R. Optimal Decision Fusion Given Sensor Rule. *Proceedings Of the International Conference on Multisource-Multisensor Information Fusion (FUSION'99)*, Sunnyvale, CA, 1999.

[103] Zhu, Y.M., Li, X.R., and You, Z.S. Unified Fusion Rule in Multisensor Netowrk Decision Systems. *Proceedings of the International Conference on Multisource-Multisensor Information Fusion (FUSION'2000)*, Paris, July 2000.

[104] Zhu, Y.M. and Li, X.R. Best Linear Unbiased Estimation Fusion. *Proceedings of the International Conference on Multisource-Multisensor Information Fusion (FUSION'99)*, 2: 1054–1061, Sunnyvale, CA, ISIF, July 1999.

[105] Zhu, Y.M., Zhang, K.S., Li, X.R., and You, Z.S. Track Fusion of Distributed EFRLS State Estimators. *Proceedings of the International Conference on Multisource-Multisensor Information Fusion (FUSION'2000)*, Paris, July 2000.

[106] Zhu, Y.M., You, Z.S., Zhao, J., Zhang, K.S., and Li, X.R. The Optimality of the Distributed Kalman Filter with Feedback. *Automatica*, 37: 1489–1493, 2001.

[107] Zhu Y.M. and Li, X.R. Recursive Least Squares with Linear Constraints. *Proceedings of the 38th IEEE Conference on Decision and Control*, 3: 2414–2419, 1999.

Index